Corrosion Characteristics, Mechanisms and Control Methods of Candidate Alloys in Sub- and Supercritical Water

Donghai Xu · Shuwei Guo

Corrosion Characteristics, Mechanisms and Control Methods of Candidate Alloys in Sub- and Supercritical Water

西安交通大学出版社
XI'AN JIAOTONG UNIVERSITY PRESS

Springer

Donghai Xu
Department of Energy and Power Engineering
Xi'an Jiaotong University
Xi'an, Shaanxi, China

Shuwei Guo
Department of Energy and Power Engineering
Xi'an Jiaotong University
Xi'an, Shaanxi, China

ISBN 978-981-16-5524-1 ISBN 978-981-16-5525-8 (eBook)
https://doi.org/10.1007/978-981-16-5525-8

Jointly published with Xi'an Jiaotong University Press
The print edition is not for sale in China (Mainland). Customers from China (Mainland) please order the print book from: Xi'an Jiaotong University Press.

This Springer imprint is published by the registered company Springer Nature Singapore Pte Ltd.
The registered company address is: 152 Beach Road, #21-01/04 Gateway East, Singapore 189721, Singapore

Preface

Supercritical water (SCW) refers to water at temperature and pressure that is above its critical point (T = 374 °C, p = 22.1 MPa), where liquid and gas phases cannot be distinguished. SCW is a homogeneous and nonpolar solvent with high diffusivity and advantageous transport properties, and can be used as an excellent reaction medium. However, high temperature, high pressure, high concentration of dissolved oxygen and/or high concentration of ions can cause severe reactor corrosion in sub- or supercritical water environment, such as supercritical water oxidation, supercritical water gasification and nuclear reactor. Chemical corrosion, stress corrosion cracking and/or hydrogen embrittlement easily appear, so severely shortening reactor lifetime.

In this book, we aim to provide readers with comprehensive and systematic introduction and summary of corrosion characteristics, mechanisms and control methods of candidate alloys in sub- and supercritical water environment. First of all, corrosion types of candidate alloys and the effects of major alloying elements on corrosion resistance of potential alloys in sub- and supercritical water are compared and analyzed. At the same time, research status of candidate materials, and development and application trends of several corrosion-resistant alloys are summarized. Then, corrosion characteristics of Ni–Cr, Ni–Cr–Mo, Ni–Fe–Cr and Ni–Fe–Cr–Mo–Cu corrosion-resistant alloys, FeCrAl alloy and Zircaloy are discussed in detail, including the corrosion rate, the structure and composition of oxide film, the effects of various surface treatment processes, etc. More specifically, we investigated corrosion behavior of Ni-based alloy, Fe–Ni-based and stainless steels in SCW. The effects of aggressive species on the corrosion behavior of Ni-based alloys are also explored in SCW. We further propose the total corrosion processes and mechanisms of typical candidate alloys in sub- and supercritical water environment. Finally, we attempt to study corrosion control methods such as ceramic coatings and passivation processes in supercritical water oxidation and in subcritical water, respectively. The challenges and development trends of corrosion research of candidate materials in sub- and supercritical water environments are identified at the end of this book. The information involved in this book is valuable for theoretically guiding material selection and design and operating parameter optimization of key equipment in the sub- and supercritical water technologies.

Generally speaking, the content of this book is suitable for students and researchers who are interested in alloy corrosion in sub- and supercritical water. People from all walks of life are also very welcome to read. At last, there are many people who have helped a lot in the process of writing this book we wish to thank. Herein, we sincerely appreciate the help and contribution from Prof. Shuzhong Wang, Yang Guo, Zefeng Jing, Yanhui Li and Jianqiao Yang in Xi'an Jiaotong University in China, and Prof. Digby D. Macdonald in University of California at Berkeley in USA. Many graduate students (including Zhijiang Ma, Guanyu Jiang, Wangpeng Yang, Yu Wang, etc.) from Xi'an Jiaotong University have made great contributions to the preparation of this book, including data treating, figure plotting, text writing, proofreading and so on.

This book was supported by the Projects from National Natural Science Foundation of China (22078258, 21576219), the National Key Research and Development Program of China (2021YFE0104900) and the Fundamental Research Funds for the Central Universities (xtr042019015, xjh012019005, xzd012020031).

Xi'an, China Donghai Xu

Contents

Chapter 1
Introduction

1.1 Supercritical Water Technology

1.1.1 Characteristics of Supercritical Water

Supercritical water (SCW) refers to water at a temperature that is above its critical point (T = 374 °C, p = 22.1 MPa), where liquid and gas phases cannot be distinguished. SCW is completely different from the water at normal temperature with respect to physical parameters such as density, solubility, dielectric constant, ionic product, viscosity and diffusion coefficient. SCW possesses transport properties that are similar to those of a gas and the solubility properties of the fluid [1, 2], and its density lies within the interval between a gas and a liquid, with these properties being sensitive to the pressure [3]. When water approaches the critical point, its dissociation constant (ionic product) are approximately higher three orders of magnitude higher ($K_w \approx 10^{-11}$) than that of water at 25 °C ($K_w \approx 10^{-14}$). However, the dissociation constant drops sharply [4], when the critical point is exceeded. Compared with normal temperature water, SCW possesses a much lower viscosity and a higher self-diffusion coefficient; both change by about an order of magnitude compared with the values at ambient temperature. The lower viscosity and higher diffusion coefficient make SCW a good transport medium [5]. When water changes from the liquid state to the supercritical state, the number of hydrogen bonds is greatly reduced, thereby resulting in a greatly reduced dielectric constant. SCW varies from a polar state to a non-polar state and its solvent power is reversed [3, 6]. Water-insoluble organic matter at normal temperature can be dissolved in SCW, and some gases (e.g., O_2 and CO_2) are also completely miscible with SCW. Thus, SCW provides an opportunity to carry out chemical reactions in a single fluid phase rather than a multiphase system under conventional conditions. The potential advantages are that higher concentration of reactants can be obtained and there is no inter-phase mass transfer that limits the reaction rate [4]. In other words, a single homogeneous phase allows reactions to

D. Xu and S. Guo, *Corrosion Characteristics, Mechanisms and Control Methods of Candidate Alloys in Sub- and Supercritical Water*,
https://doi.org/10.1007/978-981-16-5525-8_1

proceed rapidly, because the potential interfacial mass transfer limitations have been eliminated [7].

In summary, SCW is a homogeneous and non-polar solvent with high diffusivity and advantageous transport properties, and can be used as an excellent reaction medium. Chemical reactions in SCW possess a variety of advantages such as, obtaining higher concentrations of reactant gases than in a conventional gas–liquid system, eliminating inter-phase transport limitations present in a heterogeneous reaction system, easily separating products, optimizing high pressure dependence of supercritical reaction media, and owning higher diffusivity than a liquid and better heat transfer than a gas [8]. Additionally, the significant changes in physicochemical properties of SCW with temperature and pressure makes it possible to control the reactions by judiciously controlling these independent variables [9]. Therefore, the use of SCW as a reaction medium has attracted great attention.

1.1.2 Supercritical Water Oxidation

Supercritical water oxidation (SCWO) is a fast and efficient method for the destruction of toxic, organic waste [10–12], which can rapidly oxidize and decompose organic matter in a single-phase mixture by using the miscibility of organic matter and oxygen in SCW [13]. Accordingly, SCWO possesses many advantages, such as a high reaction rate, a high destruction rate (>99.9%), no NOx formation, and good reaction confinement [14]. The operating conditions of SCWO are commonly far higher than the critical point of water, typically in the ranges of 500–650 °C and 25–30 MPa. In SCWO, the reaction residence time ranges from several seconds to a few minutes, even less than 50 s is required for the complete destruction of organic contaminants at about 650 °C [2]. After SCWO, hydrocarbons and oxygenated hydrocarbons are converted into CO_2 and H_2O, and nitrogen in the feedstock is converted into N_2, while the heteroatoms such as Cl, S and P are transformed to corresponding mineral acids (i.e. HCl, H_2SO_4 and H_3PO_4) or inorganic salts [6]. Accordingly, SCWO products are mainly clean liquid water, gases, and solid inorganic salts [15].

SCWO is an efficient and environmentally friendly technology that can rapidly destroy organic matter in pollutants, and the resulting inorganic salts in the reactor effluent can be easily leached out, and the heat generated during the reactions can be recovered [16]. Therefore, it has become a good option for the commercial treatment of sewage sludge [17]. For example, SCWO can also dispose many types of organic wastewaters with organic removal rates of up to 99.99%, without incineration-related pollutant emissions [18], and other problems, such as high construction and operation costs, strict requirements on temperature, mixing, uniformity, and residence time. A high concentration of organic waste or auxiliary fuel is required to maintain the required combustion temperature, but the lack of toxic effluents negates the public opposition that has arisen in locating incineration plants [19]. Thus, SCWO is an effective substitution technology of incineration and has received extensive attention [10–12, 17]. Existing studies have demonstrated that SCWO is capable of harmlessly treating N-containing compounds (e.g., pyridine, aniline, nitrobenzene, ammonia, and nitrogen-containing phenolic compounds) and unconventional wastes (e.g., chemical weapons and radioactive waste). It can even destroy highly toxic and potentially carcinogenic polychlorinated biphenyls [20]. Greater than a 99.9% efficiency can be achieved for the destruction of polychlorinated biphenyls in SCW when a certain amount of oxidant is added according to the stoichiometric coefficient [21]. SCWO also has high potential for treating halogenated organic compounds and is competitive with other treatment methods [22, 23].

Over the past three decades, the number of large scale SCWO plants have been developed around the world to treat various organic feedstock [6, 22–24]. More and more commercial applications show a bright prospect for the SCWO technology.

1.1.3 Supercritical Water Gasification

Hydrogen as a clean fuel of high calorific value has attracted much attention. Hydrogen production methods include water electrolysis, steam reforming, autothermal reforming, aqueous-phase reforming, biomass fermentation, and supercritical water gasification (SCWG). SCWG is an efficient thermochemical gasification process of organic matters occurring in supercritical water with partial or complete absence of oxidants, which can be used to produce a hydrogen-rich flammable gas from biomass [25, 26].

The operating temperature of biomass SCWG can be divided into two types of ranges, i.e., low temperature (350–500 °C) and high temperature (500–800 °C) [27]. The raw materials of SCWG include actual biomass and organic fossil materials, such as straw, wood, sludge and coal. It is necessary to choose some model compounds as research objectives, so as to reveal the conversion mechanisms of the above-mentioned complex materials in SCWG [27]. Hence, the research on model compounds such as glucose, cellulose and phenol has attracted much attention [28, 29]. Topically gaseous products comprise H_2, CO_2, CH_4, CO and C_2 light gases, and syngas. (H_2 + CO) is a major gaseous product and can be used as a clean fuel or in the diesel fuel production by a gas-to-liquid technology [27]. Particularly, CO can react with steam to form H_2 and CO_2 through the water–gas shift reaction during SCWG, and thus the hydrogen atoms of water can be released to form H_2 [30].

Compared with conventional thermochemical gasification, SCWG does not need the drying process with high energy consumption [31, 32], so that it can avoid the high drying cost effectively [33]. SCWG can reduce char formation in the homogeneous solution, and thus obtain a higher gasification efficiency and cleaner products [34]. SCWG also has a high reaction efficiency and a high H_2 selectivity [35], and the temperature of complete gasification reaction can be as low as 600 °C, and only small amounts of tar and coke are produced under those conditions. The generated CO_2 has a high solubility in the room-temperature and high-pressure water and is easily separated from H_2; meanwhile gaseous products are clean and liquid effluents can be used as fertilizer [36]. The produced high-pressure H_2 reduces the compression cost during its storage [27]. In addition, solar energy can be used as the heat source in the hydrogen production from biomass SCWG, so the whole process can be sustainable. This conversion from solar energy to chemical energy can overcome the shortcomings of intermittent and uneven distribution of solar energy. Therefore, it is possible to realize low-cost, high-efficiency hydrogen production via solar energy [36]. Among the choices of renewable resources, the potential of SCWG being used to produce hydrogen from biomass is considered to have the lowest cost in the near- and mid-term [36]. Accordingly, biomass SCWG is a promising hydrogen production method [28, 29, 37].

1.1.4 Supercritical Water-Cooled Reactor

SCW can be used not only as a reaction medium for the harmless treatment of organic pollutant or hydrogen production by biomass gasification, but also as an excellent coolant for fossil fuels, supercritical and ultra-supercritical power plants and the potential supercritical water-cooled (nuclear) reactors (SCWR) being envisioned as one of the six Generation IV nuclear power plant technologies [38]. The goal of modern power plants is to achieve a higher thermal efficiency and lower carbon dioxide emission. SCW as a coolant has attracted more and more attention due to its potential energy saving and environmental protection performances. The basic design of the SCWR is a thermal spectrum reactor using low-enrichment uranium oxide fuel in a conventional light water reactor and utilizing water rods (containing neutron absorbers) as reaction moderators. Obviously, the SCW coolant does not undergo a phase change as it passes through the reactor core. The outlet temperature and operating pressure of SCWR can reach up to 625 °C and 25 MPa, respectively [39] yielding a high theoretical (Carnot) efficiency. The Generation IV CANDU SCWR is a new energy-saving advanced pressure tube reactor using H_2O as the coolant, and the generated SCW can be used for power generation, hydrogen production, and district heating [40]. This high-temperature (supercritical) water can be adopted to produce hydrogen by the thermochemical water-splitting cycle, which is considered to be one of the most efficient methods of producing hydrogen from water and is superior to conventional low-temperature water electrolysis [41]. The European SCWR is operated with SCW at an inlet pressure of 25 MPa and an average core outlet temperature of 500 °C, so making the power plant more economical and reliable and less nuclear waste formation simultaneously, because of the higher thermodynamic efficiency [42].

Overall, as an advanced light water reactor, SCWR can be used for basic load power production, and has a high thermodynamic efficiency, and greatly simplified plant equipment compared with the present commercial light water reactors. Unlike the boiling water reactor that uses a saturated steam cycle, the SCWR adopts a single-phase SCW coolant and thus has no the problem of burning out or drying up unless the coolant pipe is broken. The SCWR system possesses some attractive features, such as low flow rate, high enthalpy rise, and a single-phase water coolant [43]. This helps to reduce costs, achieves a high thermal efficiency, and provides a simplified, straight- steam cycle system [43, 44]. Therefore, the SCWR has a promising future as one of the key development targets for the next-generation nuclear reactor [45, 46].

1.2 Problems in Supercritical Water Technology

1.2.1 Problems in Supercritical Water Oxidation

Figure 1.1 identifies the principal difficulties and problems at different stages of a typical SCWO process stream. Obviously, a corrosion problem exists at the stages of preheating, reaction, and cooling, especially in the reactor that operates at 400–650 °C and at ~25 MPa. More specifically, chemical corrosion mainly occurs in low-density supercritical regions (likely in the preheater back end, the reactor, and the cooler front end), while electrochemical corrosion majorly appears in subcritical- and high-density supercritical regions (e.g., the front end of the preheater and the back end of the cooler). The electrochemical corrosion mechanism prevails in high-density aqueous system with a density more than 100–200 kg/m^3.

The extremely low solubility of inorganic salt in SCW leads to its precipitation, crystallization, agglomeration, and deposition [47]. This reduces the heat transfer capacity in heat exchangers and results in plugging in pipes and reactors [48], in the supercritical regions. If not controlled, the plugging can cause frequent system shutdowns and large economic loss [49]. The best option of overcoming the problem is to minimize the salt content in the feedstock [50], and to design an innovative

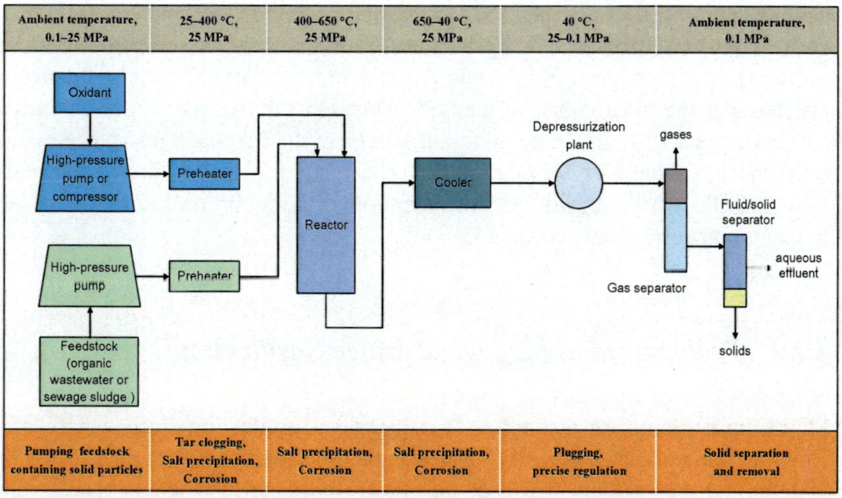

Fig. 1.1 Main problems at different stages in a typical SCWO process

reactor structure [51]. Overall, reactor corrosion and reactor plugging are the core problems in a SCWO system. They result in the difficulties in the scale-up and commercial application of SCWO devices and severely limit the development of the SCWO technology [48]. Researchers have proposed many technical solutions, among which reactor structure innovation is one of the key ways of preventing reactor corrosion and salt deposition in SCWO. Until now, various reactor structures such as counter-flow reactors, cooled wall reactors, centrifugal reactors, and transpiring wall reactor have been developed, and their work principles and methods are different. For example, the transpiring wall reactors effectively overcome the two problems above by forming a protective water film on the inner wall surface of the reactor to prevent corrosive substances and inorganic salts from contacting the inner wall [52].

High temperatures, high concentrations of dissolved oxygen, extreme pH values, high concentrations of ions and sharp pressure change [2], can cause severe reactor corrosion in SCWO, especially in acidic, high-oxidizing and high-temperature media. Corrosion is determined by solution factors (including density, temperature, pH value, electrochemical potential, the aggressiveness of aggressive anion) and material factors (including alloy composition, surface condition, material purity, and heat treatment) [53]. Corrosion is also affected by the dissociation of salt and alkali, the solubility of gases and corrosion products, and the stability of the protective oxide layer. Additionally, salt deposition is also one of the causes of equipment material corrosion, and the salts dissolved in water in the high temperature, high density and strongly oxidizing SCWO environment make reactor materials particularly susceptible to corrosion [51]. The stagnant microenvironment existing between the deposited-salt layer and the inner reactor wall can result in severe corrosion [54]. In order to prolong the service life of the SCWO reactor, researchers have developed a variety of corrosion control techniques to reduce the corrosion rate of the reactor material [55], including the use of high corrosion resistant material, liners, coatings, transpiring wall reactors, adsorption/reaction on the fluidized solid phase, vortex/circulating flow reactors, pre-neutralization, and the optimization of process operating conditions, etc. [54], while the selection of material with high corrosion resistance is the most common method. Although no one material can withstand corrosion in all systems, some materials can meet the corrosion resistance requirement under certain SCWO conditions for a considerable period of time [56]. Notably, Ni-based alloys have drawn the most attention as the SCWO reactor materials due to their good corrosion resistance [50, 57].

1.2.2 Problems in Supercritical Water Gasification

SCWG is a promising technology for hydrogen production from wet biomass. However, tar and char are easily formed in the preheater and reactor of the SCWG system, and the buildup of ash in the reactor is likely to result in reactor plugging. System shutdown caused by tar and char production has a severe influence on the energy conversion process [58]. In order to solve this problem, researchers have conducted innovative improvements of SCWG reactors. For instance, a SCW fluidized bed system for biomass SCWG has been developed to effectively avoid the plugging problem [58]. There are also the problems of slow heating rate and low gasification rate under some reaction conditions in biomass SCWG, which can be solved by adding a certain amount of oxidant (less than the stoichiometric requirement for complete oxidation) into the reactor [59]. At present, SCWG is still at the stages of experimental research and demonstration development. Before the construction of the large-scale SCWG reactor, comprehensive technical demonstration and improvement should be carried out, including the continuous delivery of high-concentration biomass, the prevention and control of reactor plugging, and material corrosion, catalyst development, the improvement of gasification efficiency and economy, etc. Certainly, detailed reaction mechanisms for biomass SCWG also need to be further explored [60]. It is worth noting that the corrosion of SCWG reactor materials under high temperature and high pressure conditions exists widely. Especially, the presence of hydrogen can induce hydrogen embrittlement and/or hydrogen attack (methane formation via the reaction of atomic hydrogen with carbides) of materials, which may result in the serious damage of the SCWG reactor. Therefore, the improvement of the corrosion resistance of reactor material is important for ensuring the serviceability of the SCWG reactor.

1.2.3 Problems in Supercritical Water-Cooled Reactor

Since the operating conditions and materials of SCWR are significantly different from those of conventional heavy water reactors and light water reactors [61], many challenges must be addressed before the construction of a SCWR power plant can be achieved. The feasibility of SCWRs depends on whether its materials can withstand thermal stress, hydrostatic stress, thermochemical stress, and radiation stress caused by the operating conditions. The material failure of the fuel cladding, which is the key component of SCWR, may be due to the following mechanisms: penetration through the tube wall caused by general corrosion, oxide accumulation thereby hindering heat transfer, stress corrosion cracking, and stress exceeding the material yield limit [62]. Herein, stress corrosion cracking is likely induced by the combined effects of chemical and/or electrochemical corrosion and stress on the material. The chemical properties of SCW under neutron and gamma photon irradiation conditions are important variables affecting the actual operation of SCWR. The chemical environment-sensitive cracking in the reactor mainly includes intergranular stress corrosion cracking, irradiation-assisted stress corrosion cracking, and corrosion fatigue [63]. Core materials should be capable of withstanding extreme environments, including high temperature and pressure, radiation and rapid chemical property change. The material choice has become one of the main obstacles to SCWR implementation [64]. Notably, the cladding material is required to withstand high temperature (600–650 °C) and neutron radiation under normal working conditions, so heat transfer during the process must be studied, simultaneously, because any reduction in heat transfer rate will raise the temperature of the fuel [65].

The long-term, reliable operation of SCWR depends on the ability of controlling and maintaining the chemical conditions of water at the supercritical state, and this can alleviate corrosion and reduce the transportation of corrosion products and radionuclides to large extent [40]. The harsh working environment of SCW coolant leads to the need to use modified stainless steels or other advanced materials [61]. Overall, the most suitable materials are austenitic stainless steels and Ni-based alloys. SS 310, Incoloy 800H and Inconel 625, all of which exhibit acceptable performance in the Canadian SCWR [62]. Until now, research on material problems has made a great progress, but the selection and identification of suitable fuel cladding alloys is still the most significant challenge [62, 64]. Therefore, from the view of material selection, the combination of high temperature, high pressure and aggressive chemical environment makes SCWR a challenging reactor [66]. Moreover, modern fossil-fueled power plants are operated at 650 °C and 34.5 MPa or higher, so as to achieve a higher thermal efficiency. In such a high-temperature and high-pressure environment, the history of using structural materials used in nuclear reactors and fossil-fueled, thermal power plants (FFTPPs) are susceptible to corrosion provide a good basis for using the same materials in SCWRs. However, caution must be exercised in simply regarding a SCR as being a FFTPP with a different heat source, as was done in selecting materials for BWRs. In that case, little recognition of the impact of water radiolysis products (H_2, O_2, H_2O_2) on raising the ECP of weld-sensitized Type 304 SS lead to extensive IGSCC that has dogged BWR over the past forty years. Even so, SCR technology still faces the challenge of material selection for fuel cladding and core components [67].

As aforementioned (in the whole Sect. 1.1.4), SCWO reactors, SCWG reactors and SCWRs employ harsh working environments, such as high temperature, high pressure, corrosive SCW, high concentration of oxygen (SCWO), hydrogen-enriched gas streams (SCWG) and/or radiation (SCWR), all of which contribute to severe corrosion problems of materials. Therefore, material corrosion has become a decisive factor affecting the safe, reliable and stable operation of the supercritical water technology system. This is clearly the most important issue in the development and application of supercritical water technologies (i.e., SCWO, SCWG and SCWR) and must be successfully addressed before these technologies become practical.

1.3 Studies on Candidate Materials for Supercritical Water Technology

1.3.1 Major Candidate Materials

Selecting high corrosion-resistant alloys as the structural materials of major SCWO equipment is a simple and basic method of corrosion prevention and control [54]. In general, some ferritic-martensitic steels [68–70], austenitic stainless steels [71–76], Ni-based alloys [77–82], Ti-based alloys, and Zr-based alloys can be used in SCW environments [83]. Four types of ferritic-martensitic steels (i.e. T91, T92, HCM12A and HT-9) have been tested in SCW and HT-9 is found to be a candidate material for the core structure of SCW supercritical water reactor [84]. Austenitic stainless steels (e.g. SS 316L, D-9) have good creep resistance and corrosion resistance in high-temperature environment. They are good material options for some SCWO equipment, due to simple manufacturing properties (e.g., weldability) and their relatively low costs. Oxide dispersion strengthened steels prepared by the mechanical alloying process exhibit good high-temperature performance and resistance to radiation-induced swelling and embrittlement, and are recommended as the structural material of the Generation IV nuclear reactor [85]. Titanium alloys (e.g., Ti-15Mo-5Zr-3Al, Ti-3Al-2.5 V, Ti-6Al-4 V and Ti-15 V-3Al-3Sn-3Cr) have better corrosion resistance in supercritical water than Ni-based alloys. Titanium alloys (e.g., Ti-Gr2, Ti-Gr9 and Ti-Gr12) exhibit good corrosion resistance in strong oxidizing environment but possess poor mechanical properties in high temperature environment, so they can be used as the coating of other materials for which high strength is not an issue [2]. Zirconium alloys are adopted as fuel cladding materials in the nuclear power industry owing to their low neutron absorption cross-section and good corrosion resistance [86]. Zircaloy-2 and Zircaloy-4 have been widely used and are regarded as standard structural materials in various fission reactors. The zirconium alloys are also potential candidate materials for supercritical water reactors [83]. Furthermore, ceramic materials can be used as reactor materials in supercritical water technologies. Al-based and Si-based ceramic materials, such as alumina, silicon carbide, and silicon nitride are suitable for use in environments of $pH < 12$ [2].

In particular, Ni-based alloys are extensively used to manufacture the reactors in many pilot- and large-scale plants in SCWO systems, due to their good performance under harsh conditions [87]. Ni-based alloys (e.g., Inconel 600, Inconel 625, Incoloy 718 and Hastelloy C-276) are commonly adopted as structural materials in the high-temperature and high-pressure equipment in SCWO systems [54]. Additionally, Ni-based alloys are also widely chosen as SCWG reactor materials, SCWR materials, and equipment materials in the supercritical fossil power industry. Overall, Ni-based alloys have attracted the most attention, have the most extensive applications, and show the most promising future in the material selection for key equipment in supercritical water technologies. This is reflected in the following 3.2 section.

1.3.2 Research Overview on Candidate Materials

Many researchers have carried out numerous studies on the corrosion behavior and corrosion control of materials in supercritical water, and several reviews have been reported. The earliest review of corrosion can be traced back to the work of Friend in 1980 [88], which systematically described the corrosion behaviors of alloys with Ni contents higher than 40 wt.% in a wide range of environments. Kritzer [89], summarized the corrosion behavior of Inconel 625 in strongly oxidizing solutions of acids (i.e., HF, HCl, HBr, H_2SO_4, and HNO_3), salts (i.e., NaCl, $NaHSO_4$, and Na_2SO_4)and alkali (i.e., NaOH) at sub- and supercritical temperatures. Mitton [90], reported the degradation characteristics of SCWO various candidate materials, including Fe-based alloys, Ni-based alloys, Ti-based alloys, ceramics, and noble metals, and described some potential methods of reducing the corrosion damage of SCWO equipment. Eliaz [91], introduced the problems relating to structural material selection and corrosion control, and mentioned the degradation characteristics of several candidate structural materials and some corrosion control approaches. Kritzer [53] also reviewed corrosion influencing factors and partial corrosion mechanisms of stainless steels and Ni-based alloys in high-temperature water and supercritical water. Sun et al. [83] summarized the behavior of alloys, ceramics, and materials used as protective coatings for SCWR components, and specially focused on material degradation characteristics, selection criteria, surface modification methods, and new processing development. Marrone and Hong [54]. presented the corrosion control methods for supercritical water applications (i.e., SCWO and SCWG), and pointed out that corrosion control could be achieved through a combination of corrosion resistant materials, chemical control, mechanical design, and process design. Guzonas et al. [92] reported the effect of changes in SCW density on corrosion mechanism and emphasized the close relationship between SCW corrosion and high temperature steam above 500 °C. Presently, Tang et al. [55] described the corrosion characteristics of Fe-based alloys, Ni-based alloys, Ti-based alloys, ceramics, niobium, tantalum, niobium, and gold under SCWO conditions.

Until now, researchers have conducted extensive experiments on the corrosion behavior of Ni-based alloys in SCW, and great progress has been made in this field. Hence, it is necessary to summarize, compare, and analyze these investigations, so as to guide the material selection and corrosion control in supercritical water technologies (i.e., SCWO, SCWG and SCWR). However, there is no any special, systematic report on the topic at the present time. Thus, in this present work, the development and application of Ni-based alloys, the types of corrosion of Ni-based alloys and the effects of major alloying elements are reviewed comprehensively. The corrosion characteristics of four types of Ni-based alloys in subcritical and supercritical water environments are summarized and compared systematically. The corrosion mechanisms of typical Ni-based alloys in supercritical water are analyzed in depth and further research directions are proposed. This information can provide significantly theoretical support for corrosion control and optimization of the design of Ni-based alloys potentially being employed in industrial implementation of various supercritical water technologies.

1.4 Development and Application of Ni-Based Corrosion-Resistant Alloy

Pure nickel is an expensive industrial raw material, of which the most important use is as an alloying element in stainless steels and in a variety of Ni-based alloys. Ni-based alloys generally possess higher strength and excellent corrosion resistance at 650–1000 °C, and can be divided into Ni-based heat-resistant alloys, Ni-based corrosion-resistant alloys, Ni-based wear-resistant alloys, Ni-based precision alloys, and Ni-based shape-memory alloys. In this work, the Ni-based alloys of interest mainly refer to the Ni-based corrosion-resistant alloys, which have good corrosion resistance, high strength, good plastic toughness, and good machinability. The most significant property of these Ni-based alloys is their corrosion resistance, which is generally dependent on chemical compositions and structure of the alloy substrate. Table 1.1 shows the classification of Ni-based alloys. According to the contents of Ni as well as Fe in the alloy, the Ni-based alloys can be divided into pure nickel, Ni-enriched corrosion-resistant alloys and Fe + Ni-based, corrosion-resistant alloys. The Ni-enriched corrosion-resistant alloys and the Fe + Ni-based corrosion-resistant alloys can further be classified into several series. The series that is suitable for subcritical and supercritical water service include Ni–Cr alloys, Ni–Cr-Fe alloys, Ni–Cr-Mo alloys, and Ni-Fe-Cr-Mo-Cu alloys. The present work will systematically summarize and analyze the corrosion resistances of these four kinds of alloys in supercritical water as well as in subcritical water.

Table 1.2 lists the Ni-based corrosion-resistant alloys commonly used in supercritical water. More specially, alloy classification, alloy abbreviation, commodity name, American brand name, and the chemical compositions can be seen here. It is worth noting that all of these alloys have high Ni (31–75 wt.%) and Cr (15–33

Table 1.1 Classification of Ni-based corrosion-resistant alloys

Classification	Matrix	Series
Pure nickel	Ni ($w_{Ni} = 100$ wt.%)	Pure Ni
Nickel-enriched corrosion-resistant alloy	Ni ($w_{Ni} \geq 50$ wt.%)	Ni-Cr alloy, Ni-Cu alloy, Ni-Mo alloy, Ni-Cr-Fe alloy, Ni-Cr-Mo alloy, and Ni-Cr-Mo-Cu alloy
Iron and Nickel based corrosion-resistant alloy	Ni + Fe ($w_{Ni+Fe} \geq 50$ wt.%)	Ni-Fe-Cr alloy, Ni-Fe-Cr-Mo alloy, Ni-Fe-Cr-Mo-Cu alloy, and Ni-Fe-Cr-Mo-Cu-N alloy

Note wt.% represents the weight content

Table 1.2 Ni-based corrosion-resistant alloys commonly used in supercritical water

Alloy classification	Alloy abbreviation	Commodity name	American brand	Main chemical composition (wt.%)
Ni-Cr alloys	600	Incoloy 600	UNS No.6600	75Ni-15Cr-8Fe-0.1C
	601	Incoloy 601	UNS No.6601	60Ni-23Cr-14Fe-1.5Al-0.1C
	X-750	Inconel X-750	–	70Ni-15Cr-3Ti-1Nb-0.7Al
	690	Inconel 690	UNS No.6690	60Ni-30Cr-10Fe-0.05C
Ni-Cr-Mo alloys	C-276	Hastelloy C-276	UNS No.10276	60Ni-16Cr-16Mo-4 W-0.08Si-0.02C
	C-22	Hastelloy C-22	UNS No.6022	60Ni-22Cr-13Mo-3 W-0.08Si-0.015C
	C-2000	Hastelloy C-2000	UNS No.6200	57Ni-23Cr-16Mo-2Cu-0.088Si-0.01C
	625	Inconel 625	UNS No.6625	60Ni-22Cr-9Mo-4Nb-0.4Al-0.4Ti-0.1C
	718	Inconel 718	UNS No.7718	52Ni-19Cr-3Mo-5Nb-0.35Ti-0.5Al-0.08C
Ni-Fe-Cr alloys	800	Incoloy 800	UNS No.8800	Fe-21Cr-32Ni-0.4Al-0.4Ti-0.10C
	800H	Incoloy 800H	UNS No.8810	Fe-21Cr-32Ni-0.4Al-0.4Ti-0.07C
	800HT	Incoloy 800HT	UNS No.8811	Fe-21Cr-32Ni-0.45Al-0.45Ti-0.06C
Ni-Fe-Cr-Mo-Cu alloys	825	Incoloy 825	UNS No.8825	42Ni-21Cr-3Mo-2Cu-0.8Ti-0.05C
	33	Nicrofer 3033	UNS No.R20033	Fe-33Cr-31Ni-1Mo-0.7Cu-0.5 N-0.15C

wt.%) contents, implying their significantly beneficial effects on corrosion resistance. Overall, the most widely studied and used alloys include Inconel 690, Hastelloy C-276, Inconel 625, Incoloy 800, and Incoloy 825 according to the current literature.

1.4.1 Development of Ni-Based Alloys

The first Ni -Cr alloy developed in 1931 was Inconel 600 with a Cr content of 15 wt.%, which has good corrosion resistance, good resistance to high-temperature oxidation and sulfidation, in which the maximum service temperature can reach up to 1200 °C. However, severe intergranular stress corrosion cracking occurs in Inconel 600 in high-temperature, high-pressure water [88], such as in the primary coolant circuits of PWRs. Inconel 690 with a higher Cr content is an improved grade of Inconel 600, with an increased Cr content and has better resistance to stress corrosion cracking. Another important development of the Ni-Cr alloys is the anti-oxidation and aging-hardened treatments by adding Al, such as Inconel 601 or Inconel X-750. The Ni-Cr-Mo alloys are the most widely used Ni-based alloys, of which the early grade is Hastelloy C introduced in the early 1930s. This alloy can withstand the corrosion from oxidizing or reducing acid, but is commonly accompanied by intergranular corrosion. The precipitation of carbides (M_6C, M represents the alloying elements) and intermetallic phases with a high Mo content is responsible for the sensitivity to intergranular corrosion. Fortunately, Hastelloy C-276, developed in 1965, can effectively alleviate the intergranular corrosion problem caused by M_6C precipitation. Later, Hastelloy C-22 developed in 1982 not only inhibits intergranular corrosion caused by the precipitation of M_6C and μ phase, but also has very good resistance to general corrosion, pitting corrosion, and crevice corrosion [93–97]. The representative grade of the Ni-Fe-Cr alloy is Incoloy 800, which has good oxidation resistance. The Ni-Fe-Cr-Mo-Cu alloy has the largest output, the most extensive application, and the most alloy grades among all Fe + Ni-based corrosion-resistant alloys. Incoloy 825, which first appeared in 1950, has good corrosion resistance in H_3PO_4, H_2SO_4, and HNO_3 solutions. Furthermore, Nicrofer 3033, developed in 1995, has not only excellent corrosion resistance in concentrated HNO_3 or H_2SO_4 solution, but also good formability and weldability.

1.4.2 Applications of Ni-Based Alloys

1.4.2.1 Applications in Supercritical Water Oxidation Reactor

The commonly used materials for SCWO systems are Ni-based alloys and austenitic stainless steels, whereas the austenitic stainless steels are only suitable for relatively mild feedstocks and low temperature conditions. Ni-based alloys are well known for their superior resistance to high-temperature oxidation, good corrosion resistance and high temperature strength over a wide range of conditions [54, 98]. Thus, Ni-based alloys are the most widely studied and used as candidate materials of fabrication of high-temperature equipment in SCWO [55]. In the supercritical region from 380 to 480 °C, Ni-based alloys can be adopted to construct supercritical water preheaters when the feedstock contains aggressive substances such as HCl, H_3PO_4 and H_2SO_4 [55]. Ni-based alloys are extensively used in SCWO reactors and play important roles in corrosion control. For example, a continuous tubular reactor has been manufactured from Hastelloy C-276, which can withstand the conditions of 600 °C and 30 MPa [99]. Apart from the tubular reactor, many other types of SCWO reactors have been developed to solve the problems regarding corrosion, salt deposition, and heat transfer in tubular reactors with a high salt organic feedstock. Transpiring wall reactors consists of a pressure-bearing wall and a porous wall which is usually a sintered stainless steel or Ni-based alloy [100]. Cool wall reactor is composed of two concentric tubes, in which the inner tube confining reactions and the other tubes are separately made of a Ni-Cr alloy or an austenitic stainless steel [101]. Veriansyah et al. [102] developed a double wall reactor, in which the inner non-porous wall and the outer pressure-bearing wall are made of titanium and Hastelloy C-276, respectively. Moreover, due to the failure to find out a super material that can withstand corrosion under all conditions in SCWO, a common method is to emplace a renewable liner on the inner surface of the reactor to avoid the direct contact of reactants with the inner wall [103]. Overall, in a complex reactor design, the Ni-based alloy can serve as not only a corrosion barrier (e.g., lining or coating), but also the substrate and pressure-bearing wall material [54].

1.4.2.2 Applications in Supercritical Water Gasification Reactor

To date, researchers have performed many SCWG experimental studies on different types of reactors, such as tubular reactors, batch reactors, continuous bed reactors, fluidized bed reactors, microchannel reactors, and packed bed reactors. These reactor employ corrosion-resistant materials, including Types 316and 304 SS, Inconel 600, Inconel 625, and Hastelloy C-276, etc., and used catalysts involve NaOH, $Ni/SiAl_2O_3$, Ru/C, Ru/Al_2O_3, and Ni/AC, etc. [104]. Taking the SCWG of glucose (one of biomass model compounds), as an example, many reactors are made of Ni-based alloys. Onwudili et al. [105] studied the effects of NaOH on hydrogen production during the SCW gasification of glucose and other biomass samples, and carried out experiments in a 75 ml batch reactor made of Hastelloy-C alloy with the highest working conditions of 600 °C and 45 MPa. Lee et al. [106] performed glucose SCWG with a Ni/activated charcoal catalyst in a packed-bed reactor made of a Hastelloy C-276 tube of 9.53 mm o.d., 6.22 mm i.d., and 670 mm length. Youssef et al. [107] conducted supercritical water gasification and partial oxidation experiments of a glucose solution under different catalyst supports and Ni loadings in a batch reactor (made of Hastelloy C-276) with the highest working conditions of 600 °C and 41.4 MPa. Lee [108], used a packed-bed reactor made of a Hastelloy C-276 tube of 9.53 mm o.d., 6.22 mm i.d., and 293 mm total length to study the catalytic activities of activated carbon supported Ni-based catalysts on glucose SCWG at 650 °C and 28 MPa. In addition to the experiments in Hastelloy alloy reactors, some tests have been implemented in reactors made of Inconel series alloys. For example, Byrd et al. [109] performed catalytic glucose SCWG in a tubular reactor made of Inconel 600 with 6.35 mm o.d., 3.05 mm i.d and 500 mm length. Zhang et al. [110] investigated the catalytic activities and stabilities of 17 heterogeneous catalysts during glucose SCWG at 600 °C and 24 MPa in a bench-scale continuous down-flow tubular reactor (made of Inconel 625 tube) with 9.55 mm o.d., 6.34 mm i.d., and 472 mm length. Overall, Hastelloy C-276, Inconel 600, and Inconel 625 are used extensively as SCWG reactor materials.

1.4.2.3 Applications in Supercritical Water-Cooled Reactor

The requirements for SCWR (supercritical water-cooled reactor) materials include good mechanical properties (e.g., strength, fracture toughness), good corrosion resistance, good neutron radiation resistance, low stress corrosion cracking sensitivity, and microstructure stability. Austenitic steels, ferritic-martensitic steels, and Ni-based alloys are considered as the candidate materials for SCWR systems, because SCW is highly corrosive to structural materials [111–113]. Incoloy 800 is a candidate material for the design of the Generation IV SCWR nuclear plant [114]. Besides, Inconel 625, Inconel 617/718, Incoloy 800H, and Hastelloy C-276 are also good candidates, due to their good corrosion resistance in high temperature water [115]. Inconel 690 has become the material of the steam generator tube and of other nuclear reactor structural components, due to its high temperature strength, good ductility, high creep rupture strength, oxidation resistance, and resistance to stress corrosion cracking. Owing to the intergranular stress corrosion cracking of Inconel 600 in a long-term operation, Inconel 690 is used as a substitute for steam generator tubing material in pressurized water reactors operating at 275–375 °C and 20 MPa [116]. Meanwhile, Inconel 690 is also a candidate material for the Generation IV SCWR nuclear plants [117]. Furthermore, Inconel 617 is also considered as the heat exchanger material for the next-generation nuclear power plant that includes the concept of a very high-temperature reactor [118].

1.4.2.4 Applications in Some Other Equipment

As an early development and application of Ni-Cr corrosion-resistant alloys, Inconel 600 is widely used in the chemical processing field, because of its simple composition, easy production, and relatively low price. It is also adopted in heat treatment equipment, nuclear fuel production equipment, and nuclear power equipment. Inconel X-750 has high strength, good corrosion resistance and radiation stability in high temperature water, so it can be used in the fabrication of bolts, springs, locating pins, and seal rings of the pressure shell in light water nuclear reactors. It can also serve as a heat-resistant alloy for producing rocket engine components, and rotor impellers of gas turbine, etc. Inconel 690 is a new generation of corrosion-resistant alloy with excellent stress corrosion resistance, which is mainly utilized in heat transfer tubes of steam generators in pressurized water reactor (PWR) nuclear power plants. Inconel 625 has high corrosion resistance and high thermal strength, and is usually used to manufacture thin-walled containers and pipe fittings, corrosion-resistant tubing, reactors, distillation columns, and heat exchangers. Hastelloy C-276 is used in phosphoric acid production, hydrofluoric acid production, nuclear fuel production, and flue gas desulfurization systems. Incoloy 800, as the representative grade of Ni-Fe-Cr alloys (see Table 1.2), has the advantages of wide applicability and good resistance to stress corrosion cracking. It is commonly used to manufacture the steam generators (boilers, evaporators) of pressurized water reactors, high temperature gas-cooled reactor, heat exchangers of sodium-cooled fast reactors, and superheated steam pipes in the nuclear power industry. Incoloy 825 has good processing performance and good resistance to general corrosion and localized corrosion and it is extensively adopted in fields such as chemical processing, energy production, petrochemical industries, hydrometallurgical industries, and deep source gas wells. Incoloy 825 can also be used as the material of heat exchangers and pipeline valves in environments dealing with hot sulfuric acid, chloride-containing solutions, and sulphurous acid, etc.

1.5 Alloys Used in Subcritical Water in Nuclear Reactors

As the core component of nuclear reactors, fuel elements release energy ceaselessly through chain nuclear fission and they are contained in cladding tubes. The fuel cladding which is regarded as the first barrier, plays an important role in preventing the leakage of fission products, hindering the fuel from being washed by coolant and effectively exporting thermal energy. Therefore, maintaining high reliability and integrity of fuel cladding is a significant prerequisite for the safe operation of nuclear reactors.

During the actual working of nuclear reactors, the cladding is subjected to complex operating conditions. The inner wall of the cladding tubes interacts with fuel pellets, which results in stress corrosion cracking, and hydrogen embrittlement occurs in the presence of hydrogen. The outer wall suffers from high temperature, high pressure and intense neutron irradiation, meanwhile, chemical corrosion easily appears due to its exposure to the coolant [119]. Since the Fukushima nuclear power plant accident, accident tolerant fuel claddings have been widely studied. In origin, Zircaloy is successful in large-scale commercialization due to good properties. However, the occurrence of both debris fretting and shadow corrosion caused by Zircaloy starves for new tolerant claddings [120]. Subsequently, many durable alloys are developed and investigated. Among them, FeCrAl alloys have become the research hotspot, due to their high strength, excellent formability and sterling oxidation resistance in high temperature steam in case of coolant loss accident or more severe accident [121].

When FeCrAl alloys are applied as fuel cladding, their corrosion resistance is connected with fuel safety and performance, and so becoming a significant benchmark in light water reactors. Nevertheless, a consensus has not been reached presently on its specific corrosion behavior under corrosive conditions although many reports have existed. Consequently, this paper focuses on the corrosion behavior of FeCrAl alloys, whose oxidation kinetics, corrosion process and influencing factors are reviewed through systematical comparison with that of Zircaloy. Challenges and developing trend of FeCrAl alloys used as fuel cladding are proposed. To the best of our knowledge, this is documented for the first time to provide a detailed review on the corrosion behavior of FeCrAl cladding in nuclear reactors. This work delivers valuable information for the further corrosion studies of FeCrAl alloys, and provides reference for the continuous development of new nuclear cladding materials and coatings.

References

1. M.C. Kutney, Thermodynamic and transport property modeling in supercritical water. Ph.D. Thesis (2005)
2. M.D. Bermejo, M.J. Cocero, Supercritical water oxidation: a technical review. AlChE J. **52**, 3933–3951 (2006)
3. T. Tassaing, Y. Danten, M. Besnard, Infrared spectroscopic study of hydrogen-bonding in water at high temperature and pressure. J. Mol. Liq. **101**, 149–158 (2002)
4. P.E. Savage, Organic chemical reactions in supercritical water. Chem. Rev. **99**, 603–622 (1999)
5. J. Tester, H. Holgate, F. Armellini, P. Webley, W. Killilea, G. Hong, H. Barner, In Emerging technologies in hazardous waste management III. ACS Symposium Series (1993), p. 25–76
6. P.A. Marrone, Supercritical water oxidation—Current status of full-scale commercial activity for waste destruction. J Supercrit. Fluids **79**, 283–288 (2013)
7. Y. Bai, Z. Wang, L. Feng, Chemical recycling of carbon fibers reinforced epoxy resin composites in oxygen in supercritical water. Mater. Design **31**, 999–1002 (2010)
8. P.E. Savage, Heterogeneous catalysis in supercritical water. Catal. Today **62**, 167–173 (2000)
9. M. Watanabe, M. Mochiduki, S. Sawamoto, T. Adschiri, K. Arai, Partial oxidation of n-hexadecane and polyethylene in supercritical water. J. Supercrit. Fluids **20**, 257–266 (2001)
10. J. Li, S. Wang, Y. Li, L. Wang, T. Xu, Y. Zhang, Z. Jiang, Supercritical water oxidation of semi-coke wastewater: effects of operating parameters, reaction mechanism and process enhancement. Sci. Total Environ. **710**, 134396 (2019)
11. J. Zhang, S.Z. Wang, Y.H. Li, J.L. Lu, S.L. Chen, X.Q. Luo, Supercritical water oxidation treatment of textile sludge. Environ. Technol. **38**(15), 1949–1960 (2017)
12. Y. Li, S. Wang, M. Ren, J. Zhang, D. Xu, L. Qian, P. Sun, Recent advances on research and application on supercritical hydrothermal combustion technology. Chem. Ind. Eng. Pro. **35**, 1942–1955 (2016)
13. P.J. Crooker, K.S. Ahluwalia, Z. Fan, Operating results from supercritical water oxidation plants. Ind. Eng. Chem. Res. **39**, 4865–4870 (2000)
14. E. Fauvel, C. Joussot-Dubien, P. Guichardon, G. Charbit, F. Charbit, S. Sarrade, A double-wall reactor for hydrothermal oxidation with supercritical water flow across the inner porous tube. J. Supercrit. Fluids **28**, 47–56 (2004)
15. M. Goto, T. Nada, A. Ogata, A. Kodama, T. Hirose, Supercritical water oxidation for the destruction of municipal excess sludge and alcohol distillery wastewater of molasses. J. Supercrit. Fluids **13**, 277–282 (1998)

16. M. Svanström, M. Fröling, M. Modell, W.A. Peters, J. Tester, Environmental assessment of supercritical water oxidation of sewage sludge. Resour. Conserv. Recycl. **41**, 321–338 (2004)
17. A. Gidner, L. Stenmark, Supercritical water oxidation of sewage sludge—State of the art. Chematur Engineering AB (2001), p. 430
18. V. Bambang, K. Jae-Duck, Supercritical water oxidation for the destruction of toxic organic wastewaters: a review. JEnvS **19**, 513–522 (2007)
19. A. Shanableh, E.F. Gloyna, Supercritical water oxidation-wastewaters and sludges. Water Sci. Technol. **23**, 389–398 (1991)
20. V. Marulanda, G. Bolaños, Supercritical water oxidation of a heavily PCB-contaminated mineral transformer oil: laboratory-scale data and economic assessment. J. Supercrit. Fluids **54**, 258–265 (2010)
21. K. Hatakeda, Y. Ikushima, O. Sato, T. Aizawa, N. Saito, Supercritical water oxidation of polychlorinated biphenyls using hydrogen peroxide. Chem. Eng. Sci. **54**, 3079–3084 (1999)
22. J. Abeln, M. Kluth, G. Petrich, H. Schmieder, Supercritical water oxidation (SCWO): a process for the treatment of industrial waste effluents. HPR **20**, 537–547 (2001)
23. J. Yang, S. Wang, Y. Li, Y. Zhang, D. Xu, Novel design concept for a commercial-scale plant for supercritical water oxidation of industrial and sewage sludge. J. Environ. Manage. **233**, 131–140 (2019)
24. P.A. Marrone, S.D. Cantwell, D.W. Dalton, SCWO system designs for waste treatment: application to chemical weapons destruction. Ind. Eng. Chem. Res. **44**, 9030–9039 (2005)
25. Y. Matsumura, T. Minowa, B. Potic, S.R.A. Kersten, W. Prins, W.P.M.v. Swaaij, B.v.d. Beld, D.C. Elliott, G.G. Neuenschwander, A. Kruse, M.J.A. Jr. Biomass gasification in near- and super-critical water: Status and prospects. Biomass Bioenergy **29**, 269–292 (2005)
26. G. Li, Y. Lu, Cyclone separation in a supercritical water circulating fluidized bed reactor for coal/biomass gasification: Structural design and numerical analysis. Particuology **39**, 55–67 (2018)
27. S.N. Reddy, S. Nanda, A.K. Dalai, J.A. Kozinski, Supercritical water gasification of biomass for hydrogen production. Int. J. Hydrogen Energy **39**, 6912–6926 (2014)
28. S. Guo, L. Guo, C. Cao, J. Yin, Y. Lu, X. Zhang, Hydrogen production from glycerol by supercritical water gasification in a continuous flow tubular reactor. Int. J. Hydrogen Energy **37**, 5559–5568 (2012)
29. R. Samiee-Zafarghandi, J. Karimi-Sabet, M.A. Abdoli, A. Karbassi, Supercritical water gasification of microalga Chlorella PTCC 6010 for hydrogen production: box-Behnken optimization and evaluating catalytic effect of MnO_2/SiO_2 and NiO/SiO_2. Renew. Energy **126**, 189–201 (2018)
30. A. Demirbas, Hydrogen production from Biomass via supercritical water gasification. Energy Source Part A **32**, 1342–1354 (2019)
31. X. Hao, L. Guo, X. Zhang, Y. Guan, Hydrogen production from catalytic gasification of cellulose in supercritical water. Chem. Eng. J. **110**, 57–65 (2005)
32. D. Hendry, A. Miller, W. Jacoby, Turbulent operation of a continuous reactor for gasification of alcohols in supercritical water. Ind. Eng. Chem. Res. **51**, 2578–2585 (2012)
33. Y. Lu, S. Li, L. Guo, X. Zhang, Hydrogen production by biomass gasification in supercritical water over $Ni/\gamma Al_2O_3$ and Ni/CeO_2-γAl_2O_3 catalysts. Int. J. Hydrogen Energy **35**, 7161–7168 (2010)
34. D. Xu, S. Wang, X. Hu, C. Chen, Q. Zhang, Y. Gong, Catalytic gasification of glycine and glycerol in supercritical water. Int. J. Hydrogen Energy **34**, 5357–5364 (2009)
35. Y. Guo, S.Z. Wang, D.H. Xu, Y.M. Gong, H.H. Ma, X.Y. Tang, Review of catalytic supercritical water gasification for hydrogen production from biomass. Renew. Sustain. Energ Rev. **14**, 334–343 (2010)
36. J. Chen, Y. Lu, L. Guo, X. Zhang, P. Xiao, Hydrogen production by biomass gasification in supercritical water using concentrated solar energy: system development and proof of concept. Int. J. Hydrogen Energy **35**, 7134–7141 (2010)
37. Y.J. Lu, L.J. Guo, C.M. Ji, X.M. Zhang, X.H. Hao, Q.H. Yan, Hydrogen production by biomass gasification in supercritical water: A parametric study. Int. J. Hydrogen Energy **31**, 822–831 (2006)

38. I. Betova, M. Bojinov, P. Kinnunen, S. Penttila, T. Saario, Surface film electrochemistry of austenitic stainless steel and its main constituents in supercritical water. J. Supercrit. Fluids **43**, 333–340 (2007)
39. M. Granovskii, I. Dincer, M.A. Rosen, I. Pioro, Performance assessment of a combined system to link a supercritical water-cooled nuclear reactor and a thermochemical water splitting cycle for hydrogen production. Energy Convers. Manage. **49**, 1873–1881 (2008)
40. D. Guzonas, F. Brosseau, P. Tremaine, J. Meesungnoen, J.P. Jay-Gerin, Water chemistry in a supercritical water-cooled pressure tube reactor. Nucl. Technol. **179**(2), 205–219 (2017)
41. M. Granovskii, I. Dincer, M.A. Rosen, I. Pioro, Thermodynamic analysis of the use a chemical heat pump to link a supercritical water-cooled nuclear reactor and a thermochemical water-splitting cycle for hydrogen production. CSEE J. Power Energy **2**, 756–767 (2008)
42. T. Schulenberg, J. Starflinger, P. Marsault, D. Bittermann, C. Maráczy, E. Laurien, J.A.L.à. Nijeholt, H. Anglart, M. Andreani, M. Ruzickova, A. Toivonen, European supercritical water cooled reactor. Nucl. Eng. Des. **241**, 3505–3513 (2011)
43. J. Yoo, Y. Ishiwatari, Y. Oka, J. Liu, Conceptual design of compact supercritical water-cooled fast reactor with thermal hydraulic coupling. Ann. Nucl. Energy **33**, 945–956 (2006)
44. X. Cheng, B. Kuang, Y.H. Yang, Numerical analysis of heat transfer in supercritical water cooled flow channels. Nucl. Eng. Des. **237**, 240–252 (2007)
45. D. Danielyan, Supercritical-water-cooled reactor system - as one of the most promising type of generation IV nuclear reactor systems. Ineel Projects Report (2003)
46. C. Fazio, A. Alamo, A. Almazouzi, S.D. Grandis, D. Gomez-Briceno, J. Henry, L. Malerba, M. Rieth, European cross-cutting research on structural materials for generation iv and transmutation systems. J. Nucl. Mater. **392**, 316–323 (2009)
47. Y. Zhang, S. Wang, W. Song, J. Yang, T. Xu, J. Li, C. Yang, Y. Li, Characteristics of sodium sulfate deposition in hydrogen production from supercritical water gasification: a review. Int. J. Hudrogen Energy **44**, 29467–29482 (2019)
48. V. Vadillo, J. Sánchez-Oneto, J.R.n. Portela, E.J.M.d.l. Ossa, Problems in supercritical water oxidation process and proposed solutions. Ind. Eng. Chem. Res. **52**, 7617–7629 (2013)
49. P.A. Marronea, M. Hodesb, K.A. Smithc, J.W. Tester, Salt precipitation and scale control in supercritical water oxidation—part B: commercial/full-scale applications. J. Supercrit. Fluids **29**, 289–312 (2004)
50. P. Kritzer, E. Dinjus, An assessment of supercritical water oxidation (SCWO) Existing problems, possible solutions and new reactor concepts. Chem. Eng. J. **83**, 207–214 (2001)
51. D. Xu, C. Huang, S. Wang, G. Lin, Y. Guo, Salt deposition problems in supercritical water oxidation. Chem. Eng. J. **279**, 1010–1022 (2015)
52. D. Xu, S. Wang, C. Huang, X. Tang, Y. Guo, Transpiring wall reactor in supercritical water oxidation. Chem. Eng. Res. Des. **92**, 2626–2639 (2014)
53. P. Kritzer, Corrosion in high-temperature and supercritical water and aqueous solutions: a review. J. Supercrit. Fluids **29**, 1–29 (2004)
54. P.A. Marrone, G.T. Hong, Corrosion control methods in supercritical water oxidation and gasification processes. J. Supercrit. Fluids **51**, 83–103 (2009)
55. X. Tang, S. Wang, L. Qian, M. Ren, P. Sun, Y. Li, J. Yang, Corrosion properties of candidate materials in supercritical water oxidation process. J. Adv. Oxid. Technol. **19**, 141–157 (2016)
56. G. Brunner, Near and supercritical water. Part II: Oxidative processes. J. Supercrit. Fluids, **47**(3), 382–390 (2009)
57. Y. Ashida, Y. Daigo, K. Sugahara, An industrial perspective on environmentally assisted cracking of some commercially used carbon steels and corrosion-resistant alloys. JOM **69**, 1381–1388 (2017)
58. Y.J. Lu, H. Jin, L.J. Guo, X.M. Zhang, C.Q. Cao, X. Guo, Hydrogen production by biomass gasification in supercritical water with a fluidized bed reactor. Int. J. Hydrogen Energy **33**, 6066–6075 (2008)
59. H. Jin, Y. Lu, L. Guo, C. Cao, X. Zhang, Hydrogen production by partial oxidative gasification of biomass and its model compounds in supercritical water. Int. J. Hydrogen Energy **35**, 3001–3010 (2010)

60. L.J. Guo, Y.J. Lu, X.M. Zhang, C.M. Ji, Y. Guan, A.X. Pei, Hydrogen production by biomass gasification in supercritical water: a systematic experimental and analytical study. Catal. Today **129**, 275–286 (2007)
61. L. Blomeley, J. Pencer, B. Hyland, F.P. Adams, Nuclear data sensitivity and uncertainty for the Canadian supercritical water-cooled reactor. Ann. Nucl. Energy **63**, 587–593 (2014)
62. D. Guzonas, M. Edwards, W. Zheng, Assessment of candidate fuel cladding alloys for the canadian supercritical water-cooled reactor concept. J. Nucl. Eng. Radiat. Sci. **2**(1), 011016 (2015)
63. S. Baindur, Materials challenges for the supercritical water-cooled reactor (SCWR). Bull. Can. Nucl. Soc. **29**, 32–38 (2008)
64. D. Guzonas, R. Novotny, Supercritical water-cooled reactor materials—summary of research and open issues. Prog. Nucl. Energy **77**, 361–372 (2014)
65. T. Reiss, G. Csom, S.F. r, S. Czifrus, The simplified supercritical water-cooled reactor (SSCWR), a new SCWR design. Prog. Nucl. Energy **52**, 177–189 (2010)
66. T.R. Allen, K. Sridharan, L. Tan, W.E. Windes, J.I. Cole, D.C. Crawford, G.S. Was, Materials challenges for generation IV nuclear energy systems. Nucl. Technol. **162**(3), 342–357 (2017)
67. K. Ehrlich, J. Konys, L. Heikinheimo, K. Ehrlich, J. Konys, L. Heikinheimo, Materials for high performance light water reactors. J. Nucl. Mater. **327** (2–3), 140–147 (2004)
68. Y. Li, T. Xu, S. Wang, J. Yang, J. Li, T. Xu, D. Xu, Characterization of oxide scales formed on heating equipment in supercritical water gasification process for producing hydrogen. Int. J. Hydrogen Energy **44**, 29508–29515 (2019)
69. Y. Li, T. Xu, S. Wang, B. Fekete, J. Yang, J. Yang, J. Qiu, A. N. Xu, J. Wang, Y. Xu, D.D. Macdonald, Modelling and analysis of the corrosion characteristics of ferritic-martensitic steels in supercritical water. Materials **12**, 409 (2019)
70. Y. Li, S. Wang, P. Sun, J. Yang, X. Tang, D. Xu, Y. Guo, J. Yang, D.D. Macdonald, Investigation on early formation and evolution of oxide scales on ferritic–martensitic steels in supercritical water. Corros. Sci. **135**, 136–146 (2018)
71. T. Xu, Y. Li, Comparative analysis on corrosion behavior of 316SS and 304SS in subcritical and supercritical aqueous systems. Key Eng. Mater. **777**, 381–385 (2018)
72. Y. Li, S. Wang, P. Sun, D. Xu, M. Ren, Y. Guo, G. Lin, Early oxidation mechanism of austenitic stainless steel TP347H in supercritical water. Corros. Sci. **128**, 241–252 (2017)
73. Y. Li, S. Wang, P. Sun, Z. Tong, D. Xu, Y. Guo, J. Yang, Early oxidation of Super304H stainless steel and its scales stability in supercritical water environments. Int. J. Hydrogen Energy **41**, 15764–15771 (2016)
74. Y. Li, S. Wang, L. Wang, P. Yu, J. Yang, Corrosion behavior of austenitic steel 304 in nearcritical aqueous solutions. Adv. Eng. Res. **45**, 189–192 (2015)
75. Y.H. Li, S.Z. Wang, X.D. Li, J.M. Lu, Corrosion of an austenitic heat-resistant steel hr3c in high-temperature steam and supercritical water. Adv Mater. Res. **908**, 67–71 (2014)
76. Y.H. Li, S. Wang, X.D. Li, M.M. Ren, Analysis on the structure and mechanism of cracking and spalling for super304H steel oxide films in high-temperature steam. Adv. Mater. Res. **908**, 72–76 (2014)
77. Y. Li, S. Wang, J. Yang, Y. Zhang, D. Xu, Y. Guo, Effect of salt deposits on corrosion behavior of Ni-based alloys in supercritical water oxidation of high salinity organic wastewater. J. Environ. Eng. **145**, 04019080 (2019)
78. J. Yang, S. Wang, X. Tang, Y. Wang, Y. Li, Effect of low oxygen concentration on the oxidation behavior of Ni-based alloys 625 and 825 in supercritical water. J. Supercrit. Fluids **131**, 1–10 (2018)
79. J. Yang, S. Wang, D. Xu, Y. Guo, C. Yang, Y. Li, Effect of ammonium chloride on corrosion behavior of ni-based alloys and stainless steel in supercritical water gasification process. Int. J. Hydrogen Energy **42**, 19788–19797 (2017)
80. Y. Li, S. Wang, J. Yang, D. Xu, Y. Guo, L. Qian, W. Song, Corrosion characteristics of a nickel-base alloy C-276 in harsh environments. Int. J. Hydrogen Energy **42**, 19829–19835 (2017)

81. Y. Li, S. Wang, X. Tang, D. Xu, Y. Guo, J. Zhang, L. Qian, Effects of sulfides on the corrosion behavior of Inconel 600 and Incoloy 825 in supercritical water. Oxid. Met. **84**, 509–526 (2015)
82. X.Y. Tang, S.Z. Wang, Y.H. Li, L.L. Qian, Influence of alloying element on corrosion behavior of Hastelloy C-276in supercritical water. Appl. Mech. Mater. **525**, 40–44 (2014)
83. C. Sun, R. Hui, W. Qu, S. Yick, Progress in corrosion resistant materials for supercritical water reactors. Corros. Sci. **51**, 2508–2523 (2009)
84. G. Gupta, P. Ampornrat, X. Ren, K. Sridharan, T.R. Allen, G.S. Was, Role of grain boundary engineering in the SCC behavior of ferritic–martensitic alloy HT-9. J. Nucl. Mater. **361**, 160–173 (2007)
85. K.L. Murty, I. Charit, Structural materials for Gen-IV nuclear reactors: challenges and opportunities. J. Nucl. Mater. **383**, 189–195 (2008)
86. W. Qin, C. Nam, H.L. Li, J.A. Szpunar, Tetragonal phase stability in ZrO_2 film formed on zirconium alloys and its effects on corrosion resistance. Acta Mater. **55**, 1695–1701 (2007)
87. D.B. Mitton, J.-H. Yoon, J.A. Cline, H.-S. Kim, N. Eliaz, R.M. Latanision, Corrosion behavior of nickel-based alloys in supercritical water oxidation systems. Ind. Eng. Chem. Res. **39**, 4689–4696 (2000)
88. W.Z. Friend, *Corrosion of Nickel and Nickel-Base Alloys*, vol. 127(8) (Wiley, New York, 1980), p. 421
89. P. Kritzer, N. Boukis, E. Dinjus, Review of the corrosion of nickel-based alloys and stainless steels in strongly oxidizing pressurized high-temperature solutions at subcritical and supercritical temperatures. NACE Int. **56**, 1093–1104 (2000)
90. D.B. Mitton, N. Eliaz, J.A. Cline, R.M. Latanision, An overview of the current understanding of corrosion in SCWO systems for the destruction of hazardous waste products. Mater. Technol. **16**, 30–53 (2001)
91. N. Eliaz, D.B. Mitton, R.M. Latanision, Review of materials issues in supercritical water oxidation systems and the need for corrosion control. Trans. Indian Inst. Met. **56**, 305–314 (2003)
92. D.A. Guzonas, W.G. Cook, Cycle chemistry and its effect on materials in a supercritical water-cooled reactor: a synthesis of current understanding. Corros. Sci. **65**, 48–66 (2012)
93. D.C. Agarwal, W.R. Herda, The "C" family of Ni-Cr-Mo alloys' partnership with the Chemical Process Industry: the last 70 years. Mater. Corros. **48**(8), 542–548 (1997)
94. R.B. Leonard, Thermal stability of hastelloy alloy C-276. Corrosion **25**(5), 222 (1969)
95. R.W. Khchner, F.G. Hodge, New Ni-Cr-Mo alloy demonstrates high-temperature structural stability with resultant increases in corrosion-resistance and mechanical properties. Mater. Corros. **24**(12), 1042–1049 (1973)
96. F.G. Hodge, Effect of aging on the anodic behavior of Ni-Cr-Mo alloys. Corrosion **29**(10), 375–383 (1973)
97. F.G. Hodge, R.W. Kirchner, An improved Ni-Cr-Mo alloy for corrosion service. Corrosion **32**(8), 332–336 (1974)
98. D. Xu, S. Wang, J. Zhang, X. Tang, Y. Guo, C. Huang, Supercritical water oxidation of a pesticide wastewater. Chem. Eng. Res. Des. **94**, 396–406 (2015)
99. D. Xu, S. Wang, C. Huang, X. Tang, Deposition properties of sodium carbonate in supercritical water in a continuous-flow tubular reactor. Adv. Mater. Res. **864–867**, 1172–1177 (2013)
100. D. Xu, S. Wang, C. Huang, X. Tang, Y. Guo, Transpiring wall reacor in supercritical water oxidation. Chem. Eng. Res. Des. **92**, 2626–2639 (2014)
101. S. Zhang, Z. Zhang, R. Zhao, J. Gu, J. Liu, O. Banu, J. Zhang, A review of challenges and recent progress in supercritical water oxidation of wastewater. Chem. Eng. Commun. **204**, 265–282 (2016)
102. B. Veriansyah, J.-D. Kim, J.-C. Lee, A double wall reactor for supercritical water oxidation: Experimental results on corrosive sulfur mustard simulant oxidation. J. Ind. Eng. Chem. **15**, 153–156 (2009)
103. Z. Chen, H. Chen, X. Liu, D. Yue, Y. Xu, An inclined plug-flow reactor design for supercritical water oxidation. Chem. Eng. J. **343**, 351–361 (2018)

104. R.F. Susanti, L.W. Dianningrum, T. Yum, Y. Kim, J. Kim, High-yield hydrogen production from glucose by supercritical water gasification without added catalyst. Int. J. Hydrogen Energy **37**, 11677–11690 (2012)
105. J.A. Onwudili, P.T. Williams, Role of sodium hydroxide in the production of hydrogen gas from the hydrothermal gasification of biomass. Int. J. Hydrogen Energy **34**, 5645–5656 (2009)
106. I.-G. Lee, S.-K. Ihm, Catalytic gasification of glucose over Ni/activated charcoal in supercritical water. Ind. Eng. Chem. Res. **48**, 1435–1442 (2009)
107. E.A. Youssef, M. Chowdhury, G. Nakhla, P. Charpentier, Effect of nickel loading on hydrogen production and chemical oxygen demand (COD) destruction from glucose oxidation and gasification in supercritical water. Int. J. Hydrogen Energy **35**, 5034–5042 (2010)
108. I.-G. Lee, Effect of metal addition to Ni/activated charcoal catalyst on gasification of glucose in supercritical water. Int. J. Hydrogen Energy **36**, 8869–8877 (2011)
109. A.J. Byrd, K.K. Pant, R.B. Gupta, Hydrogen production from glucose using ru/Al_2O_3 catalyst in supercritical water. Ind. Eng. Chem. Res. **46**, 3574–3579 (2007)
110. L. Zhang, P. Champagne, C.C. Xu, Screening of supported transition metal catalysts for hydrogen production from glucose via catalytic supercritical water gasification. Int. J. Hydrogen Energy **36**, 9591–9601 (2011)
111. S.H. Nie, Y. Chen, X. Ren, K. Sridharan, T.R. Allen, Corrosion of alumina-forming austenitic steel Fe–20Ni–14Cr–3Al–0.6Nb–0.1Ti in supercritical water. J. Nucl. Mater. **399**, 231–235 (2010)
112. M. Fulger, M. Mihalache, D. Ohai, S. Fulger, S.C. Valeca, Analyses of oxide films grown on AISI 304L stainless steel and Incoloy 800HT exposed to supercritical water environment. J. Nucl. Mater. **415**, 147–157 (2011)
113. Y. Li, T. Xu, S. Wang, J. Yang, Predictions and analyses on the growth behavior of oxide scales formed on ferritic–martensitic in supercritical water. Oxid. Met. **92**, 27–48 (2019)
114. M. Fulger, D. Ohai, M. Mihalache, M. Pantiru, V. Malinovschi, Oxidation behavior of incoloy 800 under simulated supercritical water conditions. J. Nucl. Mater. **385**, 288–293 (2009)
115. X. Zhong, X. Wu, E.-H. Han, Characteristics of Oxidation and Oxygen Penetration of alloy 690 in 600 °c aerated supercritical water. J. Mater. Sci. Technol. **34**, 561–569 (2018)
116. J.R. Crum, Stress corrosion cracking testing of Inconel alloys 600 and 690 under high-temperature caustic conditions. Corrosion **42**(6), 368–372 (1986)
117. T.R. Allen, K. Sridharan, L. Tan, W.E. Windes, J.I. Cole, D.C. Crawford, G.S. Was, Materials challenges for generation IV nuclear energy systems. Nucl. Technol. **162**, 342–357 (2017)
118. C. Sun, R. Hui, Q. Wei, S. Yick, Progress in corrosion resistant materials for supercritical water reactors. Corros. Sci. **51**, 2508–2523 (2009)
119. S. Suman, M.K. Khan, M. Pathak, R.N. Singh, J.K. Chakravartty, Rupture behaviour of nuclear fuel cladding during loss-of-coolant accident. Nucl. Eng. Des. **307**, 319–327 (2016)
120. R. B. Rebak. Chapter 5—FeCrAl–iron–chromium–aluminum monolithic alloys. Accid. Tolerant Mater. Light Water Reactor Fuels 83–141 (2020)
121. F. Ning, X. Wang, Y. Yang, J. Tan, Z. Zhang, D. Jia, X. Wu, E. Han, Uniform corrosion behavior of FeCrAl alloys in borated and lithiated high temperature water. J. Mater. Sci. Technol. **70**, 136–144 (2021)

Chapter 2
Corrosion Types and Elemental Effects of Ni-Based and FeCrAl Alloys

2.1 Corrosion Types of Ni-Based Alloys

The essence why Ni-based alloys have good corrosion resistance is that compact, protective oxide scales form on the alloy surface, in a corrosive environment, thereby inhibiting further corrosion. However, the presence of oxidant in SCWO environments leads to a higher electrochemical potential, which probably converts the trivalent chromium into hexavalent chromium during the passivation process of the Ni-based alloy. This “depassivation” dissolution process accelerates corrosion. Researchers have found that the corrosion types of Ni-based alloys in SCW primarily include general corrosion, pitting corrosion, crevice corrosion, intergranular corrosion, and environment-assisted fracture. There are also some other unusual corrosion types, such as dealloying corrosion, galvanic corrosion, and under-deposit corrosion. Pitting corrosion occurs when an aggressive anion, such as chloride, induces passivity breakdown [1–3]. When aggressive species uniformly attack the entire surface of the substrate, this process leads to general corrosion, which is highly predictable, unlike localized corrosion where theories and models that allow the prediction of accumulated damage are just now appearing [4]. Damage Function Analysis is now being used to predict general and localized corrosion damage such as pitting corrosion in a variety of practical systems [5]. Stress corrosion cracking is the result of the combination of high mechanical stress and environmental chemical factors, and generally displays no symptoms in advance, thereby frequently resulting in catastrophic failure or even in explosive failures, if it is not controlled well [6, 7]. Clearly, corrosion seriously affects the safe and reliable operation of equipment and to a certain degree hinders the commercial development of supercritical water technologies.

Ni-based corrosion-resistant alloys, to a certain extent, have exhibited the potential to solve the corrosion issues of the equipment installed in various SCW-related plants. However, present research shows that each material has special corrosion resistance and use conditions. Meanwhile, the corrosion susceptibility of Ni-based alloys in harsh environments varies with alloy type and the extent of alloying. The occurrence

D. Xu and S. Guo, *Corrosion Characteristics, Mechanisms and Control Methods of Candidate Alloys in Sub- and Supercritical Water*,
https://doi.org/10.1007/978-981-16-5525-8_2

of corrosion is closely related to the material properties and corrosive environment, and different corrosion types induce various effects and forms of damage on the material in use. In general, Hastelloy C-22 has good resistance to general corrosion, Inconel 690 and Incoloy 800 have good pitting corrosion resistance, and Inconel 625 is less prone to take place crevice corrosion than other alloys. The alloy sensitivity to intergranular corrosion is complex, depending sensitively upon the alloy composition, properties, and the environment. Although Ni-based alloys have good stress corrosion resistance in a wide range of corrosive environments, they are not immune to stress corrosion cracking. Therefore, much research is still required to overcome the stress corrosion cracking problem of Ni-based alloys and to develop enhanced resistance to stress corrosion cracking. It is also worth noting that Ni-based alloys often display a variety of corrosion types in SCW. For example, general corrosion, pitting corrosion, and nodule corrosion occur on Inconel 690 at the same time in SCW at 450–550 °C and 25 MPa [8].

2.1.1 General Corrosion

Uniform corrosion occurs on the surface of the metallic material exposed to corrosive environments via chemical and electrochemical reactions, a phenomenon that is also called “general corrosion”. Evaluation of general corrosion helps to predict the service life of equipment and components. For example, the assessment of general corrosion can determine if the candidate alloy can be used as a qualified fuel cladding material, and thus facilitates the design and analysis of fuel assemblies that are cooled by SCW in the SCWR [9]. The general corrosion rate of the thermally treated Inconel 690 exposed to high purity demineralized water containing 2.2 ppm Lithium and 1200 ppm boron (i.e., simulated primary water of a pressurized water reactor) at 330 °C, 15 MPa for 500 h significantly decreased with increasing dissolved hydrogen content, due to a shift in the corrosion potential to more negative values and to a decrease in the barrier layer oxide thickness and hence to a decrease in the passive current density (the corrosion current density). The general corrosion rate decreases rapidly by about 72% when the dissolved hydrogen content increases in the range of 5–35 $cm^3\ kg^{-1}$, while it decreases slowly in the range of 35–65 $cm^3\ kg^{-1}$ [10]. Hastelloy C-22 has good corrosion resistance, with a general corrosion rate of less than 10 $\mu m \cdot a^{-1}$ after a short-term immersion in concentrated brine (e.g., NaCl, KCl, $CaCl_2$, $NaNO_3$, and KNO_3) at 100–160 °C for 24 h [11].

Corrosion rate can be evaluated by a gravimetric method or a metal loss depth determination. The gravimetric method refers to the weight change of the sample in per unit area and per unit time, and can determine the average corrosion rate. The mathematical expression of the gravimetric method is

$$v_g = \frac{\Delta m}{St} \tag{2.1}$$

where $v_g(\mathrm{g \cdot m^{-2} \cdot h^{-1}})$ is the corrosion rate, $S(\mathrm{m^2})$ is the surface area of the sample, and $t\,(\mathrm{h})$ is the corrosion time.

If corrosion products exfoliate or are removed from the sample surface by dissolution [12], for example, the gravimetric method commonly exhibits corrosion weight loss,

$$\Delta m = m_1 - m_0 \tag{2.2}$$

where $\Delta m(\mathrm{g})$ is the weight change, $m_0(\mathrm{g})$ is the sample mass before corrosion, and $m_1(\mathrm{g})$ is the sample mass after corrosion product removal.

If corrosion products firmly adheres to the sample surface after corrosion, the gravimetric method exhibits a corrosion weight gain,

$$\Delta m = m_2 - m_0 \tag{2.3}$$

where $m_2(\mathrm{g})$ is the mass of the sample with corrosion products.

Further, the depth method is practically valuable, because the corrosion depth is reflected in the following formula. Thus, the corrosion rate, expressed by the depth method, can be estimated by

$$v_g = \frac{\Delta m}{St} = \frac{\Delta V\rho}{St} = \frac{S\Delta h\rho}{St} = \frac{v_d\rho}{t} \tag{2.4}$$

and the Equation above can be transformed into

$$v_d = \frac{v_g}{\rho} \tag{2.5}$$

where ΔV is the sample volume change due to corrosion, Δh is the sample thickness change before and after corrosion, $\rho(\mathrm{g \cdot cm^{-3}})$ is the sample density, v_d is the corrosion rate in terms of the corrosion depth. Assuming that the corrosion rate is constant, in one year v_d can be expressed in units of $\mathrm{mm \cdot a^{-1}}$.

Overall, weight change (or mass change) and corrosion rate are commonly used in the literature to evaluate general corrosion, and researchers often express $\frac{\Delta m}{S}$ as a function of t.

2.1.2 Pitting Corrosion

Localized corrosion is significantly different from general corrosion, and its corrosion rate or corrosion depth on part of alloy surface is much higher than the average value on the whole surface. Localized corrosion that occurs on Ni-based alloys includes pitting corrosion, crevice corrosion, and intergranular corrosion. In a corrosive environment, most of the metal surface does not corrode or corrodes very slightly, but some points appear as very small, etched holes or spots. These etched holes or spots continue to develop in the depth direction to form small pore shapes. This phenomenon is called as pitting corrosion. Pitting corrosion usually occurs on easily passivated metals in the case of aggressive anions (e.g., Cl^-) coexisting in the systems of interest. It is one type of significantly destructive localized corrosion, and is difficult to be predicted, although great advances have been made in recent years in developing highly deterministic methods for predicting the evolution of damage [13]. The major influencing factors in pitting corrosion include types and concentrations of aggressive anions, pH, temperature, and medium flow rate. In the chloride-containing sulphate solution ($NaCl + Na_2S_2O_3$), the pitting corrosion resistance of Inconel 600 is significantly lower than that of Inconel 690 or Incoloy 800, because Inconel 600 has a lower Cr content [14]. For pure nickel in NaOH solution, Cl^- destroys the passive film formed on the nickel electrode locally at the pitting corrosion potential or above, and subsequent chloride segregation into the acidifying pit cavity accelerates active dissolution and hence pit growth. This can be attributed to the formation of concentrated HCl in the cavity, with Cl^- being transported into the pit via the coupling current that flows from the pit cavity to the external surface where it is consumed by the prevailing partial cathodic reaction (commonly hydrogen evolution or oxygen reduction) and acidification resulting from cation hydrolysis within the cavity ($M^{n+} + nH_2O \rightarrow M(OH)_n + nH^+$) [13]. Organic amines can inhibit pitting corrosion, while urea inhibits pitting corrosion at a low concentration and promotes pitting corrosion at a high concentration [15]. The potentiodynamic polarization test in the NaCl solution shows that laser surface melting can improve the pitting corrosion resistance of Inconel 600. In addition, pitting corrosion initiates at the junction of TiN inclusions and the barrier layer matrix, or possibly from the sulfides contacting with TiN [16]. Experiments with Incoloy 800 in high temperature water at 300 °C for 100–5000 h indicate that pitting potential decreases with increasing temperature, but the preformed films in advance has no a significant effect on the pitting potential of alloys. The amount of etched pits on the surface of the prefilmed sample is much larger than that on the sample surface with the air formed oxide film [17].

Pitting resistance equivalent number (PREN) is commonly used to evaluate alloy corrosion resistance. A high PREN value means the better pitting corrosion resistance and crevice corrosion resistance of the alloy. The contents of Cr, Mo and N in an alloy have important influences on its pitting corrosion behavior [18], and alloys with high contents of Cr and Mo and a certain amount of N have large PREN values, meaning the better pitting corrosion resistance [19]. The PREN values of stainless steel and Ni-based alloy can be calculated using the following Equation [20].

$$\mathrm{PREN} = w_{\mathrm{Cr}} + 3.3 w_{\mathrm{Mo}} + 16 w_{\mathrm{N}} \tag{2.6}$$

where w is the mass fraction of an element (i.e., Cr, Mo or N). Researchers often use the electrochemical theory of corrosion to explain the pitting corrosion behaviors of an alloy, and adopt the potential dynamic method to measure the polarization curve of the alloy in a solution (usually containing Cl^-), from which the pitting potential can be obtained. When the local corrosion potential (E_{corr}) of the metal surface is higher than the pitting potential (E_{pit}), pitting corrosion develops spontaneously. The pitting corrosion resistance of an alloy decreases with increasing corrosion potential for $E_{corr} > E_{pit}$.

Most investigations focus on pitting corrosion behavior and mechanisms in normal temperature or high temperature solutions containing Cl^- at normal pressure. The differences in the pitting corrosion mechanisms in normal temperature water and in SCW are still unclear, and the effects of some corrosive anions (e.g., OH^- NO_3^-, and SO_4^{2-}) on the pitting corrosion of Ni-based alloys in SCW have not been reported extensively. Highly destructive and dangerous pitting corrosion causes not only equipment damage but also other forms of corrosions (e.g., pits can nucleate stress corrosion and corrosion fatigue cracks).

2.1.3 Crevice Corrosion

Impurities and special structures on the metal surface lead to the formation of crevices. The presence of crevices and other similarly confined spaces, to which the access of corrosion-related substances from the environment is limited, is the necessary condition for the establishment of differential aeration and the occurrence of crevice corrosion. Crevice corrosion occurs on almost all metals, especially with metals that are passive. Accordingly, almost all corrosive media can cause crevice corrosion, and the solution containing Cl^- is usually the most deleterious for crevice corrosion. Due to the geometrical factors of the structure, the cathode depolarizer (e.g., dissolved oxygen) can easily reach the metal surface outside the crevice by convection and diffusion. However, it is very difficult to reach the crevice inside, so resulting in a concentration difference of the cathode reactants between the crevice inside- and outside. Thereby, crevice corrosion also belongs to a concentration cell corrosion or "differential aeration" cell [21]. Geometric factors (e.g., the width and depth of the crevice, and the area ratio of inside and outside the crevice) and environmental factors (e.g., halogen ions and other anions, dissolved oxygen, temperature, pH, and flow rate) all affect crevice corrosion. In essence, crevice corrosion can be divided into four stages: i.e., initiation, propagation, slowing, and stagnation, and an alloy can re-passivate in some cases, at which point crevice corrosion ceases [22]. The geometry of the crevice has proven to be very important in assessing the risk of crevice corrosion [23]. Although inert oxide layers of Cr and Mo can form on alloy surface, the Ni–Cr–Mo alloy are still susceptible to crevice corrosion in the specific environments. Typical crevices may be caused by joints, gaskets, or debris deposits on the alloy surface [24].

Some researchers have reported crevice corrosion experiments on Ni–Cr–Mo alloys, such as Hastelloy C-22 and Inconel 625 in the NaCl solution. Crevice corrosion occurs on Hastelloy C-276 and Monel 400 in SCWO environments of 400,

450 and 500 °C [25]. Once crevice corrosion begins, compared with the Fe-based alloy SAM 1651 (chemical compositions (wt%) are Fe 51.3, Mo 25.7, Cr 14.9, C 3.4, Y 3.4, B 1.2), Hastelloy C-22 exhibits corrosion propagation much closer to the crevice boundary, and the localization of the crevice corrosion development is much more obvious [26]. SAM 1651 is more susceptible to passivation than is Hastelloy C-22 [26]. The Mo content in an alloy determines the distribution of corrosion in the crevice. Corrosion develops laterally on the surface of the alloy with a high Mo content, while it extends into the interior of the alloy with a low Mo content [27]. Mo also helps to suppress the generation of large-size crevices and reduce the initial propagation rate of the crevices [28]. The size order of the crevice corrosion area in 5 $mol{\cdot}L^{-1}$ NaCl solution at 150 °C is Inconel 625 > Hastelloy C22 > Hastelloy HYBRID-BC1 [29]. The ability to resist crevice corrosion in 0.1 or 1.0 $mol{\cdot}L^{-1}$ NaCl solution at 20–90 °C is in the order of Inconel 625 > Hastelloy C-22 > Hastelloy C-22HS > Hastelloy HYBRID-BC1 [30].

Experiments with several Ni-based alloys (including Inconel 625, Hastelloy G-30, Hastelloy C-22, Hastelloy C-22HS, Hastelloy HYBRID-BC1) in chloride-containing solution at 60 °C indicate that Hastelloy HYBRID-BC1 has the best crevice corrosion resistance [31]. The crevice corrosion resistance in $CaCl_2$ solutions at 20–100 °C is in the order of Hastelloy HYBRID-BC1 > Hastelloy C-22HS > Hastelloy C-22 > Inconel 625 [32]. Inhibition of crevice corrosion in Ni–Cr–Mo alloys in chloride-containing environments have been studied, and nitrates are found to be the most effective among sulfates, carbonates, bicarbonates, chromates and molybdate anions [33]. Besides, carbonates can inhibit chloride-induced crevice corrosion in Hastelloy C-22 [33], and sulfates, chromates, and molybdates are also viable corrosion inhibitors [34].

The Mo content has an important influence on the crevice corrosion resistance of an alloy. Both pitting corrosion and crevice corrosion have similar corrosion development stages, but they have two essential differences. Firstly, pitting corrosion occurs in an environment containing aggressive anions, whereas whether active anions exist in the environment does not affect the occurrence of crevice corrosion. This is because chloride induces passivity breakdown in the case of pitting corrosion, resulting in the subsequent formation of a cavity ("crevice"), whereas in the case of crevice corrosion, the crevice already exists. However, chloride may induce passivity breakdown within the crevice, as the passivity breakdown potential shifts in the negative direction in response to acidification of the crevice via differential aeration. Secondly, because a crevice already exists, crevice corrosion is more likely to occur than pitting corrosion, and can expand and create etched holes. In comparison with adding corrosion inhibitors or changing environmental conditions, avoiding crevices during design and construction is more reliable and easier to implement for crevice corrosion control. For example, it is possible to use welding instead of riveting or bolting/screwing, to improve the quality of weld seams, to apply coating to protect joints, and to enact timely clean up deposits and impurities around corner joints to eliminate crevices and hence avoid crevice corrosion.

2.1.4 Intergranular Corrosion

Intergranular corrosion occurs in a narrow region at the grain boundaries of metallic material and is also one type of localized corrosion. Almost all early Ni-based corrosion-resistant alloys suffer from intergranular corrosion. When intergranular corrosion occurs, the adhesion between the grains decreases substantially causing the grain boundaries to separate and initiating localized corrosion. Eventually, the corrosion morphology consists of many irregular polygon corrosion trenches along the grain boundaries. Alloy composition (e.g., C, N, Cr, Ni, and Mo), structure (e.g., grain size), and environmental factors (e.g., temperature, environment) are the major factors influencing intergranular corrosion.

In oxygen-containing H_2SO_4, $NaHSO_4$, Na_2SO_4 aqueous solutions and H_2SO_4 solution, intergranular attack occurs on Inconel 625 at above 150 °C, and much severe intergranular attack appears at above 250 °C. The maximum penetration depth can reach up to 20 μm in 100 h and 200 μm in 125 h, and the development of intergranular attack eventually abates somewhat within the range of 300–340 °C [2, 35]. Inconel 625 also suffers from intergranular attack in 0.2 mol·kg^{-1} HNO_3 aqueous solution containing 0.48 mol·kg^{-1}oxygen at 500 °C and 24 MPa, and thus Ni-based alloys are not suitable for the SCWO environment containing HNO_3 [36]. Importantly, intergranular corrosion occurs in Inconel 625 in pure water/oxygenated solutions at temperatures above 130 °C, and in HCl/O_2 solutions within the temperature range of 250–300 °C, and in SCW at 500 °C and 38 MPa. The largest intergranular corrosion depth can reach up 20 μm within 50 h, and unpolished samples suffer from more severe intergranular corrosion than do surface-polished samples [2]. Hastelloy C-22/276 and Inconel 625 also exhibit intergranular corrosion in SCW at 450 °C, 25 MPa, and Inconel 625 exhibits intergranular corrosion with depths of up to 0.2 mm in a solution containing H_2O_2 at 600 °C and 25 MPa [37]. The attack on Inconel 600 and Monel 400 in the NaOH solutions is intergranular, while that on a series of Ni-based alloys is transgranular under the same conditions, among which the transgranular attack (TGA) is a kind of stress corrosion in the form of transgranular cracking which has been observed and studied in austenitic stainless steels [38]. The stress corrosion cracking mode in the Ni–Cr–Fe alloys in the oxygenated NaOH environment changes from transgranular to intergranular when the Ni content is higher than 65 wt% [39]. The initial corrosion of Inconel 625 is intergranular corrosion and pitting corrosion in the oxygenated HCl solution at 350 °C and 24 MPa [40]. However, intergranular corrosion with a crack length of 30–40 μm occurs on SS 316 among several other alloys (e.g., Hastelloy C-276, Hastelloy B-22, and SS 316) at 400 °C and 60 MPa [41]. Therefore, it seems that the occurrence of intergranular corrosion on Ni-based alloys are complex and irregular. How to predict intergranular corrosion in harsh SCW environments is still a significant challenge.

Radiation-induced material changes in the center components of the light water reactor can lead to intergranular failure [42], and thus fully understanding the influences of radiation on the microstructures of material and grain boundary is important for avoiding intergranular corrosion damage in light water reactor materials. Intergranular carbide precipitation, chromium depletion, impurity segregation and local deformation characteristics are considered to be contributing factors in the intergranular stress corrosion cracking (IGSCC) of Inconel 600 in high temperature water in light water reactors [43]. In addition, the precipitation of grain boundary carbides may also play an important role in improving intergranular stress corrosion cracking by blocking the movement of dislocations.

Based on the summarization of existing studies, it is evident that research on intergranular corrosion has made a great progress. Intergranular corrosion is mostly associated with stress corrosion cracking. In the region where intergranular corrosion occurs, the corrosion develops as penetration along the grain boundary, thereby resulting in the loss of grain boundary adhesion and a significant decrease in metal strength. At this time, if the material continues to be affected by the synergistic effect of corrosive species and stress, brittle fracture (i.e., intergranular stress corrosion cracking) may occur. Accordingly, intergranular corrosion is one of causative reasons for stress corrosion cracking. There is no macroscopic change in the region where intergranular corrosion occurs, but the microstructures and compositions of the grains and grain boundaries have changed. The precipitation of harmful carbides in intermetallic phases is a generally accepted depletion theory of intergranular corrosion. Now, most studies mainly focus on the superficialphenomena of intergranular corrosion, corrosion depth, and various influencing factors, and research on intergranular corrosion mechanisms of alloys in SCW is scarce. Developing intergranular corrosion theory by exploring intergranular corrosion from a microscopic scale is clearly a fruitful area of study. As for intergranular corrosion of Ni-based alloys in SCW, future research should be focused on exploring the migration and content changes of elements and the phase behaviors of alloys, searching for non-sensitive elements for intergranular corrosion, studying the effects of carbide precipitation and carbon content on the sensitivity of intergranular corrosion, and further improving the stabilization of alloys.

2.1.5 Environment-Assisted Fracture

Environment-assisted fracture refers to the cracking and fracture of material under the effects of both environmental medium and stress (including applied stress, residual stress, and stress caused by chemical change or phase change). If the environmental medium is corrosive, the corrosion is called as "stress corrosion cracking (SCC)". The corrosion damage under the action of stress includes stress corrosion cracking (SCC), corrosion fatigue (CF), hydrogen-induced fracture (HIF), fretting corrosion (FC), and impact corrosion (IC). In a high temperature corrosive system, metallic materials are susceptible to corrosion fatigue, which is the result of the interaction between the cyclic stress and the corrosive medium. This results in the initiation and growth of cracks. The growth rate of fatigue crack of Inconel 625 in SCW increases with an increase in the test temperature [44].

2.1.5.1 Stress Corrosion Cracking (SCC)

Stress corrosion cracking refers to the phenomenon that results in the growth of a crack in a material in a specific corrosive environment. The term "stress corrosion cracking" includes one type of low stress, brittle fracture, where there is little plastic deformation before fracture. Stress corrosion cracking commonly occurs on passive metals, and commonly initiates at localized corrosion events, such as pits that act as stress raisers to yield a stress intensity factor (K_I) that exceeds the critical value for SCC (K_{ISCC}). Stress corrosion cracking can be divided into intergranular stress corrosion cracking (IGSCC), transgranular stress corrosion cracking (TGSCC) and mixed mode (IGSCC + TGSCC). The major influencing factors are stress (Mode and magnitude) and environment (e.g., special anions, pH, temperature, and interface polarization state in the medium). In condensed aqueous systems, the electrochemical corrosion potential (ECP) is usually the primary determining parameter [45, 46].

Ferritic-martensitic steels are generally resistant to stress corrosion cracking but suffer from severe general corrosion, while austenitic stainless steels and Ni-based alloys have good general corrosion resistance but are sensitive to stress corrosion cracking [47] Kim et al. [48] studied the corrosion behavior of Inconel 625 in high temperature and high pressure water (300–426 °C, 24.1 MPa and pH 2). The results show that stress corrosion cracking develops along grain boundaries under subcritical conditions and occurs at the junction of stress and defective dealloyed oxide layers. However, stress corrosion cracking is the result of the direct chemical attack of relevant HCl molecules in SCW in the presence of stress. Stress corrosion cracking does not occur under subcritical or supercritical conditions in neutral solutions, meaning that maintaining solution neutrality is beneficial to suppressing the stress corrosion cracking of Inconel 625. The stress corrosion cracking behavior of Inconel 600 and Inconel 690 in hydrogen-containing sub- and supercritical water at 320–450 °C indicate that cracks propagate along the grain boundaries close to the Cr-depleted regions, and Cr_2O_3 is found in the grain boundary before the crack tip. Intergranular cracks develop along the sample surface and are hindered by chromium carbide when entering the sample interior, and this helps to inhibit the crack propagation [49]. The effects of creep on stress corrosion cracking of cold worked Inconel 690 in SCW at 450–550 °C, 25 MPa have been investigated [50]. The results show that cold worked samples exhibit high sensitivity to intergranular cracking, and creep plays an important role in the growth of stress corrosion cracks, which contributes more than 80% of the crack growth rate. Moreover, large amounts of vacuoles are formed before the crack tip, and they weaken the bond strength of grain boundaries and accelerate the expansion of stress corrosion cracking. The attack of HCl molecules in acidic solutions probably induces the development of stress corrosion cracking [51].

Figure 2.1 displays the stress corrosion cracking of Ni-based alloys insubcritical and supercritical water. As shown in Fig. 2.1a, b, Inconel 625 exhibits 60 μm of stress corrosion cracking in SCW of pH 2, and severe stress corrosion cracking occurs on Hastelloy C-276 in the HCl solution of pH 1 at 350 °C. All cracks develop from the outer surface and penetrate to the depth of 514 μm [52]. Severe stress corrosion cracking was found to occur in a reactor that had undergone 100-loadings

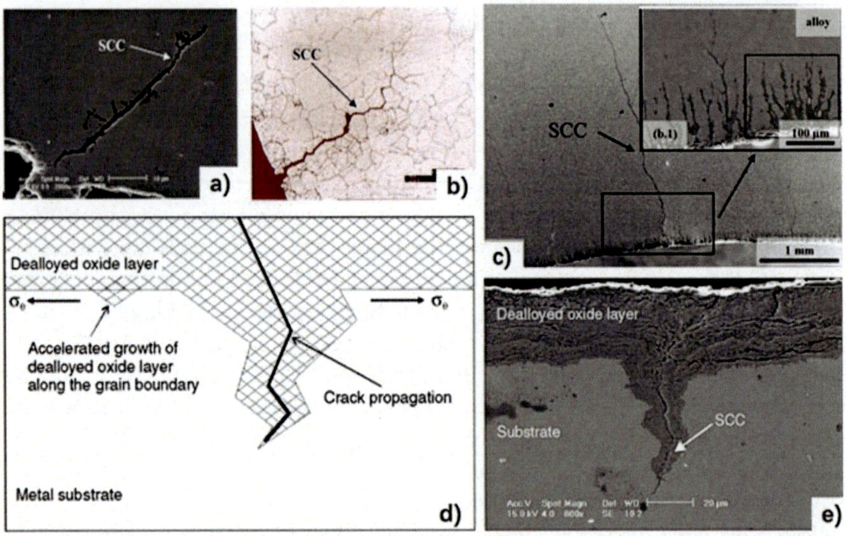

Fig. 2.1 Stress corrosion cracking (SCC) of Ni-based alloys in subcritical or supercritical water **a** the development of SCC, shown by the cross-section SEM image of Inconel 625 in the pH 2 HCl solution at 426 °C [52], **b** a laser microscope image of the etched sample, shown by the cross-section SEM analysis of Hastelloy C-276 in the pH 1 solution at 350 °C and 27.3 MPa [52], **c** SCC with a depth of up to 6 mm in a reactor used in approximately 100 ammonoacidic experiments at 600 °C, 30 MPa, and crack propagation following partly inter- and trans-granular routes [53], **d** the schematic diagram of SCC in a subcritical temperature [48], and **e** the SEM image of the dealloyed oxide layer and SCC at 351 °C and 24.1 MPa [48]

of ammonoacidic (it is not an organic matter or a compound, but an environment containing ammonia and acid) in crystal growth experiments (see Fig. 2.1c). This can be explained by the frequent mechanical loading and high stress concentration eventually inducing stress corrosion cracking, and thus the inner surface of the reactor is almost completely covered by cracks having different lengths. Additionally, stress corrosion cracking occurs in the outer oxide layer of Incoloy 625 in the aqueous solution of pH 2 at 300–426 °C and 24.1 MPa (see Fig. 2.1e). The cracking initiates and spreads along grain boundaries because they have open structures compared with most crystal lattices, thereby facilitating the diffusion of ions at the grain boundaries rather than in the lattice. As displayed in Fig. 2.1d, once the cracks are formed, the acidic solution penetrates along the grain boundary and forms oxides [48].

At present, the research objectives in studying stress corrosion cracking in Ni-based alloys mainly involve assessing microstructure size, crack growth rate, and extension direction. Environmental media primarily include high temperature and high pressure water, supercritical water, and acidic solutions. However, apart from the alloy itself and special media, tensile stress is also a necessary conditions for the development of stress corrosion cracking. The tensile stress can be the working stress caused by the applied load under the prevailing conditions or the residual stress (e.g., thermal stress and deformation stress) formed inside the material during production and installation processes, and even by the stress caused by the volume effect of corrosion products in the crack enclave (pack corrosion). In a wide range of corrosive environments, it is important to understand the source and strength of the stress for mechanistic analysis and control of stress corrosion cracking. However, the existing research on the tensile stress effects of materials is not sufficiently comprehensive to resolve many important issues, and few researchers have studied the influence of stress on stress corrosion cracking, and the measures to prevent stress corrosion cracking by stress relief, in supercritical aqueous media.

2.1.5.2 Hydrogen Embrittlement

Hydrogen embrittlement refers to the phenomenon that hydrogen enters the material interior and results in the degradation of material properties. When hydrogen embrittlement occurs, metal toughness, and various physical and chemical properties decrease due to the penetration and diffusion of hydrogen atoms into the crystal structure of alloy. According to the case that material is in a hydrogen production environment or the material has already dissolved hydrogen, hydrogen embrittlement can be divided into hydrogen environment embrittlement and internal hydrogen embrittlement [54]. Hydrogen embrittlement is an intrinsic property of Ni–Cr–Fe alloys, because the grain boundary cohesion decreases as the alloy purity increases [55].

Hydrogen embrittlement exhibits different characteristics for different Ni-based alloys. Inconel 718 has a high sensitivity to hydrogen embrittlement, and hydrogen induced cracking tends to distribute across the surface in the tensile stress concentration region. Laser surface softening can improve the elongation (fracture strain) of a material by reducing the hardness of the surface in the tensile stress concentration region, so effectively reducing the sensitivity of Inconel 718 to hydrogen embrittlement [56]. Although the sensitivity of a material to hydrogen embrittlement generally decreases with increasing temperature, Inconel 600 suffers from severe hydrogen embrittlement at 250 °C [57]. The studies regarding the hydrogen embrittlement behavior of Ni-based alloys and stainless steels, show that hydrogen embrittlement is aggravated by an increase of the Ni content, cold working, and thermal aging. The degree of hydrogen embrittlement can be measured by the hydrogen-induced ductility loss. In the case of cold working and cold work aging, standard and high-alloyed stainless steels are more resistant to hydrogen-induced ductility loss than arc Ni based alloys [58]. The addition of specific elements (such as C, P, Sn and Sb) during the heat treatment, and segregation in grain boundaries can enhance the adhesion of grain boundaries [55]. Moreover, the addition of an appropriate amount of boron can effectively reduce the sensitivity of Fe- and Ni-based alloys to hydrogen embrittlement [59].

Studies on the mechanism of hydrogen embrittlement have confirmed that the interaction of slips and grain boundaries plays an important role in hydrogen embrittlement [60]. For Fe- and Ni-based alloys containing 20–30 wt% of Ni, hydrogen embrittlement is considered to be the result of a combination of high hydrogen content near the austenite grain boundaries and the stress concentration duc to boundary disturbance caused by the impact of martensite shear [61]. Hydrogen transportation by dislocations associated with one or more cracking mechanisms is one of the mechanisms that have been postulated to be responsible for hydrogen embrittlement of some metals, which is verified by the Inconel 600 experiments [57]. For the Fe- and Ni-based alloys, hydrogen embrittlement sensitivity is also related to the microstructure of the alloy [59], and this correlation is more remarkable especially for the Ni–Fe–Cr alloys with the Cr contents of 15–20 wt% and the Ni content of higher than 40 wt% [62]. Furthermore, some researchers believe that the sensitivity of hydrogen embrittlement is significantly related to the grain boundary structure. Since boundaries possessing large gaps and structural irregularities can be affected adversely, boundaries lacking these defects are resistant to hydrogen embrittlement [63].

It is worth noting that the high-temperature and high-pressure hydrogen-rich environment formed in the hydrogen production via SCWG provides the conditions of generating hydrogen embrittlement. The inner wall material of the reactor (e.g., Ni-based alloy) may participate in hydrogen production reactions as a catalyst. The infiltration of atomic hydrogen into the reactor inner wall and thereactions between Hand the alloy can cause the embrittlement and even cracking of the reactor material. Taking the catalytic SCWG of wood biomass as an example, three important reactions of wood biomass in SCW are as shown as follows [64].

$$C + H_2O \rightarrow CO + H_2 \tag{2.7}$$

$$CO + H_2O \rightarrow CO_2 + H_2 \tag{2.8}$$

$$CO + 3H_2 \rightarrow CH_4 + H_2O \tag{2.9}$$

where Eqs. 2.7–2.9 are the water gas reaction, water–gas shift reaction, and methanation reaction, respectively. High concentrations of H_2 and CH_4 produced in the reactions accumulate near the inner wall of the reactor, and the extremely high mass transfer coefficient of SCW facilitates the absorption of these gases by the reactor material, and the high-pressure environment provides the driving force for hydrogen penetration into the material. When hydrogen atoms penetrate into the material interior and then recombine in voids to form H_2 molecules, the pressure increase inside the void results in a decrease in the external stress required for crack propagation, and then resulting in the formation of localized plastic deformation, eventually inducing material cracking.

Although hydrogen induced cracking is highly hazardous for SCWG reactor materials, the investigations concerning the characteristics, mechanisms, and control of hydrogen embrittlement in SCWG environments are still insufficient to fully understand the phenomenon of HIC. Furthermore, Ni-based alloys are important candidate materials for SCWRs, but their resistance to hydrogen embrittlement and the methods of reducing the sensitivity of hydrogen induced cracking still need further exploration.

2.2 Element Effects of Ni-Based Alloys

In the same corrosive environment, the corrosion resistance of an alloy often depends on the types and concentrations of the alloy elements [65]. The Cr content of Ni-based alloys is generally in the range of 15–30 wt% [66]. Cr can strongly improve the corrosion resistance of Ni in strong oxidizing media, imparts the high-temperature oxidation resistance to Ni. Cr is regarded as the most important alloy element for improving the corrosion resistance of alloys, especially in acidic and oxidative environments (e.g., nitric acid, hydrochloric acid, and high-temperature oxidative circumstance) and sulfur-containing gases [67]. Therefore, the Ni-based alloys with high Cr content is the first option for use in high temperature and strong oxidizing, acidic solutions. Compared with austenitic stainless steel with a low Cr content (e.g., SS 347H, and SS 316L), Ni-based alloy exhibits less weight loss in SCW. The high Cr content promotes the formation of relatively dense, protective Cr_2O_3 in the inner oxide layer, while the low Cr content results in thinner and unprotective Cr oxides in the inner layer and bubble particles in the outer layer [68].

Figure 2.2 compares the penetration rates of several Ni-based alloys in the HCl aqueous solution of pH 2 at 268–428 °C and 24.1 MPa [51]. Apparently, the highest penetration rate of each alloy is found to occur within 350–360 °C (see Fig. 2.2a), showing the severest corrosion in the high subcritical temperature region. Figure 2.2b indicates the penetration rate in the high subcritical temperature range (340–374 °C) as a function of the Cr content. The penetration rate decreases with the Cr content increasing. This is because increasing the Cr content leads to the formation of dense Cr oxide barrier layer of the passive film, which can inhibit the transport of ions (e.g., O^{2-} and Ni^{2+}) within the lattice. Notably, Inconel 671 has the highest corrosion resistance in the high subcritical temperature range among the tested alloys above. This is consistent with the lowest penetration rate and the highest Cr content in this alloy (see Fig. 2.2). Although Cr improves the corrosion resistance of the Ni-based alloy in SCW, a too high Cr content also causes some undesirable problems. For example, the increase of aging temperature and aging time aggravate the embrittlement phenomenon, and meanwhile thermal aging reduces the impact toughness and corrosion resistance of alloy, especially for the high-Cr content Ni-based alloys [69].

As one of the most important multi-functional metal elements in the modern industry, Ni has high temperature resistance, high pressure resistance and good thermal stress properties [71]. However, the dissolution of Ni is the severest among all alloy elements [65], and the synergistic effect between chloride and oxygen induces severe Ni dissolution [72]. At the same time, the dissolution of Ni oxides also cause the lack of Ni on the alloy surface and the acceleration of corrosion. Besides, Ni

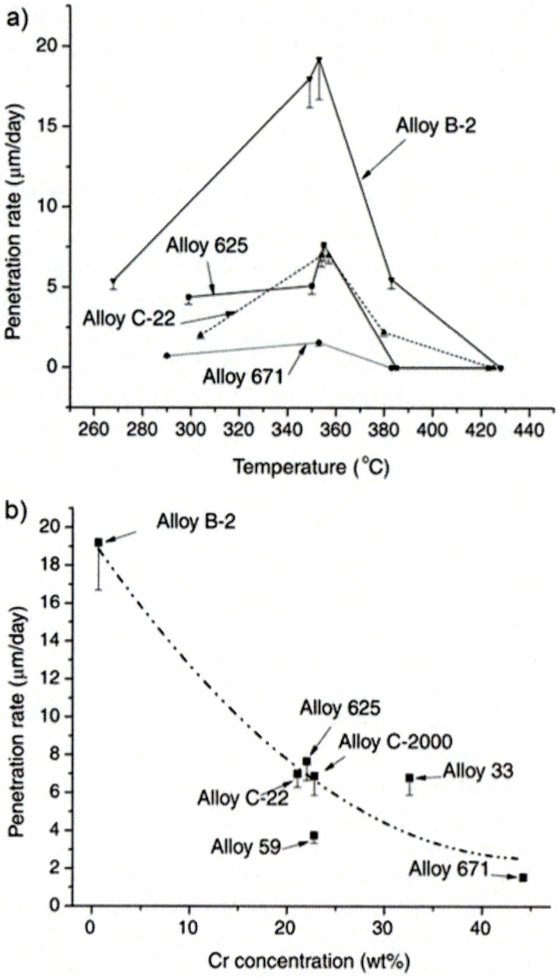

Fig. 2.2 Alloy penetration rate (i.e., dealloyed oxide growth rate) as a function of temperature or the Cr concentration of the alloy in the aqueous HCl solution of pH 2 at 268–428 °C and 24.1 MPa [51] **a** penetration rate versus temperature, **b** penetration rate versus the Cr concentration of the alloy in the high subcritical temperature range. Reprint permission from Ref. [70]

becomes unstable and then form soluble Ni ammines in high temperature water in the presence of ammonium ions [73, 74]. The addition of Fe is generally used to improve the economics of Ni-based alloys, and the increased Fe content increases the solubility of carbon in the Ni-based alloys. This can improve the resistance of alloys to intergranular corrosion and carburization performance. Under SCW conditions, the oxidation rate of Fe is the highest among Ni, Fe, Cr and Mo, and a stable oxide Fe_3O_4 is finally formed, whichis almost insoluble [72]. However, Fe weakens the corrosion resistance of Ni-based alloys in SCW [51, 52]. Nicrofer 3033 (Alloy 33) has a high Cr content, but its penetration rate is also relatively high, because a large amount of Fe (33 wt%) in the alloy rapidly dissolves from the alloy surface and thus induces an increase in the penetration rate.

Mo is an important alloy element that makes Ni-based alloys have good corrosion resistance. It can strongly improve the pitting corrosion resistance, crevice corrosion resistance, and corrosion resistance to non-oxidizing acids. When the Cr content is constant, an increase in the Mo content can reduce the corrosion rates of Ni-based alloys [73]. Hastelloy C-276, Inconel 625 and Incoloy X-750 suffer from general corrosion in SCW, and Incoloy X-750 exhibits the highest corrosion rate, likely due to the lack of Mo [75]. Although both Cr and Mo can improve corrosion resistance of Ni-based alloys in subcritical water environments, Mo cannot exhibit good corrosion resistance in absence of Cr [51, 52]. The corrosion behavior of Inconel 625 and Hastelloy C-276 in the acidic solution of pH 2 at 300–425 °C and 24.1 MPa indicate that a dealloyed oxide layer is formed at subcritical temperature while a thin oxide film is formed at supercritical temperature [51]. The selective dissolution of Ni and Fe and the oxidation of Cr and Mo are major corrosion mechanisms. Chromic oxide is soluble at pH = 1, while Mo is the only element capable of forming stable compounds under these conditions [52].

The diffusion coefficient (D) of the alloying element is an important parameter affecting the corrosion properties. It is confirmed that the diffusion coefficient of the Ni–Fe–Cr alloy has an order of $D_{Cr} > D_{Fe} > D_{Ni}$, in which $D_{Cr}/D_{Ni} \cong 2.5$ and $D_{Fe}/D_{Ni} \cong 1.8$ [76]. The Cr flux is strongly affected by the oxide microstructure. The lattice diffusion coefficient $D_L(m^2 \cdot s^{-1})$ and the grain boundary diffusion coefficient $D_G(m^2 \cdot s^{-1})$ of Cr in Incoloy 800 are separately expressed as follows [77],

$$D_L = 3.24 \times 10^{-4} e^{\frac{-287.4\,kJ/mol}{RT}} \tag{2.10}$$

$$D_G = 5.80 \times 10^{-5} e^{\frac{-184.2\,kJ/mol}{RT}} \tag{2.11}$$

where R $(J \cdot mol^{-1} \cdot K^{-1})$ is the molar gas constant, and T(K) is the thermodynamic (Kelvin) temperature. For Alloy 20Cr–25Ni, the flux of Cr ions (j_L) emerging on the surface from the interior of a single grain of the alloy is calculated by Eq. 2.12, and the flux of Cr ions (j_B) emerging from the boundary of a single grain is evaluated by Eq. 2.13, and thus the ratio of the two fluxes can be expressed by Eq. 2.14 [78].

$$j_L = \frac{d^2 D_L (C_B - C_I)}{\sqrt{\pi D_B t}} \tag{2.12}$$

$$j_B = \frac{2dw D_B (C_B - C_I)}{\sqrt{\pi D_B t}} \tag{2.13}$$

$$\frac{j_B}{j_L} = \frac{2w}{d}\sqrt{\frac{D_B}{D_L}} \tag{2.14}$$

where d is the grain size, D_L is the lattice diffusion coefficient, D_B is the grain boundary diffusion coefficient (the same as D_G), C_I is the fixed surface concentration,

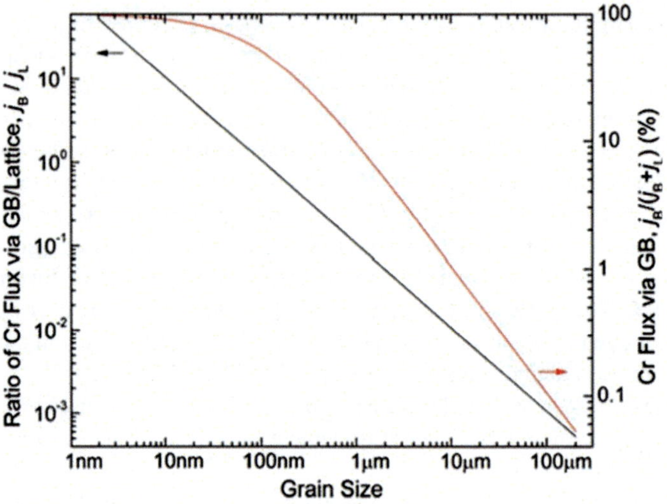

Fig. 2.3 The flux ratio of Cr via grain boundaries (G_B) versus lattice (i.e., j_B/j_L), and the fraction of the Cr flux via GB in the total flux (i.e.,$j_B/(j_B + j_L)$) as a function of the grain size for Incoloy 800H. Reprint permission from Ref. [70]

C_B is the bulk concentration, t is the oxidation time, and w is the grain boundary width. Figure 2.3 displays the flux ratio of Cr via grain boundaries versus lattice (i.e.,j_B/j_L),together with the fraction of the Cr flux via grain boundaries in the total flux (i.e.,$j_B/(j_B + j_L)$), as a function of the grain size for Incoloy 800H [70]. Apparently, the (j_B/j_L) and $j_B/(j_B + j_L)$ for Cr decrease with the increasing grain size, which is caused by grain size refinement followed by the formation of Cr-rich oxides.

The Ni-based alloys discussed above, exhibit different corrosion types and characteristics in complex and variable sub- and supercritical water. Alloying elements such as Cr, Ni, Fe, and Mo significantly affect the corrosion resistance of Ni-based alloys. The relevant investigations regarding corrosion types and alloy element effects still need to be carried out comprehensively and in depth.

2.3 Element Effects of FeCrAl Alloys

In FeCrAl alloys, in addition to the balancing Fe, other elements also widely participate in chemical reactions in the corrosion environment, which directly cause the corrosion of the alloy. Specific action of Cr, Al, Si, reactive elements (Y, Ti, Hf, Zr etc.) and other elements are clarified in the following paragraphs.

2.3.1 *Chromium (Cr)*

In low-temperature hydrogenated water chemistry, Cr participates in reactions to form an inner spinel oxide layer rich in $FeCr_2O_4$ or Cr_2O_3. When local electrochemical corrosion potential is higher than 0 mV, the probability of formation of CrO_4^{2-}, $HCrO_4^-$ and $Cr_2O_7^{2-}$due to dissolution of the inner oxide shows a close connection with the content of Cr under the irradiation in reactors [71]. In the high-temperature oxidation environment, Cr is highly associated with the transformation of the mixed oxide layer on the outer surface to Al_2O_3. Based on the oxide map for FeCrAl ternaries, Young et al. [72] concluded that 2–3 wt% Al with 20 wt% Cr in the alloy could support α-Al_2O_3 formation. Figure 2.4 depicts the phase diagram of the FeCrAl ternary alloys system. The optimum Cr content of FeCrAl alloys inducing the formation of Al_2O_3 at high temperatures was investigated by Niu et al. [73] through comparison among five types of FeCrAl alloys (P_1-P_5) with different Cr contents. For the alloy with 3 at.% Al, the critical Cr helping to form Al_2O_3 is between 5 and 10 at.%, and the higher Cr content not only produces a transition from internal to external oxidation of Al, but also decreases the growth of external FeO_x scale.

Although there is no consistent conclusion on the critical Cr content for the formation of Al_2O_3 on FeCrAl ternary alloys with constant Al under high-temperature oxygen, the Cr content in FeCrAl alloy should ensure that Cr required by external Al_2O_3 is greater than that required to avoid internal oxidation of Al. It has been reported that in order to guarantee the oxidation resistance of the alloy under high

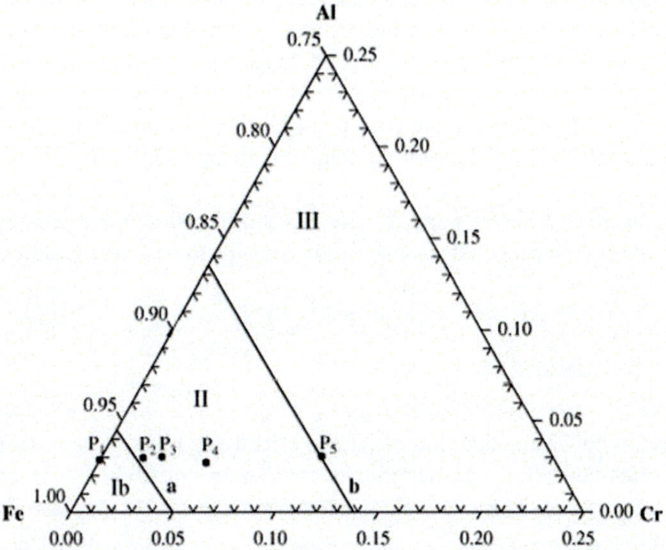

Fig. 2.4 Phase diagram of the FeCrAl ternary alloys system [73] (lines a and b represent the conditions for the occurrence of transition. Region Ib and Region III separately represent FeOx and $(Cr, Al)_2O_3$, and Region II represents FeOx and $(Cr, Al)_2O_3$)

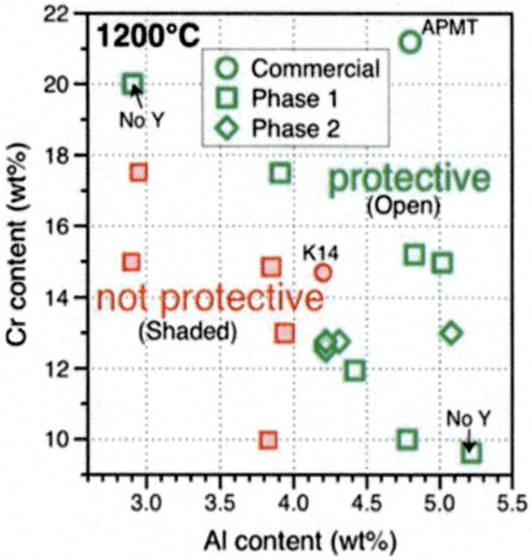

Fig. 2.5 Influence of Cr and Al content on oxidation resistance of FeCrAl alloys at 1200 °C steam [76]

temperature steam environment (>1000 °C), the content of Cr in FeCrAl alloys is supposed to reach 20% [74]. However, when Cr exceeds 8.5 wt%, neutron radiation in reactors will make α′ phase precipitate from FeCrAl through nucleation, growth or amplitude modulation decomposition, resulting in embrittlement of the alloy [75].

Figure 2.5 displays the influence of Cr and Al content on oxidation resistance of FeCrAl alloys in 1200 °C steam. Apparently, reasonable combination of Cr and Al content in steam at 1200 °C can provide good corrosion resistance for different FeCrAl alloys [76], which are candidate fuel cladding materials. Satisfactory oxidation resistance can be achieved when Cr and Al contents are in a wide composition range. A synergistic effect between Cr and Al contents to guarantee the formation of protective Al_2O_3 was confirmed by Unocic et al. [77]. However, this synergistic effect may diminish when temperature rises to 1400 °C with a high heating rate. Besides, the effect seems to be obvious on the transition to an external Al_2O_3 scale when the Al content is low (below 6–10 wt%) at 1000 °C [78]. Whether there is a quantitative relationship between Cr and Al contents at different temperatures, especially in oxygen, to provide expected anti-oxidation remains to be further studied.

2.3.2 *Aluminum (Al)*

Compared with Cr, the addition of Al can effectively raise the maximum tolerable temperature of the alloy, which highly determines the corrosion resistance of FeCrAl at high temperature. This is a big consideration for FeCrAl to replace Zr as nuclear fuel cladding [79].

Figure 2.6 illustrates the effect of Al content on oxidation behavior of FeCrAl alloys in high temperature water. The formation of Al_2O_3 is determined by the content of Al in the alloy in high-temperature steam. The increase of Al enhances its activity,

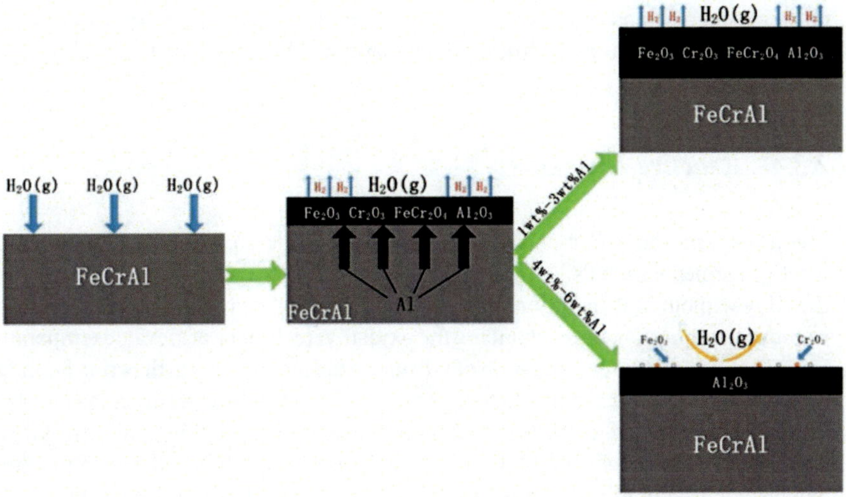

Fig. 2.6 Effect of Al content on oxidation behavior of FeCrAl alloys in high temperature water [80]

so forming Al_2O_3 through enough reduction reaction. Low Al content (1–3 wt%) is unable to completely support the conversion of Al into Al_2O_3 with Cr_2O_3, Fe_2O_3 and $FeCr_2O_4$, resulting in the appearance of porous spinel with poor oxidation resistance on the alloy surface. On the one hand, the production of dislocation loops will be inhibited if Al is up to 4 wt%, which can reduce the hardening and embrittlement of FeCrAl as claddings during the actual operation of reactors [81]. On other hand, more than 5 wt% Al increases the ductile/brittle transition temperature, reduces the ductility of weldments, and affects the machinability of the alloy [76, 82]. Therefore, the Al content should be controlled at 4–5 wt%.

2.3.3 Silicon (Si)

Specific behavior of Si in FeCrAl alloys is to form SiO_2 through absorbing oxygen, which precipitates on the matrix surface and reacts with the initial FeO to generate Fe_2SiO_4, so effectively preventing the occurrence of internal oxidation zone. During the process, the oxidation state of Si does not change with the thickness of surface oxide [83]. When FeCrAl is oxidized at high temperature, the addition of slight Si can affect the activity of Cr in the matrix, thus the evaporation of chromium into hydroxide of chromium oxide is inhibited and the shedding of protective oxide from the surface is prevented. When Cl^- exists in the environment, Si may reduce the thickness of secondary protective Fe-rich oxides, which makes FeCrAl alloy show lower corrosion weight gain. It is also reported that a positive synergistic effect takes place between Si and little Al [84]. Under medium water chemistry, Si in FeCrAl seems not greatly change hydrothermal corrosion behavior of the material [85].

2.3.4 Reactive Elements

The growth mechanism and adherence of Al_2O_3 scale are strongly determined by reactive elements such as Y, Ti, Hf, Zr, Ce, Ca, Mg and La [86–89]. Compared with the alloy without Y, its addition to FeCrAl can reduce the growth rate of alumina and improve the adhesion, thus enhancing the oxidation resistance at elevated temperature [76]. This is due to the combination of yttrium oxide precipitated at the grain boundary of alumina with external oxide, which prevents the formation of voids in the interface. The higher content of Y, the bigger doping rate and precipitation density of yttrium oxide [79]. However, it should be noted that Y easily volatilizes during arc melting, and the addition of high Y leads to internal oxidation of FeCrAl due to over-doping effect. Therefore, the composition of Y in FeCrAl should be 0–0.12 wt% [90].

In FeCrAl alloys without Ti, Al_2O_3 growth is virtually controlled by the diffusion of oxygen along oxide grain boundaries and show a columnar structure with increasing grain size in growth direction. Differently, Ti-added alloys form an outer oxide zone with equiaxed grains whose sizes get bigger with the increment of exposure time on the top of inner columnar part [87]. Hf seems to have a large impact on Al_2O_3 scale growth with a decrease in the parabolic rate constant. This means that the transport of anion and cation is influenced by the segregation of Hf. [91]. Zr can alleviate the spallation of the Al_2O_3 scale of C-containing FeCrAl alloys and change the structure of Al_2O_3 by reducing the oxide grain size as well as forming zirconia precipitates and nano-porosity [86]. The presence of Zr may make Al_2O_3 grow inwardly instead of both directions [92]. The addition of Ce exerts an effect on grain refining and inclusion modifications. When the Ce content in FeCrAl alloys increases, the average grain size of the alloy will be refined. If the content is higher than 0.05 wt%, intermetallic inclusions will appear and their distribution will alter from dispersing on the matrix to segregating on grain boundaries [88]. Thus, the amount of Ce added to FeCrAl alloys to raise corrosion resistance under high temperatures should be lower than the solubility limit at the service temperature [93]. Ca and Mg easily lose their sulfur-controlling ability by diffusing into the growing oxide scale from the matrix due to their high volatilities [89]. La appears to modify Al_2O_3 into a structure with a more stable morphology and a higher surface area, but more proof is needful to support this statement [94].

In summary, the effects of reactive elements share a big similarity on the oxidation performance of FeCrAl alloys. Reactive elements in FeCrAl play three main roles [72]. First, reactive elements help to resist scale spallation by slowing the scale growth rate and changing the mass transfer mechanism from mixed oxygen and Al transport to merely oxygen transport due to their segregation at grain boundaries of Al_2O_3 scale. The improvement in the spalling resistance is the most significant property of reactive element effects, and the scale adhesion may be attributed to desulphurization of reactive elements or other models [95]. Secondly, reactive elements change the microstructures of Cr_2O_3 and Al_2O_3 as well as the grain sizes in growth direction. Thirdly, reactive elements may strengthen the oxide-alloy interface through suppressing sulphur segregation, decreasing wrinkling and scale detachment, and making oxide intrusions or pegs grow into the alloy.

2.3.5 Other Elements

As an impurity, the increment of C content in an FeCrAl alloy may result in extensive spallation of Al_2O_3 and short limited lifetime of the alloy. This phenomenon is related to extensive formation of Cr-carbide in the alloy (grain boundaries) and at the scale/metal interface [86]. S from the alloy segregates heavily at the Al_2O_3/FeCrAl interface and may be incorporated into the Al_2O_3 scale [96]. Few reports focus on the influence of N on the corrosion performance of FeCrAl alloys. Based on existing information, it is acceptable that raising the N content can significantly enhance corrosion properties of steels, while excess nitrogen deteriorates the corrosion resistance [97]. In salt-containing environment, the existence of N can improve protective ability of passive film by reducing defect density and increasing the fraction of Cr_2O_3 and Fe_2O_3, so enhancing pitting corrosion resistance of advanced steels [98]. As a harmful component, oxygen enters into the alloy matrix automatically during alloy smelting. O is commonly added into alloys to increase the yield strength by solution strengthening [99]. However, O may exist as inclusion in oxides and its specific effect on the corrosion resistance of FeCrAl alloys requires more investigation.

Other elements, such as W, Ta, Mo, Nb, V and rare-earth elements, can not only improve the corrosion resistance of FeCrAl, but also enhance high-temperature mechanical properties. The influence of Mo on the corrosion of FeCrAl alloys seems to be connected with temperature and the concentration of Mo. Increasing the Mo content causes more precipitation of the μ phase, so reducing the mechanical strength and corrosion resistance of FeCrMo alloys [100]. In the range of 2–3 wt%, the Mo content nearly makes no differences to oxidation kinetics of different FeCrAl alloys at high temperatures [101]. Mo segregation at grain boundaries is observed when temperature reaches 500 °C, and this may be a semaphore for the change of corrosion behavior of FeCrAl alloys [102]. Mo plays the role of solution strengthening as well as hinders the recrystallization, so improving the tensile properties and maintaining the thermal stability [90]. When FeCrAl is coated on the surface of Zircaloy, the introduction of Mo as buffer layer between Zr and FeCrAl coating can significantly reduce the mutual diffusion of zirconium and iron, which makes FeCrAl/Zr show excellent oxidation resistance in high temperature steam [103]. Nb can prevent grain growth, hinder dislocation slipping and reduce the size of heat-affected zone [104]. Significant positive influences of Nb on the corrosion resistance and phase compositions of alloys have been reported in the latest research [105]. The minor addition of Nb increases the Cr availability in austenite, leads to the formation of

Cr_2O_3 in an early time, and triggers precipitation phases as the reservoir of Al for the formation of Al_2O_3 scale. High Nb content increases the weight gain rate of FeCrAl alloys and the spallation of Al_2O_3 in high-temperature steam oxidation. In order to achieve a balance between anti-corrosion ability and mechanical properties, the optimum content of Nb in FeCrAl alloys is 0.5 wt% [106].

In addition, Pt can remarkably decrease the weight gain rate of FeCrAl alloys and provide good oxide coherency [107]. FeCrAl alloys may contain Mn, Co and P, etc. The effects of these impurities are more reflected in mechanical properties other than corrosion performance. In second-generation FeCrAl alloys, the contents of Mn, Co and P, etc., are all at or below 0.01 wt% [108]. The effect of Ni on the oxidation behavior of FeCrAl alloy is mostly concentrated at high temperature. Like Cr and Al, the change of Ni has a positive influence on the rapid growth of Fe-rich oxide layer formed after the rupture of Cr/Al oxide acting as primary corrosion protection. The addition of Ni is worth taking for the corrosion protection of alloys which is incapable to improve breakaway corrosion in corrosive environment [109]. Ni in the alloy makes the austenite stable zone expand from high temperature to low temperature. Therefore, alloys always keep a face-centered cubic structure of austenite, which improves the stability of protective layer [110]. Specific percentage of different FeCrAl alloy composites is different. Although the addition of elements is helpful to improve the high temperature corrosion resistance of FeCrAl alloy, mechanical properties, radiation swelling resistance, weldability and processability demand further study. In general, the nominal composition of FeCrAl model alloy is that Fe: Bal., Cr: 10–20%, Al: 3–5%, Y: 0–0.12%, Nb: 0.5 wt% and impurities: <0.01 wt%. Tensile properties of the alloy are not sensitive to the composition, and the oxidation resistance at high temperature (about 1200 °C) is closely related to the content of Cr and Al [90].

References

1. J. Pang, D.J. Blackwood, Corrosion of titanium alloys in high temperature near anaerobic seawater. Corros. Sci. **105**, 17–24 (2016)
2. P. Kritzer, N. Boukis, E. Dinjus, Corrosion of alloy 625 in aqueous solutions containing chloride and oxygen. Corrosion, 824–834 (1998)
3. D.D. Macdonald, The point defect model for the passive state. J. Electrochem. Soc. **139**, 3434–3449 (1992)
4. D.D. Macdonald, G.R. Engelhardt, *Predictive Modeling of Corrosion. Shreir's Corrosion* (Elsevier, Oxford, 2010), pp. 1630–1679
5. D.D. Macdonald, The history of the point defect model for the passive state: a brief review of film growth aspects. Electrochim. Acta **56**, 1761–1772 (2011)
6. G. Gupta, P. Ampornrat, X. Ren, K. Sridharan, T.R. Allen, G.S. Was, Role of grain boundary engineering in the SCC behavior of ferritic–martensitic alloy HT-9. J. Nucl. Mater. **361**, 160–173 (2007)
7. R.B. Rebak, Environmentally assisted cracking of nickel alloys. Environ. Induced Cracking Mater. 435–446 (2003)
8. X. Zhong, E.-H. Han, X. Wu, Corrosion behavior of Alloy 690 in aerated supercritical water. Corros. Sci. **66**, 369–379 (2013)

9. R. Novotný,P. Janík, A. Toivonen, A. Ruiz, Z. Szaraz, L. Zhang, J. Siegl, P. Haulld, S. Penttil, J. Macak, European project "Supercritical water reactor-fuel qualification test": Summary of general corrosion tests. J. Nucl. Eng. Radiat. Sci. **2**, 031007 (2016)
10. S.H.J. Eon, E.H. Lee, D.H. Hur, Effects of dissolved hydrogen on general corrosion behavior and oxide films of alloy 690TT in PWR primary water. J. Nucl. Mater. **485**, 113–121 (2017)
11. R.B. Rebak, General corrosion behavior of n06022 in super concentrated brines at temperatures higher than 100°C. Mater. Res. Soc. Symp. Proc. **1124**, 473–480 (2009)
12. K. Wang, C. Li, Y. Li, J. Lu, X. Luo, Multi-physics coupling analysis on the time-dependent localized corrosion behavior of carbon steel in CO_2-H_2O environment. J. Electrochem. Soc. **167**, 013505 (2020)
13. J. Wang, S. Qian, Y. Li, D.D. Macdonald, Y. Jiang, J. Li, Passivity breakdown on 436 ferritic stainless steel in solutions containing chloride. J. Mater. Sci. Technol. **35**, 637–643 (2019)
14. A.A.B. Araneda, M.A. Rodríguez, R.B. Rebak, M.A. Kappes, R.M. Carranza, Effect of thiosulfate on the pitting corrosion of nickel base alloys in chloride solutions. NACE Int. (2017)
15. S.M.A. El-Haleem, S.A. El-Wanees, Chloride induced pitting corrosion of nickel in alkaline solutions and its inhibition by organic amines. Mater. Chem. Phys. **128**, 418–426 (2011)
16. Y.S. Lim, J.S. Kim, H.S. Kwon, Pitting corrosion of the laser surface melted Alloy 600. J. Nucl. Mater. **336**, 65–72 (2005)
17. B. Stellwag, Pitting Resistance of Alloy 800 as a Function of temperature and prefilming in high-temperature water. Corrosion **53**(2), 120–128 (1997)
18. R. Merello, F.J. Botana, J. Botella, M.V. Matres, M. Marcos, Influence of chemical composition on the pitting corrosion resistance of non-standard low-Ni high-Mn–N duplex stainless steels. Corros. Sci. **45**, 909–921 (2003)
19. Y. Guo, T. Sun, J. Hu, Y. Jiang, L. Jiang, J. Li, Microstructure evolution and pitting corrosion resistance of the Gleeble-simulated heat-affected zone of a newly developed lean duplex stainless steel 2002. J. Alloys. Compd. **658**, 1031–1040 (2016)
20. S.A. Al-Fozan, A.U. Malik, Effect of seawater level on corrosion behavior of different alloys. Desalination **228**, 61–67 (2008)
21. U.R. Evans, Oxygen distribution as a factor in the corrosion of metals. Ind. Eng. Chem. **17**, 363–372 (1925)
22. X. Shan, J.H. Payer, Effect of polymer and ceramic crevice formers on the crevice corrosion of Ni-Cr-Mo Alloy 22. Corrosion **66**, 105005 (2010)
23. N. Larche, P. Boillot, P. Dezerville, E. Johansson, J.M. Lardon, D. Thierry, Crevice corrosion performance of high-alloy stainless steels and Ni-based alloy in desalination industry. Desalin. Water Treat. **55**, 2491–2501 (2014)
24. B. Kobe, M. Badley, J.D. Henderson, S. Anderson, M.C. Biesinger, D. Shoesmith, Application of quantitative X-ray photoelectron spectroscopy (XPS) imaging: investigation of Ni-Cr-Mo alloys exposed to crevice corrosion solution. Surf. Interface Anal. **49**, 1345–1350 (2017)
25. E.F. Gloyna, L. Li, Supercritical Water Oxidation Model Development for Selected EPA Priority Pollutants. EPA (1995)
26. X. Shan, H. Ha, J.H. Payer, Comparison of crevice corrosion of fe-based amorphous metal and crystalline Ni-Cr-Mo alloy. Metall. Mater. Trans. A **40**, 1324–1333 (2008)
27. N. Ebrahimi, P. Jakupi, J.J. Noel, D. Shoesmith, The role of alloying elements on the crevice corrosion behavior of Ni-Cr-Mo alloys. Corrosion **71**, 1441–1451 (2015)
28. R.S. Lillard, M.P. Jurinski, J.R. Scully, Crevice corrosion of alloy 625 in chlorinated astm artificial ocean water. Corrosion **50**, 251–265 (1994)
29. N. Ebrahimi , P. Jakupi, J.J. Noël, D.W. Shoesmith, The Influence of composition on the crevice corrosion of Ni-Cr-Mo alloys. NACE Int. (2016)
30. E.C. Hornus, M.A. Rodríguez, R.M. Carranza, C.M. Giordano, R.B. Rebak, Effect of environmental variables on crevice corrosion susceptibility of Ni–Cr–Mo alloys for nuclear repositories. Procedia Mater. Sci. **8**, 11–20 (2015)
31. S.S. Haudet, M.A. Rodríguez, R.M. Carranza, R.B. Rebak, Effect of alloy composition on the crevice corrosion resistance of nickel alloys. NACE Int. (2012)

32. E.C. Hornus, M.A. Rodríguez, Effect of temperature on the crevice corrosion resistance of Ni-Cr-Mo alloys as engineered barriers of nuclear repositories. Procedia Mater. Sci. **1**, 251–258 (2012)
33. R.B. Rebak, Efficiency of inhibitors of chloride-induced crevice corrosion of a Ni-Cr-Mo alloy. NACE Int. (2011)
34. M.R. Ortíz, M.A. Rodríguez, R.M. Carranza, R.B. Rebak, Efficiency of inhibitors for chloride-induced crevice corrosion of Alloy 22. Mater. Res. Soc. Symp. Proc. **1475**, 495–500 (2012)
35. P. Kritzer, N. Boukis, E. Dinjus, Corrosion of alloy 625 in high-temperature, high-pressure sulfate solutions. Corrosion **54**, 689–699 (1998)
36. P. Kritzer, N. Boukis, E. Dinjus, The corrosion of nickel-base alloy 625 in sub- and supercritical aqueous solutions of HNO_3 in the presence of oxygen. J. Mater. Sci. Lett. 771–773 (1999)
37. S. Iimura, H. Yakuwa, M. Miyasaka, Corrosion behavior of several commercial alloys in supercritical water ncluding high concentration of chloride ions. NACE Int. (2001)
38. G.G. Chase, W.E. Glad, T.A. Knoebel, transgranular and intergranular corrosion of 18Cr-3Ni-12Mn stainless steel immersed in seawater. Corrosion **47**, 74–79 (1991)
39. J.R. Crum, L.E. Shoemaker, Corrosion resistance of nickel alloys in caustic solutions. NACE Int. (2006)
40. N. Boukis, P. Kritzer, M. Schacht, P. Kritzer, The initiation of material corrosion in semicritical and upercritical aqueous hydrochloric acid solutions. NACE Int. (1999)
41. Y. Watanabe, T. Kobayashi, T. Adschiri, Significance of water density on corrosion behavior of alloys in supercritical water. NACE Int. (2001)
42. S.M. Bruemmer, E.P. Simonen, P.M. Scott, P.L. Andreson, G.S. Was, J.L. Nelson, Radiation-induced material changes and susceptibility to intergranular failure of light-water-reactor core internals. J. Nucl. Mater. **274**, 299–314 (1999)
43. S.M. Bruemmer, Microstructural and microchemical mechanisms controlling intergranular stress corrosion cracking in light-water-reactor systems. J. Nucl. Mater. **216**, 348–363 (1994)
44. H.I. Khan, N. Zhang, H. Xu, Z. Zhu, D. Jiang, Effect of temperature on corrosion fatigue cracking behavior of inconel 625 in steam and supercritical water. Int. J. Chem. Mater. Eng. **12** (2018)
45. J. Shi, J. Wang, D.D. Macdonald, Prediction of crack growth rate in Type 304 stainless steel using artificial neural networks and the coupled environment fracture model. Corros. Sci. **89**, 69–80 (2014)
46. J. Yang, Y. Li, A. Xu, B. Fekete, D.D. Macdonald, The electrochemical properties of alloy 690 in simulated pressurized water reactor primary water: Effect of temperature. J. Nucl. Mater. **518**, 305–315 (2019)
47. G.S. Was, P. Ampornrat, G. Gupta, S. Teysseyre, C. Pister, Corrosion and stress corrosion cracking in supercritical water. J. Nucl. Mater. **371**, 176–201 (2007)
48. H. Kim, D.B. Mitton, R.M. Latanision, Stress corrosion cracking of alloy 625 in pH 2 aqueous solution at high temperature and pressure. Corrosion, **67**, 035002 (2011)
49. T. Moss, G.S. Was, accelerated stress corrosion crack initiation of alloys 600 and 690 in hydrogenated supercritical water. MMTA **48**, 1613–1628 (2017)
50. L. Zhang, K. Chen, D. Du, W. Gao, P.L. Andresen, X. Guo, Characterizing the effect of creep on stress corrosion cracking of cold worked Alloy 690 in supercritical water environment. J. Nucl. Mater. **492**, 32–40 (2017)
51. H. Kim, D.B. Mitton, R.M. Latanision, Corrosion behavior of Ni-base alloys in aqueous HCl solution of pH 2 at high temperature and pressure. Corros. Sci. **52**, 801–809 (2010)
52. H. Kim, D.B. Mitton, R.M. Latanision, Effect of pH and temperature on corrosion of Nickel-Base alloys in high temperature and pressure aqueous solutions. J. Electrochem. Soc. **157**, 194–199 (2010)
53. B. Hertweck, T.G. Steigerwald, N.S.A. Alt, E. Schluecker, Different corrosion behaviour of autoclaves made of nickel base alloy 718 in ammonobasic and ammonoacidic environments. J. Supercrit. Fluids **95**, 158–166 (2014)
54. D.M. Symons, A comparison of internal hydrogen embrittlement and hydrogen environment embrittlement of X-750. Eng. Fract. Mech. **68**, 751–771 (2001)

55. M. Cornet, C. Bertrand, M.D.C. Belo, Hydrogen embrittlement of ultra-pure alloys of the inconel 600 type: Influence of the additions of elements (C, P, Sn, Sb). Metall. Trans. A **13A**, 141–144 (1982)
56. A. Hirose, Y. Arita, Y. Nakanishi et al., Decrease in hydrogen embrittlement sensitivity of INCONEL 718 by laser surface softening. Mater. Sci. Eng. A **219**, 71–79 (1996)
57. F. Lecoester, J. Chene, D. Noel, Hydrogen embrittlement of the Ni-base alloy 600 correlated with hydrogen transport by dislocations. Mater. Sci. Eng. A **262**, 173–183 (1999)
58. O.A.E. Kebir, A. Szummer, Comparison of hydrogen embrittlement of stainless steels and nickel-base alloys. Int. J. Hydrogen Energy **27**, 793–800 (2002)
59. Z. Guo, H. Liang, M. Zhao et al., Effect of boron addition on hydrogen embrittlement sensitivity in Fe–Ni based alloys. Mater. Sci. Eng. A **527**, 6620–6625 (2010)
60. K. Miyata, M. Igarashi, Effect of ordering on susceptibility to hydrogen embrittlement of a ni-base superalloy. Metall. Trans. A **23A**, 953–961 (1992)
61. M.L. Wayman, G.C. Smith, The hydrogen embrittlement of Fe-Ni martensites. Metall. Trans. **1**, 1189–1193 (1970)
62. T. Michler, J.R. Naumann, M.P. Balogh, Hydrogen environment embrittlement of solution treated Fe–Cr–Ni super alloys. Mater. Sci. Eng. A **607**, 71–80 (2014)
63. R.W. Smith, G.S. Was, Application of molecular dynamics to the study of hydrogen embrittlement in Ni-Cr-Fe alloys. PhRvB **40**, 10322–10336 (1989)
64. M. Osada, T. Sato, M. Watanabe, M. Shirai, K. Arai, Catalytic gasification of wood biomass in subcritical and supercritical water. Combust. Sci. Technol. **178**, 537–552 (2006)
65. X. Tang, S. Wang, D. Xu, Y. Guo, J. Zhang, Y. Wang, Corrosion behavior of Ni-based alloys in supercritical water containing high concentrations of salt and oxygen. Ind. Eng. Chem. Res. **52**, 18241–18250 (2013)
66. G.A. Young, T.E. Capobianco, M.A. Penik, B.W. Morris, J.J. McGee, The mechanism of ductility dip cracking in nickel-chromium alloys. Weld. Res. **87**(2), 31S-43S (2008)
67. M.G. Pujar, R.K. Dayal, A.S. Khanna, E.W. Kreutz, Effect of laser surface melting on the corrosion resistance of chromium-plated 9Cr-1 Mo ferritic steel in an acidic medium. JMatS **28**, 3089–3096 (1993)
68. Y. Behnamian, A. Mostafaei, A. Kohandehghan, B.S. Amirkhiz, L.L. Jing, A comparative study of oxide scales grown on stainless steel and nickel-based superalloys in ultra-high temperature supercritical water at 800°C. Corros. Sci. **106**, 188–207 (2016)
69. Y. Daigo, Y. Watanabe, K. Sugahara, T. Isobe, Compatibility of nickel-based alloys with supercritical water applications: aging effects on corrosion resistance and mechanical properties. Corrosion **62**(2), 174–181 (2006)
70. L. Tan, X. Ren, K. Sridharan, T.R. Allen, Effect of shot-peening on the oxidation of alloy 800H exposed to supercritical water and cyclic oxidation. Corros. Sci. **50**, 2040–2046 (2008)
71. P. Wang, S. Grdanovska, D.M. Bartels, G.S. Was, Effect of radiation damage and water radiolysis on corrosion of FeCrAl alloys in hydrogenated water. J. Nucl. Mater. **533**, 152108 (2020)
72. D.J. Young, in *Chapter 7—Alloy Oxidation III: Multiphase Scales*. High Temperature Oxidation and Corrosion of Metals (Elsevier, 2016), pp. 335–392
73. Y. Niu, S. Wang, F. Gao, Z.G. Zhang, F. Gesmundo, The Nature of the third-element effect in the oxidation of Fe–xCr–3 at.% Al alloys in 1atm O_2 at 1000°C. Corros. Sci. **50**, 345–356 (2008)
74. D. Pan, R. Zhang, H. Wang, C. Lu, Y. Liu, Formation and stability of oxide layer in FeCrAl fuel cladding material under high-temperature steam. J. Alloys Compd. **684**, 549–555 (2016)
75. K.G. Field, K.C. Littrell, S.A. Briggs, Precipitation of α' in neutron irradiated commercial FeCrAl alloys. Scr. Mater. **142**, 41–45 (2018)
76. Y. Yamamoto, B.A. Pint, K.A. Terrani, K.G. Field, Y. Yang, L.L. Snead, Development and property evaluation of nuclear grade wrought FeCrAl fuel cladding for light water reactors. J. Nucl. Mater. **467**, 703–716 (2015)
77. K.A. Unocic, Y. Yamamoto, B.A. Pint, Effect of Al and Cr content on air and steam oxidation of FeCrAl Alloys and commercial APMT alloy. Oxid. Met. **87**, 431–441 (2017)

78. S. Yoneda, S. Hayashi, S. Ukai, The Transition from transient oxide to protective Al_2O_3 Scale on Fe–Cr–Al alloys during heating to 1000 °C. Oxid. Met. **89**, 81–97 (2018)
79. C. Tang, A. Jianu, M. Steinbrueck, M. Grosse, A. Weisenburger, H.J. Seifert, Influence of composition and heating schedules on compatibility of FeCrAl alloys with high-temperature steam. J. Nucl. Mater. **511**, 496–507 (2018)
80. Y. Qiao, P. Wang, W. Qi, S. Du, Z. Liu, F. Meng, X. Zhang, K. Wang, Q. Liu, Z. Yao, C. Bai, X. Wang, Mechanism of Al on FeCrAl steam oxidation behavior and molecular dynamics simulations. J. Alloys Compd. **828**, 154310 (2020)
81. X. Zhou, L. Guo, Y. Wei, H. Wang, S. Mo, Effect of aluminum content on dislocation loops in model FeCrAl alloys. Nucl. Mater. Energy **21**, 100718 (2019)
82. M.N. Gussev, K.G. Field, Y. Yamamoto, Design, properties, and weldability of advanced oxidation-resistant FeCrAl alloys. Mater. Des. **129**, 227–238 (2017)
83. R. Qiang, A. Leong, J. Zhang, M.P. Short, Corrosion behavior of Fe–Cr–Si alloys in simulated PWR primary water environment. J. Nucl. Mater. **526**, 151735 (2019)
84. J. Eklund, B. Jönsson, A. Persdotter, J. Liske, J.E. Svensson, T. Jonsson, The influence of silicon on the corrosion properties of FeCrAl model alloys in oxidizing environments at 600 °C. Corros. Sci. **144**, 266–276 (2018)
85. W. Wu, G. Ran, Y. Li, S. Cong, Y. Sun, Early corrosion behaviour of irradiated FeCrAl alloy in a simulated pressurized water reactor environment. Corros. Sci. **174**, 108824 (2020)
86. V. Kochubey, H. Al-Badairy, G. Tatlock, J. Le-Coze, W.J. Quadakkers, Effects of minor additions and impurities on oxidation behaviour of FeCrAl alloys. Development of novel surface coatings compositions. Mater. Corros. **56**, 848–853 (2005)
87. T. Huang, D. Naumenko, P. Song, J. Lu, Q.W. Joseph, Effect of titanium addition on alumina growth mechanism on yttria-containing FeCrAl-base alloy. Oxid. Met. **90**, 671–690 (2018)
88. N. Liu, Z. Zhang, Y. Zhang,Y. Cui, D. Chen, Y. Zhao, S.S. Xu, H. Gao, Effects of Ce addition on the microstructure and mechanical properties of accident-tolerance Fe-Cr-Al fuel cladding materials. TMS 2018 147th Annual Meeting & Exhibition Supplemental Proceedings (2018), pp. 61–71
89. D.R. Sigler, Adherence behavior of oxide grown in air and synthetic exhaust gas on Fe-Cr-Al alloys containing strong sulfide-forming elements: Ca, Mg, Y, Ce, La, Ti, and Zr. Oxid. Met. **40**, 555–583 (1993)
90. B. Qiu, J. Wang, Y. Deng, M. Wang, S.Z. Qiu, A review on thermohydraulic and mechanical-physical properties of SiC, FeCrAl and Ti_3SiC_2 for ATF cladding. Nucl. Eng. Technol. **52**, 1–13 (2020)
91. D. Naumenko, B.A. Pint, W.J. Quadakkers, Current thoughts on reactive element effects in alumina-forming systems: in memory of John stringer. Oxid. Met. **86**, 1–43 (2016)
92. S. Hayashi, T. Maeda, Effect of Zr on initial oxidation behavior of FeCrAl alloys. Oxid. Met. **93**, 573–586 (2020)
93. Y. Okanda, M. Fukusumi, S. Nenno, J.B. Newkirk, The effects of Ce additions on the high temperature corrosion behavior of Fe-30Cr-5Al alloy in 80 Pet Na_2SO_4-20 Pct NaCI molten salt. Metall. Trans. A **14**, 2131–2142 (1983)
94. M. Ozawa, K.-I. Araki, Effect of La modification on the stability of coating alumina layer on FeCrAl alloy substrate. Surf. Coat Technol. **271**, 80–86 (2015)
95. J. Stringer, The reactive element effect in high-temperature corrosion. Mater. Sci. Eng. A **120–121**, 129–137 (1989)
96. P.Y. Hou, X.F. Zhang, R.M. Cannon, Impurity distribution in Al_2O_3formed on an FeCrAl alloy. Scr. Mater. **50**, 45–49 (2004)
97. H. Feng, Z. Jiang, H. Li, P. Lu, S. Zhang, H. Zhu, B. Zhang, T. Zhang, D. Xu, Z. Chen, Influence of nitrogen on corrosion behaviour of high nitrogen martensitic stainless steels manufactured by pressurized metallurgy. Corros. Sci. **144**, 288–300 (2018)
98. J. Dai, H. Feng, H.B. Li, Z.H. Jiang, T. Zhang, Nitrogen significantly enhances corrosion resistance of 316L Stainless steel in thiosulfate-chloride solution. Corros. Sci. **174**, 108792 (2020)

99. Z. Duan, H. Yang, Y. Satoh, K. Murakami, S. Kano, Z. Zhao, J. Shen, H. Abe, Current status of materials development of nuclear fuel cladding tubes for light water reactors. Nucl. Eng. Des. **316**, 131–150 (2017)
100. M. Silva, L. Herculano, A. Urcezino, W. Araujo, H. Abreu, P. Lima-Noto, Influence of Mo content on the phase evolution and corrosion behavior of model Fe–9Cr–xMo (x = 5, 7, and 9 wt%) alloys. J. Mater. Res. **30**, 1999–2007 (2015)
101. R.B. Rebak, V.K. Gupta, M. Larsen, Oxidation characteristics of two FeCrAl alloys in air and steam from 800°c to 1300°C. JOM **70**, 1484–1492 (2018)
102. N. Li, S.S. Parker, E.S. Wood, A.T. Nelson, Oxide morphology of a FeCrAl Alloy, kanthal APMT, following extended aging in air at 300 °C to 600 °C. Metall. Mater. Trans. A Phys. Metall. Mater. Sci. **49**, 2940–2950 (2018)
103. X. Han, Y. Wang, S. Peng, H. Zhang, Oxidation behavior of FeCrAl coated Zry-4 under high temperature steam environment. Corros. Sci. **149**, 45–53 (2019)
104. X. Huang, X.Y. Li, X.D. Fang, Research progress in FeCrAl alloys for accident-tolerant fuel cladding. Cailiao Gongcheng-J. Mater. Eng. **48**, 19–33 (2020)
105. H. Shi, R. Fetzer, C. Tang, D.W.V. Szabo, G. Mueller, The influence of y and nb addition on the corrosion resistance of Fe-Cr-Al-Ni model alloys exposed to oxygen-containing molten Pb. Corros. Sci. **179**, 109152 (2021)
106. G. Chen, H. Wang, H. Sun, Y. Zhang, P. Cao, J. Wang, Effects of Nb-doping on the mechanical properties and high-temperature steam oxidation of annealing FeCrAl fuel cladding alloys. Mater. Sci. Eng. A **803**, 140500 (2020)
107. T. Amano, High-temperature oxidation of FeCrAl(Y, Pt) alloys in oxygen–water vapour. Mater. High Temp. **28**, 342–348 (2011)
108. M.N. Gussev, K.G. Field, E. Cakmak, Y. Yamamoto, in *Mechanical Behavior and Structure of Advanced Fe-Cr-Al Alloy Weldments*. Proceedings of the 18th International Conference on Environmental Degradation of Materials in Nuclear Power Systems—Water Reactors, vol. 2 (2018), pp. 201–214
109. A. Persdotter, J. Eklund, J. Liske, T. Jonsson, Beyond breakaway corrosion—influence of chromium, nickel and aluminum on corrosion of iron-based alloys at 600 °C. Corros. Sci. **177**, 108961 (2020)
110. M.S. Elbakhshwan, S.K. Gill, A.K. Rumaiz, J. Bai, S. Ghose, R.B. Rebek, L.E. Ecker, High-temperature oxidation of advanced FeCrNi alloy in steam environments. Appl. Surf. Sci. **426**, 562–571 (2017)

Chapter 3
Corrosion Characteristics of Candidate Alloys

3.1 Corrosion Characteristics of Ni–Cr Alloys

The addition of Cr can improve the corrosion resistance of Ni in the oxidizing media. When the alloy contains 10 wt% Cr or more, the corrosion resistance and oxidation resistance of the Ni–Cr alloy is significantly enhanced. As shown in Table 3.1, the grades and chemical compositions of Ni–Cr corrosion-resistant alloys mainly include Inconel 600, Inconel 601, Inconel 690 and Inconel X-750. These Ni-based alloys commonly contain a Cr content as high as 14–31 wt% and 79–90 wt% of combined content of Ni and Cr.

3.1.1 *Corrosion Weight Change*

Various researchers have found that the corrosion weight change of Inconel 690 in SCW first increases and then decreases with increasing exposure time, in most cases. The corrosion weight gain of the as-received sample and the grain boundary engineering (GBE) sample of Inconel 690 are less than 0.12 mg cm^{-2} after 10 weeks exposure in SCW. The as-received sample exhibits a higher corrosion rate than the GBE sample, which proves that the oxide layer of the former cannot be firmly attached on the sample surface [1]. The corrosion weight gain of Inconel 690 at 550 °C is greater than that at 450 °C, and it increases first and then decreases with increasing exposure time, while there are some fluctuations at 450 °C [2]. These fluctuation can be attributed to the oxidation and mass loss caused by pitting corrosion and oxide dissolution. The corrosion weight gains of Inconel 690 at the two temperatures above are less than 0.12 mg cm^{-2} at 25 MPa for 500 h.

Figure 3.1 summarizes the corrosion weight changes with exposure time increasing for typical Ni–Cr alloys exposed to SCW for 0–1700 h. Overall, the weight changes of these alloy samples are reflected by corrosion weight gains after exposure

D. Xu and S. Guo, *Corrosion Characteristics, Mechanisms and Control Methods of Candidate Alloys in Sub- and Supercritical Water*,
https://doi.org/10.1007/978-981-16-5525-8_3

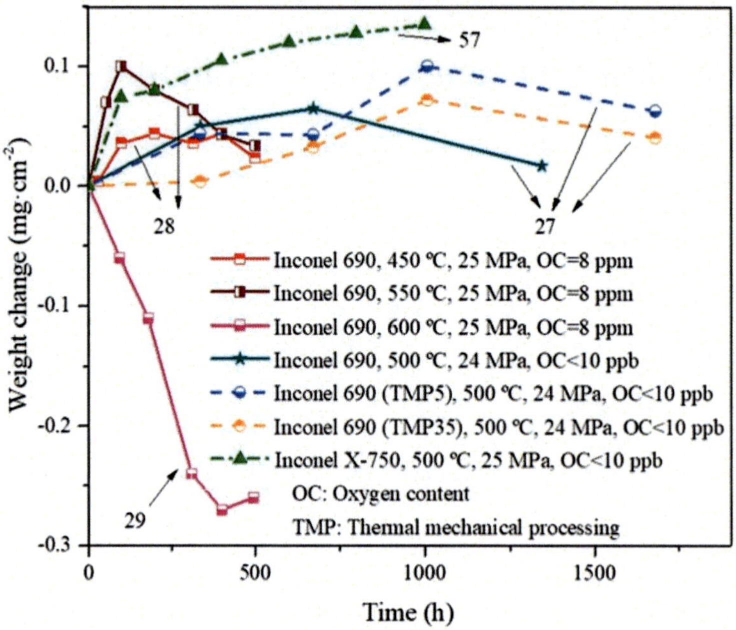

Fig. 3.1 Weight changes of typical Ni–Cr alloys exposed to supercritical water with time increasing

to SCW at different conditions. They are in the range of 0–0.15 mg cm^{-2}, and increase first and then decrease with increasing exposure time. Notably, the corrosion weight gain of Inconel X-750 increases over the whole period, likely due to the relatively short test time. Only Inconel 690 exhibits corrosion weight loss over the entire time range in SCW at 600 °C, 25 MPa and 8 ppm of oxygen content (OC). Its corrosion weight loss increases with increasing time and this change is much more rapid at the beginning of oxidation but tends to be moderate after 400 h. This weight loss could be due to exfoliation of the corrosion layer, but specific reasons need further investigation. The value of the corrosion weight change of Inconel 690 increases from 450 to 550 °C but becomes negative at 600 °C under the same time conditions, meaning that increasing temperature promotes the corrosion of Inconel 690 in SCW. Based on the corrosion weight gain change in Fig. 3.1, it can be postulated that thermal mechanical processing (TMP) can improve the corrosion resistance of Inconel 690 within 700 h of exposure time, but is ineffective after 800 h. Moreover, the maximum corrosion weight gain occurs on Inconel X-750 in SCW at 500 °C and 25 MPa with an oxygen content of below 10 ppb, showing that the corrosion resistance of Inconel X-750 is lower than those of Inconel 690 (TMP5) and Inconel 690 (TMP35) under the same conditions. This can be ascribed to the difference of the contents of Ni and Cr (see Table 3.1), especially for a significantly lower Cr content in Inconel X-750 than Inconel 690.

Table 3.1 Chemical compositions of main Ni–Cr alloys

Alloy	Chemical composition (wt%)								
	Ni	Cr	Fe	Mn	Si	C	S	P	Others
Inconel 600	Bal	14–17	6–10	≤ 1.0	≤ 0.5	≤ 0.10	≤ 0.015	≤ 0.03	Co $\le$ 0.04
Inconel 601	58–63	21–25	Bal	≤ 1.0	≤ 0.5	≤ 0.10	≤ 0.015	≤ 0.03	0–1.7
Inconel 690	Bal	27–31	7–11	≤ 0.5	≤ 0.5	≤ 0.05	≤ 0.015	≤ 0.02	Cu $\le$ 0.50
Inconel X-750	Bal	14–17	5–9	≤ 1.0	≤ 0.5	≤ 0.10	≤ 0.010	≤ 0.02	Al 0.4–1.0 Ti 2.25–2.75 Nb 0.70–1.20

Note Bal. represents balance

3.1.2 Oxides Formed on Alloy Surface

For the corrosion behavior of Ni–Cr alloys in SCW, it is found that Inconel 690 produces a double-layer oxide film structure in SCW, i.e., a dense inner (barrier) layer rich in Cr and a porous, loose outer layer rich in Ni [3]. Specifically, the inner oxide layer is composed of Cr oxides including Cr_2O_3 and $NiCr_2O_4$ spinel, while the outer oxide layer consists Fe oxides and Ni oxides including $NiFe_2O_4$ spinel, NiO and $Ni(OH)_2$ [2, 4]. We should note that because the point defective chromia lattice contains Ni and Fe interstitials, "$NiCr_2O_4$" may, in fact, be Cr_2O_3 containing a high concentration of Ni^{2+} interstitials. It is believed that the inner oxide layer is formed by the transport of anions, or conversely, of oxygen vacancies, while the outer layer is formed by the diffusion of cations but continuously dissolved [4]. Inconel 690 forms a multi-layer oxide film in hydrogen-containing SCW, and the outer layer is composed of granular grains NiO and $NiFe_2O_4$ rich in Ni but poor in Cr, and the dense inner layer is composed of spinel and Cr_2O_3 rich in Cr and Fe but poor in Ni [5]. A thin outer layer of spinel is found on Inconel 182 in high-temperature and high-pressure environment, while the inner oxide layer consists of spinel and Cr_2O_3 [6]. Furthermore, a duplex-structure corrosion layer is also observed on Inconel 600 in SCW containing organic sulfides. The outer layer contains NiS, while the inner layer is composed of Cr-rich spinel oxides and the fast channels for outward transport of metal cations [7]. It should be noted that the severe surface plastic deformation, which is caused by surface treatment process such as work-hardened treatment, shot peening and other mechanical working, results in the formation of a thinner Cr-rich oxide layer and the occurrence of a deeper international oxidation [8]. Besides, work hardening induced by surface finishes may lead to a considerable thinning of the oxide scale due to the formation of a Cr-rich internal oxide layer against ionic species responsible for its growth and Fe cations implied in the external layer growth [9].

Table 3.2 summarizes the studies on the oxide film structures and compositions of Ni-based alloys in SCW as given in different reports. It can be observed that the

Table 3.2 Summary of oxide film structure and composition on Ni-based alloys tested in sub- and supercritical water

Alloy	Pressure (MPa)	Temperature (°C)	Test environment	Outer oxide layer	Medium oxide layer or oxide layer sequence	Inner oxide layer	Ref
Hastelloy C-276	25	500	SCWG (COD = 52,000 ppm)	Isolated Ni/Fe-rich spinel particles	–	Cr-rich oxides (Cr-rich spinels and Cr_2O_3) and little MoO_3	[10]
	25	500	SCW	Ni-rich layer	–	Cr_2O_3/$NiCr_2O_4$ mixed layer	[11]
	25	500–600	SCW (OC < 10 ppb)	NiO layer	–	Cr_2O_3/$NiCr_2O_4$ mixed layer	[12]
	25	500	SCW (OC = 100,000 ppb)	Coarse Fe-rich outer layer ($NiFe_2O_4$)	–	Fine Cr rich inner layer $NiCr_2O_4$	[13]
Inconel 625					–		
	24	400–500	SCW (2 wt% H_2O_2)	$Ni(OH)_2$/NiO outer layer	$Ni(OH)_2$-NiO-$NiCr_2O_4$-Cr_2O_3-Inconel 625	Cr_2O_3 inner layer	[14]
	24.8	700	SCW (DO = 150 ppb)	A spinel oxide $Ni(Cr, Fe)_2O_4$	–	Continuous Cr_2O_3	[15]
Inconel 625/718	25	500–600	SCW (DO = 2 ppm)	A Ni/Fe-rich outer layer	NiO-spinel-$(Cr_xFe_{1-x})_2O_3$-(Fe, Ni)	A Cr-rich spinel layer	[16]
Inconel 718	27	527.5, 600	Sub/super/ultrasuper-critical water (OC < 50 ppb)	Nb_2O_5	$NiFe_2O_4$	Cr_2O_3	[17]
		325		$NiFe_xCr_{2-x}O_4$ spinel	–	$Ni(OH)_2$ and CrO_4^{2-} film	
Inconel 625		527.5, 600		Nb_2O_5	$NiFe_xCr_{2-x}O_4$	Cr_2O_3	
		325		$NiFe_xCr_{2-x}O_4$ spinel	–	$Ni(OH)_2$ and CrO_4^{2-} film	

(continued)

Table 3.2 (continued)

Alloy	Pressure (MPa)	Temperature (°C)	Test environment	Outer oxide layer	Medium oxide layer or oxide layer sequence	Inner oxide layer	Ref
	24	500	SCW (DO = 5000 ppm)	Fe oxides, spinel structures, and $Ni(OH)_2$, loose, bar shaped particles	–	A compact layer of acicular-like NiO oxide	[18]
Inconel 690	25	450–550	SCW (DO = 8 ppm)	Ni and Fe rich porous layer: $NiFe_2O_4$ spinel, NiO and $Ni(OH)_2$	–	Cr-rich compact inner layer: Cr_2O_3 and $NiCr_2O_4$ spinel	[2]
	25	400	Hydrogenated SCW	Granular crystallites (NiO and $NiFe_2O_4$)	NiO/$NiFe_2O_4$-Ni(Fe, Cr)$_2O_4$-Cr_2O_3-alloy	Compact layer (spinel and Cr_2O_3), Fe-rich spinel on top of continuous Cr_2O_3 layer	[5]
	25	600	SCW (DO = 8 ppm)	Outer layer rich in Ni and Fe	–	Inner layer rich in Cr	[3]
Inconel X-750	25	500	SCW (OC = 100,000 ppb)	Coarse Fe rich outer layer ($NiFe_2O_4$)	–	Fine Cr rich inner layer $NiCr_2O_4$	[13]

Note DO is dissolved oxygen, and OC is oxygen content

inner layer of the double-layer oxide film is mainly composed of Cr oxides such as Cr_2O_3 and $NiCr_2O_4$, and the outer layer is primarily composed of Fe oxides and Ni oxides such as NiO, $Ni(OH)_2$ and $NiFe_2O_4$. The oxides of Ni, Fe and Cr (e.g., $NiFe_2O_4$ and $NiCr_2O_4$) can be detected between the inner and outer oxide layers due to the presence of hydrogen, which may be attributed to the outward diffusion of Ni or Fe through the short-cut paths is suppressed by the outer oxide layer, resulting in their oxidation to form an intermediate layer. In corrosive SCW environments as employed in most investigations, the concentration of dissolved oxygen is generally small. In the future, it will be necessary to study the structure and composition of oxide layers on the alloy surfaces in the presence of added oxidants and other aggressive ions, especially for SCWO, in order to fully capture the impact of redox conditions. Certainly, similar exploration in SCW with a large amount of H_2 is sufficient for the development of SCWG. Although there have been many reports on the oxide film structure and composition on Ni–Cr alloys in SCW, few studies check the effects and properties of various oxidation products. It is important to fully understand the chemical stability of each oxide so as to evaluate the alloy corrosion resistance in SCW.

Figure 3.2 illustrates the oxide film structure of Inconel 690 at SCW conditions. As shown in Fig. 3.2a, the electron probe micro-analyzer images of Inconel 690 exposed to SCW at 550 °C for 500 h indicates that the outer oxide layer is rich in

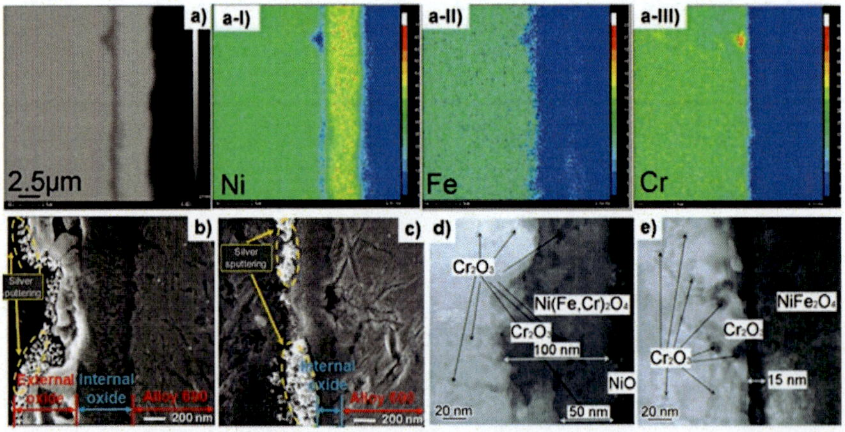

Fig. 3.2 Oxide film structures of Inconel 690 at SCW conditions **a** the distributions of Fe, Ni and Cr in the oxide film formed on the Inconel 690 surface exposed to SCW at 550 °C, 25 MPa for 500 h [2], **b** the cross-section SEM image of the Inconel 690 sample exposed to hydrogenated SCW at 600 °C, 25 MPa for 335 h [4], **c** the cross-section SEM image of the Inconel 690 sample exposed to hydrogenated SCW at 600 °C, 25 MPa for 1740 h [4], **d** the high-magnification scanning transmission electron microscope (STEM) image for the thick inner layer under a small amount of NiO in the oxide/alloy interface region of Inconel 690 exposed to hydrogenated SCW at 400 °C, 25 MPa for 1000 h [5], **e** the high-magnification STEM image for the thin layer under a large amount of $NiFe_2O_4$ in the oxide/alloy interface region of Inconel 690 exposed to hydrogenated SCW at 400 °C, 25 MPa for 1000 h [5]

Ni and Fe but Cr is enriched in the inner oxide layer [2], in accordance with other researcher's findings on the oxide structure. Herein, the Ni and Cr images also exhibit a complementary relationship that the Cr-rich region is poor in Ni and the Ni-rich region is poor in Cr. The oxide layer produced on Inconel 690 in the hydrogenated SCW at 600 °C, 25 MPa for 335 h is mainly divided into an inner layer and an outer layer (see Fig. 3.2b, c). The outer oxide layer exhibits an irregular thickness and its surface is discontinuous, and some oxide layers have exfoliated or show decreased adhesion, while the inner layer is uniform in thickness (about 500 nm). The outer oxide layer gradually disappears when the exposure time increases from 335 to 1740 h [4]. Small Cr_2O_3 particlesappears in the inner oxide layer on Inconel 690 and in the metal matrix after exposure to the hydrogenated SCW at 400 °C, 25 MPa for 1000 h (see Fig. 3.2d, e) [5]. This shows that corrosion does not stop at the metal/oxide interface, but continues to penetrate into the metal substrate. Therefore, it can be inferred that the corrosion of the Ni–Cr corrosion-resistant alloy in SCW originates from the outer surface of the alloy, and that fewer defects in the outer oxide layer are better for its protective properties. It can be seen from Figs. 3.2b–e that there is no obvious boundary between the inner and outer oxide layers, which are mainly distinguished according to the microstructure of the oxide layers. As the exposure time increases, the parallel and original neat oxide interface gradually appears to exhibit a serrated shape, showing that exposure time has great influence on the structure of the oxide film within 335–1740 h.

3.1.3 Comparison of Corrosion Characteristics

In spite of having a high price, Niobium exhibits the optimum corrosion resistance in SCW at 380–390 °C and 25 MPa among Inconel 625, SS 316L, Ni-20Cr and Niobium [19]. Schroer et al. [20] found that NiO inhibited the dissolution of Ni-containing oxide layers, so effectively protecting metal from further corrosion. The corrosion behaviors of binary Ni–Cr alloys with different Cr contents in SCW indicate that the Ni–Cr alloys have the best corrosion resistance when the Cr content is about 25 wt%. The Cr content is a decisive factor for whether the Ni–Cr alloys can be effective candidate materials of SCWO reactors for treating chloride-containing wastewater. Grain boundary engineering technology can change the grain boundary properties of Inconel 690, and thus change its oxidation behavior in SCW and hence obtain the required corrosion resistance of the material. Xu et al. [1] studied the oxidation behaviors of as-received Inconel 690 and treated with grain boundary engineering in SCW, and found that there were three distinct regions on the alloy surface, i.e., a flat region where flaky oxides arranged parallel to the surface, a rough region randomly distributed with polygonal oxide particles on the surface, and a region where flaky oxides were perpendicular to the surface. Study on the stress corrosion cracking of cold worked (CW) Inconel 690 proves that corrosion induced cracking controls the crack growth rate of CW Inconel 690 in subcritical water, while in supercritical water creep is the major factor that controls the cracking [20]. Intergranular carbides are detrimental to stress-corrosion cracking in CW Inconel 690. A recent study shows removing intergranular carbides by prior solution annealed treatment can decrease the crack growth rates by 4X in subcritical water of 360 °C [22].

Table 3.3 summarizes the experimental conditions and principal findings related to the corrosion of Ni–Cr corrosion-resistant alloys in subcritical and supercritical water, as reported in some studies. It is observed that the research on Ni–Cr alloys is mainly focused on Inconel 690, and some researchers have carried out experimental comparisons on a series of alloys with different Ni and Cr contents. The Ni–Cr alloys generally form double-layer oxide film structures when the corrosive environment is supercritical water, and, in a few cases in hydrogen-containing solution, organic solution, and hydrochloric acid. In general, the ranges of temperature, pressure, and test time cover 100–600 °C, 13–40 MPa, and 0–1740 h, respectively. The oxidant is generally dissolved oxygen but sometimes H_2O_2 is employed. Experiments in the HCl solution at 500 °C and 29.4 MPa shows that for Ni–Cr alloys with 10–30 wt% of Cr content, the Cr content increase is beneficial to the improvement of corrosion resistance of the alloys. Inconel 690 has excellent corrosion resistance among the commonly used Ni–Cr alloys, because it has a higher Cr content (27–31 wt%) than the other Ni–Cr alloys, including Inconel 600, Inconel 601, and Inconel X-750 (see Table 3.1), This is the reason why Inconel 690 becomes the representative grade of Ni–Cr corrosion-resistant alloys and is the focus of experimental studies. Notably, NiCr25 performs best in the CH_2Cl_2 solution at 100–415 °C and 40 MPa other than NiCr45 with the highest Cr content among NiCr0, NiCr25, NiCr30, NiCr35 and NiCr45 [20]. Accordingly, the influence of the Cr content on the corrosion resistance and corrosion rate of alloys with a Cr content higher than 30 wt% in SCW requires further discussion. It can be inferred from Table 3.3 that the alloy grain size is a decisive factor in corrosion mechanism, because oxygen can diffuse into the metal matrix through grain boundaries to internally oxidize the metal. Thereby, improving alloy properties via changing grain boundary characteristics is an important method of improving the corrosion behaviors of alloys in SCW.

3.2 Corrosion Characteristics of Ni–Cr–Mo Alloys

Ni–Cr–Mo alloys have good corrosion resistance in a wide range of corrosive environments, which can withstand not only the oxidizing media (e.g., oxidizing acids containing F^- and Cl^-) but also the reducing media. Compared with Ni–Cr alloys, the higher Mo content in Ni–Cr–Mo alloys can strongly improve the pitting corrosion resistance, crevice corrosion resistance, and corrosion resistance to non-oxidizing acids. When the Cr content is constant, an increase in the Mo content can reduce the corrosion rates of Ni-based alloys [24]. Hastelloy C-276, Inconel 625 and Incoloy X-750 suffer from general corrosion in SCW, and Incoloy X-750 exhibits the highest corrosion rate, likely due to the lack of Mo [13]. Table 3.4 lists the chemical compositions of major Ni–Cr–Mo alloys, including Hastelloy C-276, Hastelloy C-22, Hastelloy C-2000, Inconel 625, and Inconel 718. These alloys contain complex

Table 3.3 Corrosion results of Ni–Cr alloys in sub- and supercritical water

Ref.	Alloy	Test solution	Oxidant	Temperature (°C)	Pressure (MPa)	Exposure time (h)	Main conclusions
[49]	Ni–Cr alloys (10–30 wt% Cr)	1 mmol·L^{-1} HCl	0.6 mol/L H_2O_2	500	24	0–50	The corrosion resistance of the Ni–Cr alloy increases significantly with the Cr content increasing up to 20 wt%
[46]	NiCr0, NiCr25, NiCr30, NiCr35, NiCr45	CH_2Cl_2 aqueous solution	H_2O_2	100–415	40	48, 120, 480	NiCr25 performs best in the experiments for 120 h, and NiO retards the degradation of the Cr-rich oxide
[27]	Inconel 690 (AR and GBE samples)	Water	OC < 10 ppb	500	24	336–1680	The required corrosion performance in SCW can be achieved by grain boundary engineering
[28]	Inconel 690	Aerated pure water	OC = 8 ppm	450, 550	25	500	General, pitting, and nodule corrosions are observed, and the corrosion of TiN inclusions is faster than that of the alloy matrix

(continued)

Table 3.3 (continued)

Ref.	Alloy	Test solution	Oxidant	Temperature (°C)	Pressure (MPa)	Exposure time (h)	Main conclusions
[30]	Inconel 690 (polished and as-machined samples)	Ultrapure water	–	600	25	335, 840, 1740	The corrosion mechanism is dependent on the alloy grain size, and the diffusion within grain boundaries is accelerated at high temperatures
[31]	Inconel 690	Hydrogenated SCW	DH = 107 $mL \cdot kg^{-1}$	400	25	1000	A double-layer oxide structure is formed on the surface of Inconel 690
[29]	Inconel 690	Aerated SCW	OC = 8 ppm	600	25	494	The oxide penetration depth along grain boundaries increases with exposure time increasing
[32]	Inconel 182	Water	–	320	13	1440	There are different types of duplex-structural oxide layers on Inconel 182

Note AR is as-received, GBE is grain boundary engineering, OC is oxygen content, and DH is dissolved hydrogen

Table 3.4 Chemical compositions of major Ni–Cr–Mo alloys

Alloy	Chemical composition (wt%)														
	Ni	Cr	Mo	W	Fe	C	Si	Mn	P	S	Nb	Al	Ti	Cu	Co	B
Hastelloy C-276	Bal	14.5–16.5	15.0–17.0	3.0–4.5	4.0–7.0	≤0.02	≤0.08	≤1.0	≤0.04	≤0.03	–	–	–	–	–	–
Hastelloy C-22	Bal	20.0–22.5	12.5–14.5	2.5–3.5	2.0–6.0	≤0.015	≤0.08	≤0.5	≤0.04	≤0.02	–	–	–	–	–	–
Hastelloy C-2000	Bal	20.0–24.0	15.0–17.0	–	≤3.0	≤0.010	≤0.08	≤0.5	≤0.04	≤0.02	–	≤ 0.5	–	1.3–1.9	–	–
Inconel 625	≥58	20.0–23.0	8.0–10.0	–	≤5.0	≤0.10	≤0.5	≤0.5	≤0.015	≤0.015	3.5–4.2	≤ 0.4	≤ 0.4	–	–	–
Inconel 718	50–55	17–21	2.8–3.3	–	Bal	≤0.08	≤0.35	≤0.35	≤0.01	≤0.01	4.8–5.5	0.2–0.8	0.65–1.15	≤0.3	≤1.0	≤0.006

Note Bal. represents balance

elemental compositions, but are enriched in Ni, Cr and Mo. The two most commonly used Ni–Cr–Mo alloys are Hastelloy C-276 and Inconel 625. Hastelloy C-276 and Hastelloy C-2000 have high Mo contents (15.0–17.0 wt%), Hastelloy C-22, and Inconel 625 have medium Mo contents (8.0–14.5 wt%), and Inconel 718 has a low Mo content (~3.0 wt%). The Cr contents of Hastelloy C-22, Hastelloy C-2000 and Inconel 625 are higher than 20 wt%. The high Cr and Mo contents are responsible for the good corrosion resistance of these alloys.

3.2.1 Corrosion Weight Change

Table 3.5 illustrates the corrosion weight gain of Ni–Cr–Mo alloys in SCW as a function of exposure time. The corrosion weight gain of Hastelloy C-276 at 500 °C or Inconel 625 at 700 °C follows a parabolic law. Similarly, the corrosion weight gain of Hastelloy C-276 in a SCWG environment at 500 °C, 25 MPa and Incoloy 625 in SCW at 700 °C, 24.8 MPa with different oxygen contents (i.e., 150 ppb, 1 and 8.3 ppm) follow a near-parabolic law with increasing exposure time [10, 15]. However, it is necessary to verify the inconsistency with respect to the rate constant and the reaction order for Hastelloy C-276 in SCW at 500 °C and 25 MPa in Table 3.5. Sun et al. [14] found that the corrosion weight gain of Incoloy 625 in SCW increased with increasing exposure time, and the maximum and minimum weight gains separately occurred at 400 and 450 °C within the temperature range of 400–500 °C. Temperature and pressure affect the corrosion behavior of Ni–Cr–Mo alloys under SCW conditions. Zhang et al. [12] reported that an increase in temperature within 500–600 °C promoted oxidation and thus resulted in a thicker oxide layer, a larger corrosion weight gain, and the more remarkable tendency of corrosion layer exfoliation. The increase in temperature and pressure also slightly increase the corrosion rate of Inconel 740H at 600–700 °C and 60–100 MPa [43]. Ni and Cr are the two principal alloying elements, and the oxygen concentration or pressure does not have a significant effect on the Ni ion concentration, whereas the concentration of Cr ions increases as the oxygen concentration and pressure increase [26].

Figure 3.3 elucidates the corrosion weight changes of different Ni–Cr–Mo alloys exposed to SCW with the exposure time increasing from 0 to more than 1000 h. Overall, in most cases, the corrosion weight change is reflected by weight gain, and first increases and then decreases with increasing exposure time over a wide range of time. This may mainly be explained by that they initially form adherent oxides on the alloy surface but then suffer oxide exfoliation after a certain period of time. The negative weight gain is attributed to the exfoliation of the corrosion layer and the dissolution of metal elements and oxides. The weight gain of Inconel 625 increases rapidly at short times (0–200 h) in SCW containing 2 wt% H_2O_2 at 400 or 500 °C. Their almost highest values in the same time range (see Fig. 3.3) shows that the weight gain of Inconel 625 under oxygen-containing conditions is greater than that in the oxygen-free solution and the solution with a low oxygen content, so the presence of oxygen promotes the corrosion of Ni–Cr–Mo alloys in

Table 3.5 Equations used to describe the relationships between weight changes and exposure times of Ni–Cr–Mo and Ni–Fe–Cr alloys exposed to supercritical water

Equation	Alloy	Oxygen content	k	Temperature (°C)	Pressure (MPa)	n	R^2	Kinetics	Ref
$w^n = kt$	Inconel 617	1–25 ppb	2.10×10^{-6}	500	25	2.2	–	Parabolic	[27]
$w^n = kt$	Inconel 625	150 ppb	2.6×10^{-5}	700	24.8	2.10	–	Parabolic	[15]
$\Delta w = kt^n$	Hastelloy C-276	–	0.00178	500	25	0.49	0.99	Parabolic	[10]
$\Delta w = kt^n$	Hastelloy C-276	–	–	500	25	0.57	0.95	Parabolic	[11]

Note w is weight change, Δw is weight gain, t is exposure time, n is reaction order, k is rate constant or reaction constant, and the fitting parameter R^2 with a value approaching 1.00 denotes the increasing consistence of the data with the fitting

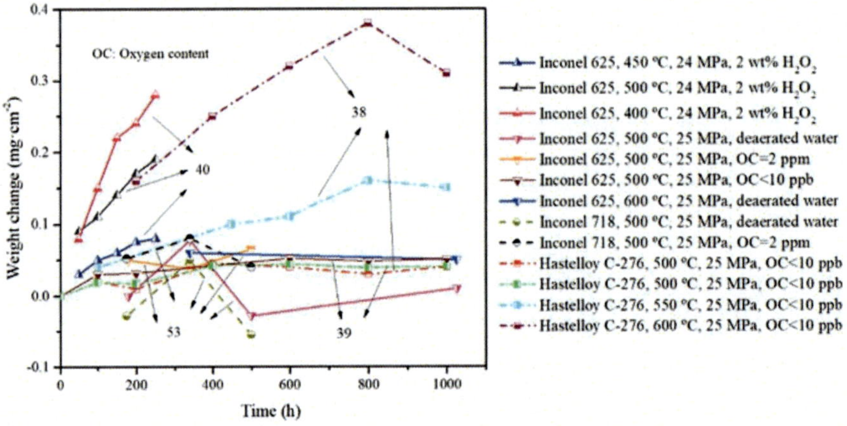

Fig. 3.3 Weight changes of Ni–Cr–Mo alloys exposed to supercritical water with time increasing

SCW. Apparently, further experimental studies over longer exposure times are still required here. The corrosion weight gain of Hastelloy C-276 at 600 °C is much higher than that of Inconel 625, suggesting that the corrosion resistance of the former may be worse than that of the latter, in accordance with the conclusions of Tang et al. [24]. As the temperature increases from 500 to 600 °C, the corrosion weight gain of Hastelloy C-276 gradually increases as a result of more oxide formation reflecting an enhancement of the transport of elements in the alloy and cations and oxygen ions in the oxide in SCW at the higher temperature. As shown in Fig. 3.3, the temperature effect noted above is mainly validated in SCW with a low content of dissolved oxygen in the corrosion tests on Hastelloy C-276 and Inconel 625. However, this trend does not seem to hold in the case of Inconel 625 in SCW with a high content of H_2O_2 because the severest corrosion appears at 400 °C rather than at 450 or 500 °C. Therefore, the coupling effects of temperature and oxidant content probably play a significant role in the corrosion behavior of Ni–Cr–Mo alloys. Moreover, Inconel 718 and Inconel 625 have similar corrosion weight gain curves under the same conditions, so they may have similar corrosion resistance in SCW under similar conditions. Overall, according to the findings in Fig. 3.3, it is clear that alloy type, temperature, time, oxidant content, and combinations of these factors significantly affect the final corrosion characteristics of Ni–Cr–Mo alloys in SCW.

3.2.2 *Oxides Formed on Alloy Surface*

As noted previously, the oxide layers formed on the Ni–Cr–Mo corrosion-resistant alloys in SCW have double-layer structures. After exposure to SCW at 600 °C, 25 MPa for 1000 h, a double-layer oxide film is generated on Hastelloy C-276, and the outer and inner layers are separately composed of NiO and the Cr_2O_3/$NiCr_2O_4$ mixture [12]. Similarly, Hastelloy C-276 develops a double-layer oxide structure in a SCWG environment at 500 °C and 25 MPa, and the outer layer comprises dispersed Ni/Fe-rich spinel particles, and the inner layer is composed of Cr oxides (i.e., Cr-rich spinel, Cr_2O_3) and a small amount of MoO_3. The oxide layer lacks NiO and $Ni(OH)_2$ in comparison to that in the SCW mentioned above, and this can be attributed to the higher hydrogen molar ratio in the SCWG environment [10]. The corrosion behavior of Inconel 625 and Inconel 718 in subcritical, supercritical and ultra-supercritical water (USCW) at 325, 425, 527.5 °C and 27 MPa confirms that the oxides in SCW are mainly composed of $NiFe_xCr_{2-x}O_4$ spinel. Nonetheless, the study on fully developed oxide layer characterization shows that the oxide layer of Inconel 625 contains $NiFe_xCr_{2-x}O_4$, Nb_2O_5, and Cr_2O_3, while that of Inconel 718 comprises Nb_2O_5 and Cr_2O_3 [17]. The double-layer oxide film formed on Inconel 625 inSCW containing 2 wt% H_2O_2 at 400 °C, 450, or 500 °C and 24 MPa is composed of $Ni(OH)_2$, NiO, $NiCr_2O_4$, Cr_2O_3 and Inconel 625 substrate from the outside to the inside of the alloy. As the temperature increases, the crystal size of the oxides on the metal surface exhibits an increase trend [14]. The dissolved oxygen content has a certain influence on the structure of the oxide film. Chang et al. [15] found that the oxide film of Incoloy 625 in SCW at 700 °C and 24.8 MPa varied with the dissolved oxygen content. A low oxygen content leads to a thickness reduction of the double-layer oxide film, with an outer layer composed of spinel oxide $Ni(Cr, Fe)_2O_4$ and an inner layer composed of Cr_2O_3, and suppresses the formation of Ni-containing oxides such as NiO and spinel oxide $Ni(Cr, Fe)_2O_4$ as well.

Table 3.2 alsosummarizes the structures and components of oxide films formed on Ni–Cr–Mo alloys in SCW as reported in different studies. It is seen that the outer oxide layer is mainly composed of Fe and Ni oxides such as Fe_2O_3, $NiFe_2O_4$, $NiCr_2O_4$, NiO and $Ni(OH)_2$, and the inner layer primarily consists of Cr oxides such as Cr_2O_3 and $NiCr_2O_4$. $NiFe_2O_4$ and $NiCr_2O_4$ are detected between the inner and outer oxide layers in some cases. Although experimental conditions are not identical, almost all researchers have demonstrated the presence of Cr_2O_3 in the inner layer. Thus, it can be inferred that Cr_2O_3 has a high stability in SCW, and that it adheres to the alloy surface to form a dense film once formed. The change of experimental conditions has a great influence on the composition of the outer oxide layer. After the initial formation of the outer oxide layer, chemical reactions may continue to occur with increasing exposure time to form other substances. The outer layer oxides are relatively unstable and only weakly protect the alloy substrate.

3.2.3 Comparison of Corrosion Characteristics

Some surface treatment methods can improve the corrosion resistance of Ni–Cr–Mo alloys in SCW. The corrosion resistance of Hastelloy C-2000, which is processed by grain boundary engineering, is improved by optimizing the arrangement of grain boundary atoms to increase small amounts of special low-energy grain boundaries [28]. The surface of Incoloy 625 can be covered with a NiCrAlY coating and then heated by the plasma spray method. Selvig et al. [29] reported that the NiCrAlY coating could effectively inhibit the intergranular corrosion and pitting corrosion of the sample in SCW. Daigo et al. [48] used a reactor made of Ni-44Cr-1Mo to study the corrosion behavior of the Ni-based alloy SCW, and found that its corrosion rate was lower than that in the reactor made of titanium alloy under the same conditions. This is due to the dense chromium oxide formed on the alloy surface, which helps to inhibit the corrosion in SCW. Son et al. studied the corrosion properties of Hastelloy C-276, Titanium Gr2, and Zirconium 702 in sub- and supercritical water containing 2-chlorophenol in a continuous experimental plant [49] and confirmed that the metal loss and the oxide film formation dominated in subcritical water and SCW, respectively. The weight loss in subcritical water was higher than that in SCW, and Titanium Gr2 exhibited the best corrosion resistance among these three alloys. Inconel 718 has a greater corrosion weight gain than Inconel 625 in subcritical, supercritical, and ultra-supercritical water [17], and Inconel 625 has better corrosion resistance than Inconel 718 [16]. This can be ascribed to the lower Cr content and the higher Fe content in Inconel 718. A proper pH range (6.5–10.5) can also improve the corrosion resistance of Ni-based alloys in reducing subcritical aqueous environments, of which the upper limit decreases with increasing oxidicability of solutions [50].

Table 3.6 lists the corrosion test conditions and principal conclusions from the study of some Ni–Cr–Mo alloys in subcritical or supercritical water over the past two decades. The test pressure range was 22.3–100 MPa but typically was around 25 MPa. Corrosion media, including deionized water, aqueous, organic, acidic, alkaline, and salt-containing solutions, with dissolved oxygen or H_2O_2 as oxidant, the exposure time range was 0.05–1026 h, and the test temperature varied from 300 to 600 °C, spanning the subcritical and supercritical regions of water. Overall, it was found that Hastelloy G-30, Hastelloy C-276, Ni-19Cr-19Mo and Inconel 625 show better corrosion resistance than does Inconel 718 in the Ni–Cr–Mo alloys studied, and the corrosion resistance of the Ni–Cr–Mo alloys is significantly better than that of austenitic stainless steel (e.g., SS 316). Titanium alloys (e.g., Titanium Gr2) have much better corrosion resistance than the Ni–Cr–Mo alloys, however, they are difficult to be adopted in large scale SCW reactors due to their relatively high prices and poor mechanical properties. Ti-containing coating on the Ni-based alloy surface may help to improve the corrosion resistance of the Ni-based alloy in subcritical or supercritical water. The corrosion behaviors of Ni–Cr–Mo alloys are primarily affected by temperature, time, alloying elements, alloy type, oxide film stability, coatings, aggressive salts, and oxygen content. Some researchers believe that high temperature promotes alloy oxidation in SCW or enlarges oxide particles,

Table 3.6 Corrosion experimental results of Ni–Cr–Mo alloys in sub- and supercritical water in recent two decades

Ref.	Alloy	Test solution	Oxidant	Temperature (°C)	Pressure (MPa)	Exposure time (h)	Major conclusions
[33]	Inconel 625 Inconel 686 Hastelloy G-30	CH_2Cl_2 with Cl_2	H_2O_2	420	40	24	Alloy G-30 exhibits good corrosion resistance due to the formation of a protective oxide layer
[34]	Inconel 625	5 wt% methanol	–	600	25	1011	The severe ductility loss of Inconel 625 occurs at the higher temperature section of the reactor, whereas no severe corrosion is observed
[35]	Hastelloy C-276, Inconel 625, Ni-45Cr-1Mo, Ni-19Cr-19Mo, SS 316	Deionized water and HCl solution (0.005–0.02 $mol \cdot L^{-1}$)	–	350–550	25	50, 100	In reducing SCW, Hastelloy C-276 and Ni-19Cr-19Mo are more resistant to corrosion, and the corrosion rates of Ni-based alloys are lower than that of SS 316

(continued)

Table 3.6 (continued)

Ref.	Alloy	Test solution	Oxidant	Temperature (°C)	Pressure (MPa)	Exposure time (h)	Major conclusions
[36]	Inconel 625 Hastelloy C-276 Ni-45Cr-Mo Titanium alloys	Organic wastes generated from nuclear power plants	H_2O_2	300, 400, 450	28.5	100	Ti alloys are selected as candidate materials for organic waste SCWO processes due to their good corrosion resistance
[30]	13 kinds of Ni-based alloys	Aqueous solution (0.01 mol·kg^{-1} H_2SO_4)	OC = 800 ppm	360, 400	30–60	50–100	The Cr-rich oxide layer, originating from the Ni-44Cr-1Mo autoclave material, provides a protective function against corrosion
[16]	Inconel 625, Inconel 718	Deaerated water	DO = 2 ppm	500, 600	25	1026	Alloy 625 has a slightly better corrosion resistance than Alloy 718 due to its higher Cr content
[31]	Hastelloy C-276, Titanium Gr2, Zirconium 702	Aqueous solution (3.89 mmol·L^{-1} C_6H_5ClO, 50.58 mmol·L^{-1} NaOH)	H_2O_2 (3.89 mmol·L^{-1})	340	22.3	0.05	Titanium Gr2 shows the strongest corrosion resistance, and the metal depletion and the formation of metal oxide are separately dominant at sub- and supercritical conditions
				400	25.3	0.04	

(continued)

Table 3.6 (continued)

Ref.	Alloy	Test solution	Oxidant	Temperature (°C)	Pressure (MPa)	Exposure time (h)	Major conclusions
[27]	Inconel 625 and 718, AR and GBE Inconel 617 samples	Deionized water	OC < 25 ppb	500	25	0–1000	GBE 617 and Inconel 718 samples show the best and the worst corrosion resistance, respectively
[14]	Inconel 625	Deionized water	2.0 wt% H_2O_2	400, 450, 500	24	50–250	The size of oxide crystals on the alloy surface gradually grows with temperature increasing, and the mass gain of Inconel 625 in SCWO increases with exposure time
[12]	Hastelloy C-276	Neutral water	OC < 10 ppb	500, 550, 600	25	1000	High temperature promotes oxidation, and the oxide growth mechanism of Hastelloy C-276 can be explained by the solid-state growth mechanism

(continued)

Table 3.6 (continued)

Ref.	Alloy	Test solution	Oxidant	Temperature (°C)	Pressure (MPa)	Exposure time (h)	Major conclusions
[37]	Inconel 625, Hastelloy C-276	Aqueous solutions (pH 1, 2 and 7)	–	300–426	24.1	53–204.5	The dealloying of elements and high corrosion rate at a high subcritical temperature can be explained by the changes of water properties and thermodynamics
[38]	Hastelloy C-22/2000 Hastelloy B-2 Alloy 33/59/671 Inconel 625	Aqueous HCl solution (pH 2)	–	300–425	24.1	20–24	Aggressive and minimal corrosions are separately observed at high subcritical and supercritical temperatures, and Cr and Mo can improve the corrosion resistance
[39]	Inconel 625	HCl aqueous solution (pH 2)	–	300–426	24.1	204.5	The direct chemical attack of hydrochloric acid molecules causes the stress corrosion cracking of samples at a supercritical temperature

(continued)

Table 3.6 (continued)

Ref.	Alloy	Test solution	Oxidant	Temperature (°C)	Pressure (MPa)	Exposure time (h)	Major conclusions
[29]	Inconel 625 and NiCrAlY coated Inconel 625	High density SCW	DO < 1 ppm	381	27	500	Air heat treated NiCrAlY coating avoids intergranular attack by forming a continuous oxide layer under the high density SCW condition
		Low density SCW		500, 470	27.6, 27.7	189	
[15]	Inconel 625	High-purity water	DO = 150 ppb, 1 ppm, 8.3 ppm	700	24.8	70, 160, 300, 600, 1000	Oxygen content affects the composition and structure of oxide film of Inconel 625 in SCWO, and the thickness of the duplex oxide layer decreases with decreasing DO
[40]	Inconel 718	Ammonobasic and ammonoacidic	–	600	300	–	The corrosion rate at the ammonoacidic condition is far higher than that at the ammonobasic condition, and stress corrosion cracking is a critical issue

(continued)

Table 3.6 (continued)

Ref.	Alloy	Test solution	Oxidant	Temperature (°C)	Pressure (MPa)	Exposure time (h)	Major conclusions
[26]	Inconel 625	5 mmol·kg^{-1} CH_3COOH or 0.5 mmol·kg^{-1} HNO_3	2.5–250 mmol·kg^{-1} O_2	400	30–100	100	The concentration of dissolved ions (i.e., Ni, Cr) increases with pressure increasing
[10]	Hastelloy C-276	Waste water (COD = 52,000 ppm)	–	500	25	10–120	The corrosion rate of Hastelloy C-276 exposed to SCWG is 2.5–3 times higher than that in high temperature air
		High temperature air	–	450–550	0.1	10–1000	
[17]	Inconel 625, Inconel 718	Deionized water	OC<50 ppb	325, 425, 527.5, 600	27	2–200	Inconel 718 exhibits a greater mass gain than Inconel 625 due to its lower Cr content and increased Fe content
[11]	Hastelloy C-276	SCW	–	500	25	50, 250, 500	Higher temperature promotes the oxidation rate and results in thermal stress which is responsible for the corrosion layer exfoliation

Note DO is dissolved oxygen, and OC is oxygen content. COD means chemical oxygen demand. AR is as-received, and GBE denotes grain boundary engineering thereby increasing thermal stress and eventually inducing corrosion layer exfoliation. Currently, it is evident that the severe corrosion that occurs in the high subcritical temperature region indicates that the influences of temperature on alloy corrosion behavior are quite complex. Therefore, the effects of the factors identified above on the corrosion behavior of Ni–Cr–Mo alloys in subcritical and supercritical water needs be further explored in detail.

3.3 Corrosion Characteristics of Ni–Fe–Cr Alloys

Based on the industrial application of Ni–Cr alloys, Ni–Fe–Cr alloys were developed by adding Fe to partiallyreplace Ni. The alloys exhibit comparable or superior performance in comparison with the Ni–Cr alloys. Table 3.7 displays the chemical compositions of typical Ni–Fe–Cr corrosion-resistant alloys. These alloys primarily contain 19–23 wt% of Cr, 30–35 wt% of Ni, and small amounts of Al and Ti, etc. Ni–Fe–Cr alloys used in SCW typically include Incoloy 800, Incoloy 800H, and Incoloy 800HT. As a representative grade of the Ni–Fe–Cr alloys, Incoloy 800 has a C content of less than 0.1 wt%, a Cr content of about 21 wt%, a Ni content of around 32 wt%, and other elements in trace. Significant differences between Incoloy 800 and Incoloy 825 rest with the addition of Mo and Cu in Incoloy 825, which belongs to one of Ni–Cr–Fe–Mo–Cu alloys.

3.3.1 *Corrosion Weight Change*

Corrosion weight change, usually evaluated by a gravimetric method, refers to the weight change of the sample in per unit area and per unit time. If corrosion products exfoliate or are removed from the sample surface by dissolution, the gravimetric method commonly exhibits corrosion weight loss. On the contrary, if corrosion products firmly adheres to the sample surface after corrosion, the gravimetric method exhibits a corrosion weight gain. The relationships between corrosion weight change and exposure time for Ni-based alloys under different temperature conditions follow parabolic rate kinetics, linear rate kinetics, or cubic rate kinetics. Equation 3.1 is usually used to fit experimental data,

$$\Delta w = kt^n \tag{3.1}$$

where Δw is the weight change, and k and n are constants. The oxidation kineticsdepends on the value of n, it follows the linear rate law when n = 1, the parabolic law when n = $\frac{1}{2}$, and the cubic law when n = $\frac{1}{3}$. However, the oxidation process may follow the parabolic lawin many experiments that have been reported so far [41]. Apparently, the corrosion weight gain that follows a linear law increases with a constant rate, while that

Table 3.7 Chemical compositions of main Ni–Fe–Cr alloys, Ni–Cr–Fe–Mo–Cu alloy, and other alloys involved in paper

Alloy	Chemical composition (wt%)														
	Fe	Ni	Cr	Al	Nb	Ti	C	Si	Mn	P	S	Cu	N	W	Co	Mo
Incoloy 800[a]	Bal	30.0–35.0	19.0–23.0	0.15–0.6	–	0.15–0.6	≤0.1	≤1.0	≤1.5	≤0.02	–	≤ 0.75	–	–	–	–
Incoloy 800H[a]					–		0.05–0.1			–	–	–	–	–	–	–
Incoloy 800HT[a]					–		0.06–0.1			≤0.03	–	–	–	–	–	–
Incoloy 825[b]	Bal	34.0–46.0	19.5–23.5	≤0.2	0.6–1.2	–	≤0.05	≤0.5	≤1.0	≤0.04	≤0.03	1.5–3.0	0.6–1.2	–	–	2.5–3.5
AFA steel	Bal	20.0	14.0	3.0	0.6	0.1	–	–	–	–	–	–	–	–	–	–
Inconel 182	≤ 10.0	Bal	13.0–17.0	–	1.0–2.5	≤1.0	≤0.1	≤1.0	5.0–9.5	≤0.03	≤0.015	≤0.5	–	–	–	–
316 SS	Bal	10.0–14.0	16.0–18.0	–	–	–	≤0.03	≤1.0	≤2.0	≤0.045	≤0.03	–	–	–	–	2.0–3.0
Alloy 214	3.0	Bal	16.0	4.5	–	–	0.05	0.17	0.37	0.01	0.01	–	–	–	–	1
Alloy 33	Bal	31.9	33.4	–	–	–	0.01	0.2	0.6	–	–	0.5	0.4	–	–	1.5
Hastelloy C-276	4.0–7.0	Bal	14.5–16.5	–	–	–	≤0.02	≤0.08	≤1.0	≤0.04	≤0.03	–	–	3.0–4.5	–	15.0–17.0
Inconel 625	≤5.0	Bal	20.0–23.0	≤0.4	3.5–4.2	≤0.4	≤0.1	≤0.5	≤0.5	≤0.015	≤0.015	–	–	–	–	8.0–10.0
Inconel 718	Bal	50.0–55.0	17.0–21.0	0.2–0.8	4.8–5.5	0.65–1.15	≤0.08	≤0.35	≤0.35	≤0.01	≤0.01	≤0.3	–	–	≤1.0	2.8–3.3
316L SS	Bal	12.0–15.0	16.0–18.0	–	–	–	≤0.03	≤1.0	≤2.0	≤0.035	≤0.03	–	–	–	–	2.0–3.0
304L SS	Bal	9.0–12.0	18.0–20.0	–	–	–	≤0.03	≤1.0	≤2.0	≤0.045	≤0.03	–	–	–	–	–

(continued)

Table 3.7 (continued)

Alloy	Chemical composition (wt%)															
	Fe	Ni	Cr	Al	Nb	Ti	C	Si	Mn	P	S	Cu	N	W	Co	Mo
Inconel 600	6.0–10.0	Bal	14.0–17.0	–	–	–	≤0.1	≤0.5	≤1.0	≤0.03	≤0.015	–	–	–	≤0.04	–
Inconel 690	7.0–11.0	Bal	27.0–31.0	–	–	–	≤0.05	≤0.5	≤0.5	≤0.02	≤0.015	≤0.5	–	–	–	–
Inconel X-750	5.0–9.0	Bal	14.0–17.0	0.4–1.0	–	2.25–2.75	≤0.1	≤0.5	≤1.0	≤0.02	≤0.01	–	–	–	–	–

And Bal. represents balance, [a]is a Ni–Fe–Cr alloy, [b]is a Ni–Fe–Cr–Mo–Cu alloy following a cubic law gradually decreases with time increasing. It can be concluded that the corrosion of the Ni-based alloy following a parabolic law or a cubic law in SCW can reach a relatively stable state with time increasing.

The Arrhenius relation is commonly adopted to express the relationship between the rate constant k and temperature T [42].

$$k = k_0 e^{-\frac{E_a}{RT}} \tag{3.2}$$

where E_a is the activation energy of oxidation reaction, R is the universal gas constant, and k_0 is a constant. Equation 3.2 can be used to calculate the activation energy of the oxidation occurred on the alloys in SCW environment.

The oxidation kinetics that approximately follows the parabolic law shows that the corrosion process is driven by transport processes [41]. Fulger et al. [42] found that Incoloy 800 exposed to SCW at 600 °C for 1440 h formed a thicker zone of intergranular corrosion and a more porous oxide film, and had weaker corrosion resistance, and its the oxidation kinetics follows the parabolic law above 500 °C. The oxidation kinetics of Incoloy 800HT follows the parabolic rate law within 500–600 °C but displays a linear rate at 450 °C. Table 3.8 indicates mathematical models and values of oxidation kinetics of typical Ni–Fe–Cr corrosion-resistant alloys (e.g., Incoloy 800, Incoloy 800H, and Incoloy 800HT). Corrosion weight gains of most Ni–Fe–Cr alloys in SCW approximately follow the parabolic law with increasing exposure time. As the temperature increases from 500, 550 to 600 °C, the rate constant increases, as expected from Eq. 3.2 However, the reaction order decreases, as observed in the corrosion tests on Incoloy 800HT in SCW at 25 MPa. Notably, the reaction order rises for Incoloy 800 under the same conditions noted above. According to the findings in Table 3.8, it can be concluded that temperature, oxygen content and grain boundary engineering affect the kinetics (rate constant) and reaction order in the corrosion of Ni–Fe–Cr alloys in SCW.

In existing studies, corrosion times for Ni-based alloys in SCW are usually hundreds of hours or up to thousands of hours in few reports. However, the exposure time of materials in actual SCW reactors is much higher than thousands of hours, and these materials also suffer from frequent startup and shutdown and working condition adjustment. Therefore, it is still necessary to verify whether the obtained weight gain curve can predict the corrosion weight change over a longer exposure time. Furthermore, the research on corrosion weight change with exposure time mainly focuses on oxidation kinetics laws. Further investigation concerning the fundamental origins of these rate laws is still required in subsequent work.

Ni–Fe–Cr alloys in SCW always show weight increase with temperature increasing [43]. An Incoloy 800H sample treated by grain boundary engineering (GBE) generally has a larger corrosion weight gain than the as-received sample. However, the sample treated by shot peening (SP) displays a much lower corrosion weight gain than the as-received sample, and its oxidation kinetics approximately follows the linear rate law [45]. In addition, the as-received sample has a greater sensitivity to changes in reaction conditions than does the GBE sample. Compared with

Table 3.8 Equations used to describe relationships between weight changes and exposure times of Ni–Cr–Mo and Ni–Fe–Cr alloys exposed to supercritical water

Equation	Alloy	Oxygen content	k	Temperature (°C)	Pressure (MPa)	n	R^{2f}	Kinetics	Ref
$w = kt^n$	Incoloy 800H (AR)	<25 ppb	–	500	25	1.05	0.96	Linear	[41]
	Incoloy 800H (GBE)		–	500		0.84	0.88	–	
	Incoloy 800H (AR)		–	600		0.34	0.47	Cubic	
	Incoloy 800H (AR)	<2 ppm	–	500		0.62	0.49	Parabolic	
$\Delta w = kt^n$	Incoloy 800	<2 ppm	–	450	25	0.32	0.97	Cubic	[43]
			–	500		0.40	0.91	Parabolic	
			–	550		0.56	0.98	Parabolic	
			–	600		0.65	0.97	Parabolic	
$\Delta w = kt^n$	Fe20Ni14Cr3Al0.6Nb0.1Ti	25 ppb	0.00197	500	25	0.680	–	Parabolic	[44]
	Incoloy 800H		0.00285	500		0.693	–	Parabolic	
$\Delta w = kt^n$	Incoloy 800HT	<2 ppm	0.01	450	25	0.81	0.99	Linear	[42]
			0.066	500		0.62	0.96	Parabolic	
			0.233	550		0.49	0.99	Parabolic	
			0.492	600		0.47	0.98	Parabolic	

And w is weight change, Δw is weight gain, t is exposure time, n is reaction order, k is rate constant or reaction constant, fitting parameter R^2 with a value approaching 1.00 denotes increasing consistence of data with fitting, AR is as-received, and GBE denotes grain boundary engineering the GBE sample, the annealed sample exhibits considerable fluctuation of corrosion weight gain with increasing exposure time, and its corrosion weight gain is also significantly lower than that of the GBE sample after a certain period of time [46]. However, thermo-mechanically processed Incoloy 800HT exhibits a weight loss after 100 h of exposure time in SCW at 600 °C and 25 MPa [47]. An increase in the oxygen content (from 25 ppb to 2 ppm) increases corrosion weight gain when other conditions are constant [41].

Figure 3.4 indicates that the corrosion weight changes of Ni–Fe–Cr alloys in SCW as exposure time increases within the first 1000 h of exposure. Like Ni–Cr alloys and Ni–Cr–Mo alloys, Ni–Fe–Cr alloys also exhibit corrosion weight gains, which initially increase and then decrease with increasing time in most cases. Although Alloy 33 shows a weight loss during the entire exposure time, its lower corrosion weight loss indicates that it is more stable in SCW, due to its better corrosion resistance. Notably, the maximum corrosion weight gain occurs on the annealed Incoloy 800H sample, which has the largest weight change fluctuation with increasing time. Nonetheless, the weight gain curve of Incoloy 800H treated by GBE (grain boundary engineering) or SP (shot peening) is relatively monotonic. Compared with the as-received sample and the GBE sample of Incoloy 800H at the same temperature, the shot peened sample of Incoloy 800H has the minimum weight gain in SCW at

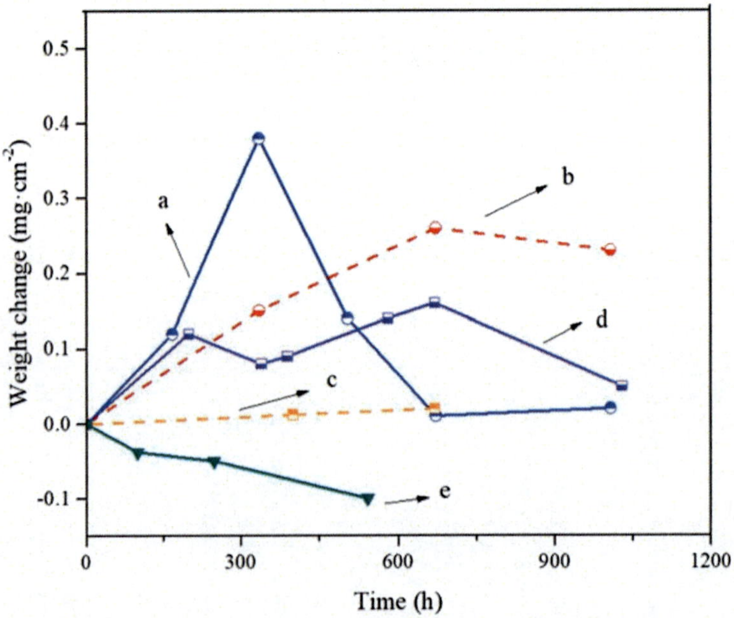

Fig. 3.4 Weight changes of Ni–Fe–Cr alloys exposed to supercritical water with time increasing. **a** Incoloy 800H (annealed), 500 °C, 25 MPa, oxygen content = 25 ppb [46]. **b** Incoloy 800H (grain boundary engineering), 500 °C, 25 MPa, oxygen content = 25 ppb [46]. **c** Incoloy 800H (shot peening), 500 °C, 25 MPa, oxygen content <25 ppb [45]. **d** Incoloy 800H (as-received), 500 °C, 25 MPa, oxygen content <25 ppb [45]. **e** Nicrlfer 3033 (Alloy 33), 550 °C, 25 MPa [59]

500 °C and 25 MPa with the oxygen content of <25 ppb. Apparently, shot peening can reduce the corrosion weight gain of Incoloy 800H in SCW, proving that shot peening is able to improve the corrosion resistance of Ni-based alloys. Therefore, it is important to choose a suitable surface treatment process in order to improve the corrosion resistance of Ni–Fe–Cr alloys, but relevant feasibility and economics factors requires comprehensive evaluation.

3.3.2 Oxides Formed on Alloy Surface

Researchers have found that Ni–Fe–Cr alloys typically form a double-layer oxide film structure in a SCW environment, with the outer layer containing Fe oxides and the inner layer comprising Cr and Ni oxides (including mixed oxides). For example, the oxide film formed on Incoloy 800 in SCW has a double-layer structure, with an outer oxide layer rich in Fe and an inner layer rich in Cr and Ni [43]. Similarly, Incoloy 800HT has a double-layer oxide structure in SCW, with an outer oxide layer composed of Fe oxide/Fe–Ni spinel oxides and an inner layer of Cr oxides [42]. Some researchers have discovered not only a double-layer oxide film structure but also other substances between the inner and outer oxide layers or at the interface between the oxide layer and the metal matrix. Incoloy 800H forms a double-layer oxide structure in oxygen-containing SCW, and its outer layer is composed of magnetite (Fe_3O_4) and hematite (Fe_2O_3) while the inner layer consists of Cr-rich Fe–Cr oxide $(Cr, Fe)_2O_3$ and a small amount of dissolved Ni [48]. At the same time, a Fe_3O_4 oxide layer with a small amount of dissolved Mn is found to exist between the inner and outer oxide layers. Incoloy 800H forms a multi-layer oxide structure in SCW at 500 and 660 °C, with the outer layer consisting of Fe_3O_4 and $(Fe, Cr)_2O_3$ and the inner layer comprising $(Fe, Cr)_3O_4$ and $FeCr_2O_4$. Also, semi-continuous Ni and a small amount of Cr_2O_3 are found at the interface of the inner and outer oxide layers and at the interface of $FeCr_2O_4$ and the metal matrix at 600 °C, respectively [41]. Alumina-forming austenitic steel (i.e., Fe-20Ni-14Cr-3Al-0.6Nb-0.1Ti) generates a multi-layer oxide film structure in SCW, that is, an outer oxide layer rich in Fe_2O_3, an inner oxide layer rich in Al–Cr–Fe and a small amount of Ni, and a thin Ni-rich oxide layer at the interface of the oxide layer and the metal substrate [44]. A three-layer oxide film has also been found in SCW for corrosion tests of other alloys such as Inconel 625 and Inconel 718 (see Table 3.9 for details).

Surface-treated alloys exhibit different surface oxide properties. For example, the Incoloy 800H sample treated by shot peening has two microstructure regions, i.e., an ultra-fine-grained region on the surface and a sequential transition region on the sub-surface [45]. It has a simpler oxide composition (including Cr_2O_3, $(Cr, Ti)_2O_3$, and $FeCr_2O_4$) compared with the as-received sample. The process of forming a stable oxide film on the surface of Incoloy 800H can be divided into five stages. At the beginning, Fe, Al, Ni and Mn are released into SCW at very low concentrations but no release of Cr is observed. The dissolution of Mn can cause a decrease in the protective property of the Cr oxide film and the generation of the outer layer of Fe

Table 3.9 Summary of oxide film structure and composition on Ni-based alloys tested in sub- and supercritical water. And DO is dissolved oxygen, OC is oxygen content

Alloy	Pressure (MPa)	Temperature (°C)	OC	DO	Outer oxide layer	Medium oxide layer or oxide layer sequence	Inner oxide layer	Ref
Incoloy 800	25	450–600	–	<2 ppm	Oxides enriched in Fe	–	Cr and Ni-rich oxides	[43]
Incoloy 800H	25	500	–	25 ppb	Hematite and magnetite	–	A mixture of austenite (substrate) and spinel	[46]
	25	500	<25 ppb	–	Fe_3O_4 with some $(Fe, Cr)_2O_3$	–	$(Fe, Cr)_3O_4$ with some $FeCr_2O_4$	[41]
		600	<25 ppb	–	Fe_3O_4 with some $(Fe, Cr)_2O_3$	A thin layer of semi-continuous pure Ni and a trace amount of Cr_2O_3 at the interfaces of the inner-outer layers and the $FeCr_2O_4$-metal matrix, respectively	$(Fe, Cr)_3O_4$ with some $FeCr_2O_4$	
	25	650	<10 ppb	–	Magnetite (Fe_3O_4) and hematite (Fe_2O_3)	A very fine Fe_3O_4 layer with a small amount of dissolved Mn	Cr-rich $(Cr, Fe)_2O_3$ with a small amount of dissolved Ni	[48]
Incoloy 800HT	25	450–600	<2 ppm	–	A mixture of Fe oxides/Fe–Ni spinel oxides	–	Cr oxide	[42]
	25	550	–	<8 ppm	Discontinuous Fe_3O_4	–	Fe–Cr oxide, $(Cr, Fe)_2O_3$	[50]

(continued)

Table 3.9 (continued)

Alloy	Pressure (MPa)	Temperature (°C)	OC	DO	Outer oxide layer	Medium oxide layer or oxide layer sequence	Inner oxide layer	Ref
Inconel 718	27	527.5, 600	<50 ppb	–	Nb_2O_5	$NiFe_2O_4$	Cr_2O_3	[17]
		325	<50 ppb	–	$NiFe_xCr_{2-x}O_4$ spinel	–	$Ni(OH)_2$ and CrO_4^{2-} film	
Inconel 625		527.5, 600	<50 ppb	–	Nb_2O_5	$NiFe_xCr_{2-x}O_4$	Cr_2O_3	
		325	<50 ppb	–	$NiFe_xCr_{2-x}O_4$ spinel	–	$Ni(OH)_2$ and CrO_4^{2-} film	
	24	500	–	5000 ppm	Fe oxides, spinel structures, and $Ni(OH)_2$, porous, bar shaped particles	–	A compact layer of acicular-like NiO oxide	[18]
Inconel 825			–	5000 ppm				
Inconel 182	13	320	–	–	A component of flaky and spinel oxides	–	Spinel oxide and Cr_2O_3	[6]

Table 3.10 Corrosion experimental results for Ni–Fe–Cr alloys in supercritical water

Alloy	Test solution	Oxidant	Temperature (°C)	Pressure (MPa)	Exposure time (h)	Main conclusions	Ref.
Incoloy 800H (Control and GBE samples)	SCW	DO = 25 ppb	500	25	1026	GBE improves protective oxidation behaviors by increasing spallation resistance and reducing oxidation rate	[46]
Incoloy 800H (AR and SP samples)	SCW	DO < 25 ppb	500	25	0–1100	Shot-peening processing improves oxidation resistance and alleviates the oxide exfoliation of Incoloy 800H	[45]
Incoloy 800	Deionized water	OC < 2 ppm	450, 500, 550, 600	25	1440	A faster mass gain is observed with an increase in temperature after1000 h, and corrosion is driven by a transport process	[43]
AFA steel	SCW	DO = 250,000 ppb	500	25	1300, 1700, 3000	The AFA steel exhibits a superior corrosion resistance than 316 SS, Incoloy 800H and D9 at the same conditions	[44]

(continued)

Table 3.10 (continued)

Alloy	Test solution	Oxidant	Temperature (°C)	Pressure (MPa)	Exposure time (h)	Main conclusions	Ref.
Incoloy 800HT, SS 304L	Deionized water	OC < 2 ppm	450, 500, 550, 600	25	1680	The corrosion resistance of oxides grown on the Incoloy 800HT surface is better than that on 304L SS	[42]
Incoloy 800H (AR and GBE samples)	Deionized water	OC < 25 ppb or 2 ppm	500, 600	25	1–1000	Oxide exfoliation is alleviated on the GBE sample compared with the AR sample, and the oxidation is influenced by the preferential transport of specific species and the short-cut diffusion path	[55]
Incoloy 800HT AR and TMP samples	SCW	DO = 150 ppb	600	25	100, 300, 1000	TMP improves oxidation resistance and alleviates oxide scale exfoliation, and grain size has a strong effect on the oxidation resistance	[47]

(continued)

Table 3.10 (continued)

Alloy	Test solution	Oxidant	Temperature (°C)	Pressure (MPa)	Exposure time (h)	Main conclusions	Ref.
Incoloy 800H, SS 316	Deoxygenated water	OC < 10 ppb	650	25	300–775	Numerous cracks are formed in the oxide layer after oxygenated treatment, and Incoloy 800H is more resistant to surface oxidation than SS 316	[48]
	Oxygenated water	OC = 20–40 ppm					
Incoloy 800H, Alloy 33	Deionized water	OC = 8 ppm	550	25	500	Alloy 33 exhibits a higher corrosion resistance than Alloy 800HT	[50]
Incoloy 800H	SCW O_2/H_2 = 0.56	DO < 20 ppm	500 – 750	25	550	The incorporation of Mn in the oxides results in no Cr release during the initial stage, whereas Mn dissolution induces the Cr release and the formation of an outer Fe oxide	[49]

And DO is dissolved oxygen, OC is oxygen content. GBE, AR and SP refer to grain boundary engineering, as-received, and shot peening, respectively. AFA steel denotesAlumina-forming austenitic steel (Fe-20Ni-14Cr-3Al-0.6Nb-0.1Ti). TMP represents thermo-mechanical processing, which contains a series of unidirectional rolling and cross rolling

oxides (i.e., magnetite and hematite). At the finally stable stage, Cr and Al become main release elements, and relevant data indicate that it takes thousands of hours to form a balanced oxide film [49].

Table 3.9 summarizesthe structures and components of oxide films of Ni–Fe–Cr alloys in SCW, as reported in different studies. Unlike Ni–Cr and Ni–Cr–Mo alloys, the outer oxide layer of the Ni–Fe–Cr alloy is mainly Fe oxides (i.e., Fe_2O_3 and

Fe_3O_4), and the inner oxide layer comprises Cr oxides (i.e., Cr_2O_3, $FeCr_2O_4$ as well as Fe_3O_4 and Fe_2O_3). It is worth noting that Ni oxide is almost undetectable in the oxide film formed on the Ni–Fe–Cr alloy in SCW. A small amount of pure Ni is found between the inner and outer oxide layers on Incoloy 800H, but only in SCW at 600 °C and 25 MPa with oxygen content less than 25 ppb. One possible reason is that the Ni content of the Ni–Fe–Cr alloy is usually within 30–45 wt%, while that of the Ni–Cr alloy or the Ni–Cr–Mo alloy is commonly above 50 wt%, and the lower Ni content makes it difficult to form stable compounds which adhere to the alloy surface in SCW. Another explanation is that Fe probably preferentially reacts with oxygen to form oxides, because the Ni–Fe–Cr alloy has reduced the Ni content by increasing the Fe content compared with the Ni–Cr alloy. According to present research, this substitution of Ni by Fe has little effect on alloy corrosion resistance in SCW. However, this can improve the economics of Ni–Cr alloys when ensuring its corrosion resistance.

The microstructures of as-received Incoloy 800 sample, together with Incoloy 800, Incoloy 800H and Incoloy 800HT samples in SCW at different conditions are illustrated in Fig. 3.5 High temperature corrosion in SCW can change the microstructures of Incoloy 800, thereby affecting its corrosion behavior. The fine grain structure contributes to the formation of protective Cr oxides, so the fine grained austenitic structure which represents a fine and uniform structure has better corrosion resistance than the coarse grained structure [43]. From the cross-section image of Incoloy 800H exposed to deoxygenated SCW at 650 °C, 25 MPa for 650 h, it can be seen that the oxide film is about 20 μm thickness and can be divided into two layers (see Fig. 3.5d),

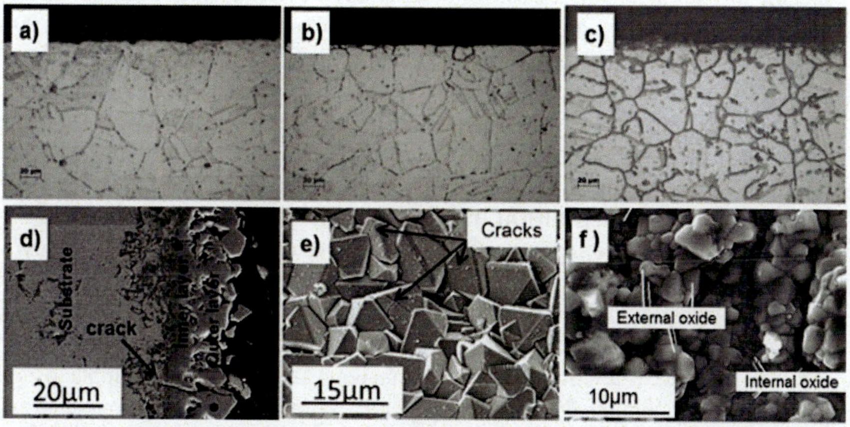

Fig. 3.5 Grain boundaries and oxide film structures of Ni–Fe–Cr alloys **a–c** grain boundary structures of Incoloy 800: **a** as-received sample, **b** corroded at 500 °C, 25 MPa for 1440 h, **c** corroded at 600 °C, 25 MPa for 1440 h [43], **d** cross-section image of oxide layer formed on Incoloy 800H in SCW (oxygen content of <10 ppb) at 650 °C, 25 MPa for 650 h [48], **e** surface morphology of oxide scale grown on Incoloy 800H in SCW (oxygen content of <10 ppb) at 650 °C, 25 MPa for 650 h [48], **f** SEM image of oxide layer developed on Incoloy 800HT in SCW at 600 °C, 25 MPa for 1680 h [42]

and several cracks extend from the oxide layer to the alloy substrate (see Fig. 3.5e). The outer oxide layer of Incoloy 800HT consists of particles in different sizes while the inner oxide layer is more uniform (see Fig. 3.5f). The growth process of the oxide film can be divided into two steps, that is, first oxide nucleation at selected sites and then followed by the uniform growth of oxide grains until they combine to form a continuous, dense oxide layer [42]. Notably, the grain boundary size of an alloy in SCW changes with changing temperature, and the fine grain structure helps to improve the corrosion resistance of the alloy. Therefore, increasing the amount of fine grains or maintaining the grains at the fine particle level is beneficial to improving the corrosion resistance of the Ni–Fe–Cr alloy in SCW. Surface treatment processes can be adopted to improve the grain boundary microstructure of material to enhance its corrosion resistance.

3.3.3 *Surface Treatment Processes*

Some surface treatment processes, such as shot peening and thermo-mechanical processing, can improve the corrosion resistance of the Ni–Fe–Cr alloy in SCW. The ultrafine grains formed by shot peening increase the amount of grain boundaries. This can accelerate the diffusion of Cr and the formation of Cr oxides, and meanwhile the high content of Ti in $(Cr, Ti)_2O_3$ and TiO_2 on the shot peening sample can reduce the volatilization of Cr. Hence, shot peening is able to suppress the exfoliation of the oxide film formed on alloy surface [45] and is an effective anti-corrosion method. After specific thermo-mechanical treatment, especially a series of cold-rolling treatments after annealing, the oxide exfoliation and corrosion resistance of Incoloy 800HT can be effectively reduced and improved, respectively [48]. NiCrAlY coatings form a dense, adherent oxide layer of Cr_2O_3 and spinel $(Ni(Cr, Al)_2O_4)$ on the alloy surface, which can enhance the corrosion resistance of Alloy 214 in SCW [51].

Tan et al. [46] conducted grain boundary engineering on Incoloy 800H by thermo-mechanical processing, and found that its corrosion resistance in SCW was greatly improved compared with that of the sample before treatment. They proposed that grain boundary engineering improved the protective oxide film by increasing exfoliation resistance and reducing the corrosion rate. The cyclic oxidation test of Incoloy 800H shows that grain boundary engineering promotes the formation of Cr_2O_3 and Cr-rich spinel [45]. The corrosion behavior of two Incoloy 800H samples in SCW indicates that the oxide exfoliation of the grain boundary engineering sample is significantly reduced compared with the as-received sample [41]. Grain boundary engineering can also increase the strength of Incoloy 800H by 10% at room temperature without a significant toughness reduction of the material [52, 53].

Figure 3.6 displays the surface morphologies and oxide layer structure of Incoloy 800H and its processed samples after exposure to SCW. As shown in Fig. 3.6a, the as-received sample of Incoloy 800, after exposure to oxygen-containing SCW at 500 °C for 503 h, the oxide film comprises three layers with approximately parallel interfaces. However, the interface between the outer two layers on the GBE (grain boundary

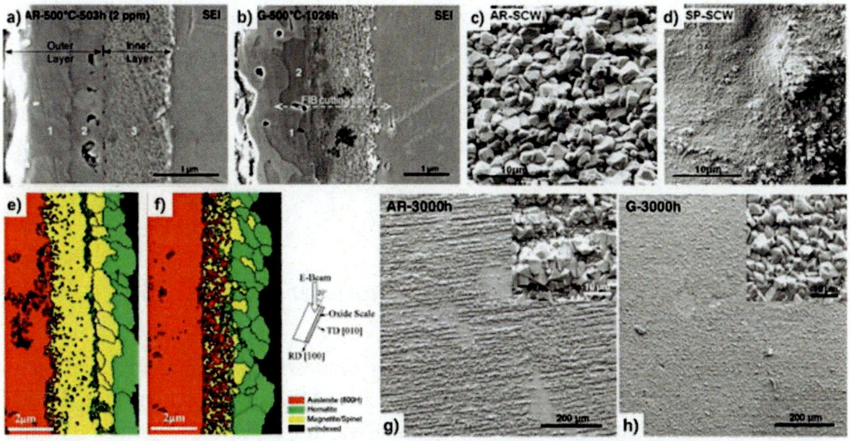

Fig. 3.6 Surface morphologies and oxide layer structures of Incoloy 800H and its processed samples exposed to SCW. **a** oxide layer structure of Incoloy 800H AR (as-received) sample exposed to SCW (oxygen content of 2 ppm) at 500 °C, 25 MPa for 503 h [41], **b** oxide layer structure of Incoloy 800H GBE (grain boundary engineering) sample exposed to SCW (oxygen content of <25 ppb) at 500 °C for 1026 h [41], **c** and **d** surface morphologies of Incoloy 800H AR sample and Incoloy 800H SP (shot peening) sample after exposure to SCW at 500 °C, 25 MPa for 667 h, respectively [45], **e** electron back-scattered diffraction (EBSD) scanning map of three-week SCW-exposed Incoloy 800H control (annealed) sample at 500 °C, 25 MPa with 25 ppb of dissolved oxygen [46], **f** EBSD scanning map of four-week SCW-exposed Incoloy 800H GBE sample at 500 °C, 25 MPa with 25 ppb of dissolved oxygen (in left side) and schematic diagram of specimen layout in EBSD analysis (in right side)[46], **g** and **h** surface morphologies of Incoloy 800H AR and GBE samples exposed to SCW (oxygen content of <25 ppb) at 500 °C, 25 MP for 3000 h, respectively [41]

engineered) sample is serrated (see Fig. 3.6b), which helps to maintain the integrity of the oxide structure and inhibit the exfoliation of oxides [41]. As indicated in Fig. 3.6c, d, the as-received sample surface is entirely covered by large particle oxides, while the shot peened (SP) sample of Incoloy 800H forms a dense oxide layer with small amounts of larger particles dispersed in the vicinity of the crack after exposure to SCW at 500 °C for 667 h [45]. Accordingly, the shot peening treatment is also helpful for the improvement of the corrosion resistance of Incoloy 800H. Figure 3.6e, f separately elucidate EBSD (Electron Back-Scattered Diffraction) scanning images of the annealed sample (for three weeks) and the GBE sample of Incoloy 800H after four weeks of exposure to SCW. It can be observed that the crystal structures of the two samples are very similar, and mainly contain austenite, magnetite/spinel (Fe_3O_4) and hematite (Fe_2O_3) [41]. It is worth noting that the mixture of austenite and spinel in the inner oxide layer of the GBE sample is more uniform than that of the annealed sample, thereby improving the adhesion between substrate and oxides [46]. After an exposure time of 3000 h, the as-received sample has a loose surface with dispersed grooves and exfoliation traces, whereas the GBE sample has a smooth and compact surface (see Fig. 3.6 g, h). In general, the surface-treated sample has a denser and smoother oxide layer than the as-received sample in SCW, and the more compact oxide layer is postulated to reduce the generation paths of oxidation reactions. Therefore, it can be concluded that the short-cut diffusion via grain boundary networks and dislocation network clusters is a key factor causing oxidation to occur. Micro-cracks on the material surface and the gaps between grain boundaries become the major paths of oxygen diffusion, thereby reducing alloy surface defects is an important method of overcoming corrosion.

3.3.4 Comparison of Corrosion Characteristics

Ni–Fe–Cr corrosion-resistant alloys have good corrosion resistance in SCW. Incoloy 800HT has better corrosion resistance than does SS 316L [42], because SS 316L has lower Cr content, coarser grain boundaries, more inter-crystalline and intra-crystalline carbide precipitates, and more porous outer oxidation layers. Inter-crystalline carbides promote the appearance of Cr-depleted region adjacent to boundaries, thereby resulting in intergranular attack. Choudhry et al. [48] studied the corrosion behavior of Incoloy 800H and SS 316 in a continuous SCW reactor, and found that the hematite generated in the oxide layer on Incoloy 800H was protective for the sample substrate. The oxide film on the Incoloy 800H surface is thinner and more stable than that on SS 316 [54], so leading to better corrosion resistance. A new type of alumina-forming austenitic steel (i.e., Fe-20Ni-14Cr-3Al-0.6Nb-0.1Ti) exhibits better corrosion resistance than do SS 316, Incoloy 800H, and D9 (austenitic stainless steel 14Cr-15Ni-2.25Mo-1.78Mn-0.61Si-0.26Ti-0.04C) in SCW [44], due to the Al–Cr–Fe oxide layer formed on the alloy surface. However, Ni–Fe–Cr alloys may have a strong tendency towards stress corrosion cracking in an ammonobasic or ammonoacidic environment [40]. Increasing the Ni content in Incoloy 800H can reduce the surface hydrogen evolution and corrosion rate of the alloy [55].

Table 3.9 summarizes some experimental conclusions on the corrosion tests of Ni–Fe–Cr alloys in SCW. Test temperature and time ranges are 450–750 °C and 0–1700 h, respectively, and pressure is controlled at 25 MPa, and corrosion medium is SCW with a trace of dissolved oxygen. The Ni–Fe–Cr alloy has a relatively low Cr content and hardly contains Mo, so its corrosion resistance may be lower than that of the Ni–Cr–Mo alloy or the Ni–Fe–Cr–Mo–Cu alloy. Many researchers have used surface treatment processes (e.g., grain boundary engineering, shot peening, and thermo-mechanical treatment) to modify the alloy surface, and thus to improve the corrosion resistance of the Ni–Fe–Cr alloy in SCW. Most of these processes can increase the protective properties of the oxide layer and inhibit the diffusion of corrosive species and the exfoliation of oxides. The integrity of the oxide layer on alloy surface is very important for inhibiting further corrosion damage. In the process of corrosion oxidation, oxygen can diffuse through grain boundaries to alloy interior, and meanwhile oxides can also be formed by the diffusion of metal cations through grain boundaries to the corrosive environment. Overall, corrosion is driven by the migration processes of some species, such as oxygen, metal cations, and corrosive anions. Thermo-mechanical processing (e.g., grain boundary engineering or shot peening) is an important method of improving alloy corrosion resistance. It can alter grain sizes and induce special structures, and adjust the crystal lattice, and thus significantly improve material properties relating to grain boundaries. Therefore, systematical investigations concerning the surface treatment processes above are also of significance for extensive application of some Ni-based alloys.

Grain boundary engineering aims to increase the proportion of low-ΣCSLBs (coincidence site lattice boundaries) grain boundaries by thermomechanical processing. Herein, Σ is a value defined as the reciprocal density of coincident sites at the grain boundary between ΣCSLBs two adjoining grains [56]. Low-ΣCSLBs contains more coincident sites of lattice on grain boundaries, and Σ3 is the most frequently used for improving material properties among all low-. In existing research on grain boundary engineering, the improvement of corrosion resistance of Ni–Fe–Cr alloys by Σ3, Σ9 and Σ27 is the most significant. Now, the research and improvement of corrosion resistance of Ni-based alloys in SCW is progressing toward microscopic scale, and desired material properties can be obtained by changing grain boundary structure characteristics. This has become an important developmental trend in the studies regarding the property modification of SCW candidate materials.

3.4 Corrosion Characteristics of Ni–Fe–Cr–Mo–Cu Alloys and Other Ni-Based Alloys

Ni–Fe–Cr–Mo-Cu alloys are a series of alloys obtained by adding an appropriate amount of Cu to Ni–Fe–Cr–Mo alloys, so they contain a large amount of Cu. The addition of Cu can improve the corrosion resistance of Ni–Fe–Cr–Mo alloys in reducing acid, especially in hot sulfuric acid. Ni–Fe–Cr–Mo–Cu alloys have good resistance to oxidizing media, stress corrosion cracking, pitting corrosion, and crevice corrosion. Compared with commonly used Ni–Cr–Mo–Cu alloys, the Ni–Fe–Cr–Mo–Cu alloy exhibits comparable or superior corrosion resistance in some media, and the replacement of parts of Ni and Cu with Fe makes the Ni–Cr–Mo–Cu alloy cheaper. Incoloy 825 is typically adopted in SCWO, and it is a Ti-stabilized Ni–Fe–Cr–Mo–Cu corrosion-resistant alloy and its chemical composition can be seen in Table 3.7.

3.4.1 Oxides Formed on Alloy Surface

Hastelloy C-276, Inconel 625 or Incoloy X-750 forms a double-layer oxide film in SCW, and its outer layer is loose Fe-rich $NiFe_2O_4$ while the inner layer is adherent Cr-rich $NiCr_2O_4$ [60]. Both Inconel 625 and Incoloy 825 exhibit phenomena like pitting corrosion in oxygen-containing SCW and have a similar oxide film structure, that is, the outer layer is loose and consists of Fe oxide, spinel and rod-like $Ni(OH)_2$

particles, while the inner layer is dense needle-like NiO. Cr_2O_3 is easily oxidized to soluble tetravalent Cr oxide and to chromate ion (CrO_4^{2-}) and thus loses its protective effect [18]. Similarly, Inconel 600 and Incoloy 825 have a multi-layer oxide structure in sulfide-containing SCW, and the outer oxide layer of Inconel 600 is Ni_3S_2 and its inner layer is mainly Cr-rich spinel and reticulated Ni_3S_2. However, from the outside to the inside layer, the oxides of Incoloy 825 are Ni_3S_2 and (Ni, Fe)$_3S_4$, Cr oxides (e.g., Cr_2O_3)and a small amount of Ni_3S_2 [56]. The oxidation products of Inconel 625/600 and Incoloy 800 in the SCWG environment containing ammonium chloride are mainly Fe oxides, spinel and trace NiO. Herein, Fe_3O_4 is found due to the accumulation of Fe_3O_4 formed by the inner surface of the SS 316 reactor on the sample surface [57]. The oxidation products of Incoloy 800, Incoloy 825, Inconel 625 and Hastelloy C-276 in sub- and supercritical water with high concentrations of salt and oxygen are mainly NiO, $NiCr_2O_4$ and Cr_2O_3 [24].

Figure 3.7a–c are the cross-section TEM (Transmission Electron Microscope) images of Incoloy 800H, Alloy 214 and Hastelloy C-2000, respectively, after exposure to SCW at 800 °C for 12 h. Incoloy 800H has a three-layer oxide film structure (see Fig. 3.7a), and its outer layer contains (Cr, Fe)$_3O_4$, the continuous intermediate layer is rich in SiO_2, and the discontinuous inner layer has Al_2O_3. Thus, the order of oxide formation is consistent with the degree of thermodynamic stability, i.e., Al_2O_3

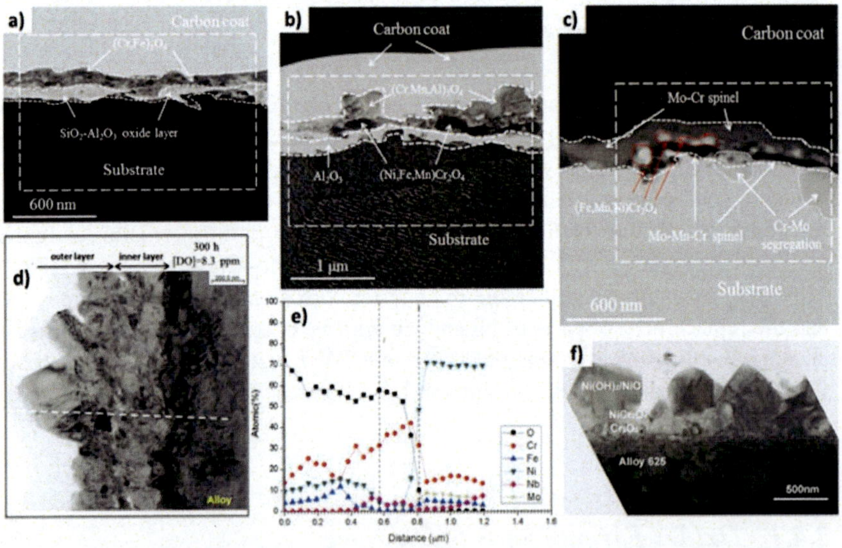

Fig. 3.7 Cross-section TEM images of different Ni-based alloys exposed to SCW **a–c** cross-section TEM (Transmission Electron Microscope) images of Incoloy 800H, Alloy 214, and Hastelloy C-2000 after exposure to SCW at 800 °C, 25 MPa for 12 h, respectively [58], **d** cross-section TEM image of Inconel 625 sample exposed to SCW of 8.3 ppm of dissolved oxygen at 700 °C, 24.8 MPa for 300 h [15], **e** EDS profiles for line-scanning across oxide layer corresponding to dash line location in (**d**) [15], **f** TEM cross-section image of equiaxed crystals of outer oxide film grown on Inconel 625 in oxidizing SCW at 450 °C, 24 MPa for 250 h [14]

> SiO_2 > Cr_2O_3. Alloy 214 has a double-layer oxide structure (see Fig. 3.7b), in which the inner layer contains Al_2O_3 while the outer layer may be (Ni, Fe, Mn)Cr_2O_4, (Ni, Fe)(Cr, Mn)$_2O_4$, and (Cr, Al)$_2O_3$, etc. Hastelloy C-2000 also has a double-layer oxide structure (see Fig. 3.7c), in which the inner layer contains Mo–Mn–Cr spinel and the outer layer includes Mo–Cr spinel. Behnamian et al. [58] reported that the thermodynamic stability of oxides in high temperature SCW had the following sequence of Al_2O_3 > TiO_2 > SiO_2 > MnO > Cr_2O_3 > $FeCr_2O_4$ > Fe_3O_4 > MoO_2 > NiO. Figure 3.7f shows the cross-section TEM image of Incoloy 625 after exposure to SCW at 450 °C for 250 h, and its oxide layer also has a double-layer structure. Similarly, Incoloy 625 forms a double-layer oxide structure after exposure to SCW at 700 °C for 300 h, and sub-micron crystal oxide particles are distributed in its outer oxide layer, while dense nano-scale crystal oxide particles are in its inner layer (see Fig. 3.7d). As shown in Fig. 3.7e, the EDS (Energy Dispersive Spectrometer) profiles for the line-scanning across the oxide layer corresponding to the dash line location in Fig. 3.7d show that Ni, Cr and Fe are concentrated in the outer oxide layer but just Cr is detected in the inner layer, except for oxygen. On the whole, if oxides in the multi-layer oxide film of the Ni-based alloy is closer to the alloy/water interface, they are coarser and more unstable. However, if the oxides are closer to the alloy matrix, they are denser and more stable. The reason for the looseness and discontinuity of the outer oxide layer may be due to the decomposition of oxides in the layer, the scouring of external fluids, or growth via electro-crystallization.

3.4.2 Comparison of Corrosion Characteristics

Many researchers have compared the corrosion behavior of various Ni-based alloys in SCW and provided relevant explanations for any differences. Li et al. [56] investigated the corrosion behavior of Inconel 600 and Incoloy 825 in sulfide-containing SCW [containing 12,800 ppm of sulfide, bivalent sulfur content and 4800 ppm COD (chemical oxygen demand)] at 520 °C, 25 MPa, and found that Incoloy 825 exhibited slightly better corrosion resistance than Inconel 600 due to its higher Cr content. Inconel 625 shows better corrosion resistance than does Incoloy 825 in SCW containing 5000 ppm of dissolved oxygen at 500 °C, 24 MPa [18]. Alloy 33 has better corrosion resistance than does SS Type 310S in SCW at 550 °C and 25 MPa, because Cr_2O_3 formed on the Alloy 33 surface is more stable [59]. By studying the oxidation behavior of stainless steels and a series of Ni-based alloys in high-temperature SCW at 800 °C and 25 MPa for 12 h, Behnamian et al. [58] provided the order of oxidation resistance of these alloys: Alloy 214 > Incoloy 800H = Inconel 625 > Hastelloy C-2000 > SS 310S > SS 347H > SS 316L, and the order of pitting corrosion resistance: Alloy 214 > SS 310S > SS 316L > Incoloy 800H > Hastelloy C-2000 > SS 347H > Alloy 62.

Yang et al. [57] studied the corrosion behavior of stainless steel and Ni-based alloys (i.e. Inconel 625, Inconel 600, and Incoloy 800) in the SCWG environment containing 6000 ppm NH_4Cl at 400 °C, 24 MPa, and found that SS 316 suffered from the most severe corrosion, including general corrosion and intergranular corrosion, while Inconel 625 exhibited the best corrosion resistance due to its high Cr content and protective Cr oxide film. By comparing the corrosion behavior of Incoloy 800, Incoloy 825, Inconel 625 and Hastelloy C-276 in sub- (350 °C, 25 MPa) and supercritical (450 °C, 25 MPa) water environments with high contents of salt and oxygen, Tang et al. [24] reported that Incoloy 800 exhibited the severest pitting corrosion under subcritical conditions while Inconel 625 and Hastelloy C-276 indicated the best corrosion resistance. Inconel 625 shows better corrosion resistance than does Hastelloy C-276 due to its higher Cr content, and Incoloy 825 has the worst corrosion resistance, with the strong, synergistic effect between salt and oxygen causing severe attack. The corrosion behavior of SS 316, Inconel 600, Inconel 625, Hastelloy C-276 and TA10 (titanium alloy Ti-0.8Ni-0.3Mo) in SCW containing oxygen, Cl^-, PO_4^{3-} at 400 °C, 25 MPa, confirms that SS 316 suffers from the severest oxide film exfoliation, Inconel 600 exhibits the severest pitting corrosion due to the lack of Mo, TA10 displays the lowest corrosion rate and the most smooth corrosion morphology, because formed TiO_2, Fe_3O_4, Cr_2O_3 and MoO_2 are hardly soluble in SCW and have good stability under SCWO conditions [60].

The corrosion rate is always expressed by the gravimetric method and depth method, which can be estimated by

$$v_g = \frac{\Delta m}{St} = \frac{\Delta V\rho}{St} = \frac{S\Delta h\rho}{St} = \frac{v_d\rho}{t} \tag{3.3}$$

and the Equation above can be transformed into

$$v_d = \frac{v_g}{\rho} \tag{3.4}$$

where v_g is the corrosion rate in terms of the gravimetric method, Δm(g) is the sample weight change before and after corrosion, S (mm^2) is the surface area of the sample, t (h) is the corrosion time, ΔV (mm^3) is the sample volume change due to corrosion, ρ (g/cm^3) is the sample density, Δh (mm) is the sample thickness change before and after corrosion, v_d is the corrosion rate in terms of the corrosion depth. Assuming that the corrosion rate is constant, in one year v_d can be expressed in units of $mm \cdot a^{-1}$, where a is the abbreviation of annual.

Figure 3.8 provides the corrosion rates of different Ni-based alloys in SCW. Inconel 600, Incoloy 800H, and Incoloy 825 represent Ni–Cr alloys, Ni–Fe-Cr alloys, and Ni–Fe–Cr–Mo–Cu alloys, respectively, that are commonly used in SCW. Apparently, Inconel 625 and Hastelloy C-276 generally have low corrosion rates, while Incoloy 800H has a relatively high corrosion rate in most of these studies. Similarly, they two also show relatively low corrosion rates compared with Inconel 600 and Incoloy 825 in most cases. Therefore, Inconel 625 and Hastelloy C-276 possess good corrosion resistance over a wide range of test environments, in terms of current investigation data regarding Ni-based alloys. Now, researchers commonly use corrosion

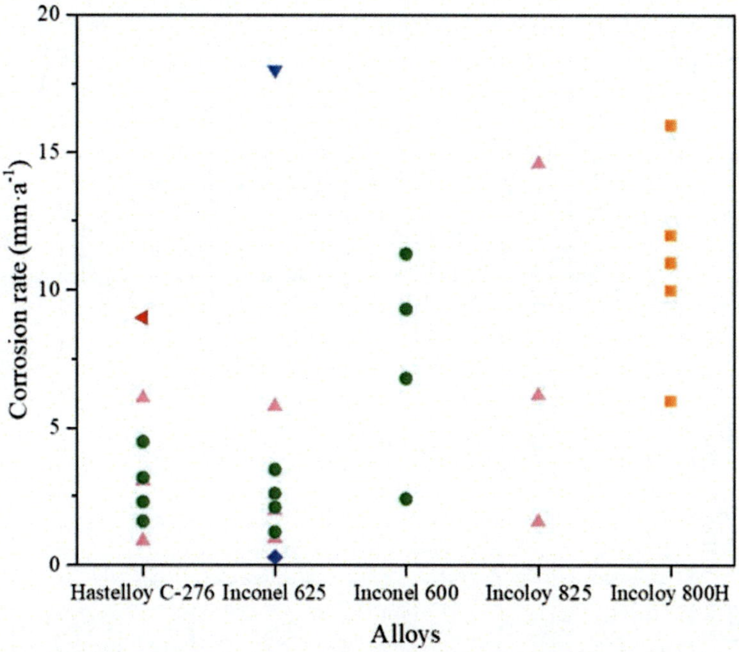

Fig. 3.8 Corrosion rates of Ni-based alloys exposed to supercritical water. Hastelloy C-276 exposed to SCW at 600 °C, 25 MPa for 66.2 h (◀) [21], Incoloy 800H exposed to SCW at 650, 700 and 750 °C, 25 MPa (■) [55], Incoloy 825, Inconel 625, and Hastelloy C-276 exposed to subcritical water (350 °C, 25 MPa) and SCW (450 °C, 25 MPa) containing high concentrations of chloride and oxygen for 60 h (▲) [24], Inconel 600, Inconel 625, Hastelloy C-276 exposed to SCW containing sodium chloride, sodium triphosphate and hydrogen peroxide at 400 °C, 25 MPa for 80 h (●) [60], Inconel 625 exposed to oxygenated ammoniacal sulphate solution at 390 °C, 25 MPa (▼) [70],Inconel 625 exposed to SCW at 426 °C, 24.1 MPa for 204.5 h (◆) [38]

weight change as an indicator of corrosion performance. However, the corrosion rate can directly reflect the corrosion thickness variation of material within a certain time range, and it should be extensively used as a crucial evaluation index in investigating the corrosion characteristics of candidate materials used in SCW in the future.

Table 3.11 displays experimental conditions and main conclusions concerning corrosion of some Ni–Fe–Cr–Mo–Cu alloys and other Ni-based alloys in subcritical or supercritical water. Temperature, pressure, and exposure time are commonly controlled within 350–800 °C, ~25 MPa, and within 1000 h, respectively. The major test environment is pure water containing dissolved oxygen or added H_2O_2, together with some inorganic salt ions in some cases. Researchers usually compare corrosion behavior of several alloys at the same conditions. Overall, Ni-based alloys have better corrosion resistance than does SS 316, while the corrosion resistance of titanium alloys (e.g., TA 10) are superior to that of Ni-based alloys. Hastelloy C-276 and Inconel 625 have better corrosion resistance than do Inconel 825 and Inconel 600. In addition to SCW, the corrosive environments involve ammonia-containing solutions,

Table 3.11 Experimentalcorrosion results for Ni–Cr-Fe-Mo-Cu alloys together with other Ni-based alloys in sub- and supercritical water

Alloy	Test solution	Oxidant	Temperature (°C)	Pressure (MPa)	Exposure time (h)	Major conclusions	Ref
Hastelloy C-276, Incoloy 625, a series of Ni–Cr alloys	SCW (0.01 mol·kg^{-1} H_2SO_4)	OC = 8 ppm	400	30	100	Thermal aging effects on the corrosion resistance and mechanical properties of alloys in SCW are very important	[61]
Hastelloy C-276, Inconel 625, Incoloy X-750	SCW	OC < 10 ppb	500	25	100–1000	Incoloy X-750 exhibits the highest corrosion rate, and the mass gains of all three alloys are small	[13]
Ni-20Cr alloy, Inconel 625, SS 316L and Niobium	2.4 wt% NH_3 and 2 wt% H_2SO_4	4.8 wt% H_2O_2	380–390	–	≈2	Niobium is resistant to corrosion under tested conditions, and the high corrosion rate of SS 316L is due to fluid mechanical effects associated with buoyancy changes	[19]
Incoloy 800/825, Inconel 625, Hastelloy C-276	Pesticide wastewater (containingCl$^-$, Br$^-$ and SO_4^{2-})	H_2O_2	350 450	25	60	Incoloy 825 shows the weakest corrosion resistance, and salt and oxygen have a synergism effect causing severe corrosion	[24]

(continued)

Table 3.11 (continued)

Alloy	Test solution	Oxidant	Temperature (°C)	Pressure (MPa)	Exposure time (h)	Major conclusions	Ref
Inconel 600, Incoloy 825	Aqueous solution (COD = 4800 ppm)	–	520	25	1–60	Sulfides deteriorate the alloy corrosion resistance in SCW, and Incoloy 825 shows slightly better corrosion resistance than Inconel 600	[43]
SS 316, TA10 Inconel 600/625, Hastelloy C-276,	SCW (containing Cl^- and PO_4^{3-})	H_2O_2	400	25	80	SS 316 and TA10 display the worst and best corrosion resistances, respectively, and phosphate shows a significant corrosion inhibition effect on alloys	[50]
SS 347H/316L/310S, Incoloy 800H, Inconel 625, Alloy 214, Hastelloy C-2000	SCW	OC = 8 ppm	800	25	12	The order of oxidation resistances is: Alloy 214 > Incoloy 800H = Inconel 625 > C-2000 > SS 310S > SS 347H > SS 316L	[46]
Inconel 600/625, Incoloy 800, SS 316	6000 ppm NH_4Cl solution	–	400	24	70	Inconel 625 shows the best corrosion resistance while SS 316 suffers from general corrosion and intergranular attack	[44]

(continued)

Table 3.11 (continued)

Alloy	Test solution	Oxidant	Temperature (°C)	Pressure (MPa)	Exposure time (h)	Major conclusions	Ref
Inconel 625, Incoloy 825	SCW	DO = 5000 ppm	500	24	24–500	Inconel 625 and Incoloy 825 exhibit pitting-like phenomena after SCWO, while Inconel 625 shows a better corrosion resistance than Incoloy 825	[18]

And OC, DO represent oxygen content and dissolved oxygen, respectively. COD refers to chemical oxygen demand halogen ion-containing solutions, and solutions containing SO_4^{2-}or PO_4^{3-}. Sometimes, the purpose of adding different ions is to simulate the investigations regarding corrosion behavior of Ni-based alloys in actual SCW environments. It is concluded that the corrosion resistance of Ni-based alloys are primarily affected by thermal aging, inorganic salts, oxidant type and concentration, and alloy elements.

Actual feedstocks treated in SCWO and SCWG may be sewage sludge, landfill leachate, phenol-containing wastewater, polychlorinated biphenyls wastewater, and wet biomass, etc., which commonly have complex compositions and properties. As far as current research results are concerned, most corrosion experiments of Ni-based alloys are limited to relatively simple SCW environments in laboratories, and corresponding corrosion characteristics and mechanisms are still insufficiently resolved, so relevant conclusions are difficult to apply in actual processes. Therefore, it is necessary to further explore the corrosion behavior of Ni-based alloys under actual operating conditions, so as to provide strong support information for the selection of candidate materials for supercritical water technologies.

3.5 Corrosion Characteristics of FeCrAl Alloys

In typical water chemistry of nuclear reactors, FeCrAl alloys can form multi-layer oxides on the surface of the matrix. With the extension of exposure time, oxides become compact. Because oxygen diffusing into the matrix is blocked due to the obstruction of multilayer protective layers, corrosion of the matrix is greatly reduced. Compared with Zircaloy, FeCrAl cladding can withstand higher temperature (Zr: 1200 K while FeCrAl: 1500 K) [62]. The rate of occurrence at high temperature steam caused by loss of coolant accident is lower than that of Zircaloy-4, which effectively slows down accident process and fuel degradation [63]. Therefore, FeCrAl alloys are expected to be a substitute for Zircaloy fuel cladding tubes.

3.5.1 *Oxidation Kinetics*

The behavior of FeCrAl in different corrosive environments has been widely investigated. Figure 3.9 presents weight changes of FeCrAl alloys exposed to various circumstances. Zircaloy gains weight in different corrosion environments in the form of parabola first and then linear, and the weight change is obvious at high temperature. In contrast, FeCrAl alloys show different oxidation kinetics. In the presence of dissolved hydrogen, FeCrAl exhibits weight loss, which is attributed to the dissolution of metals. However, in the presence of dissolved oxygen, it shows weight gain. This can be explained that the presence of dissolved oxygen promotes the formation of oxides on the metal surface. In addition, It is also reported that when the exposure time is long enough (5000 days), FeCrAl alloys will present a combination of weight gain and weight loss, and the changing rate is significantly different [64].

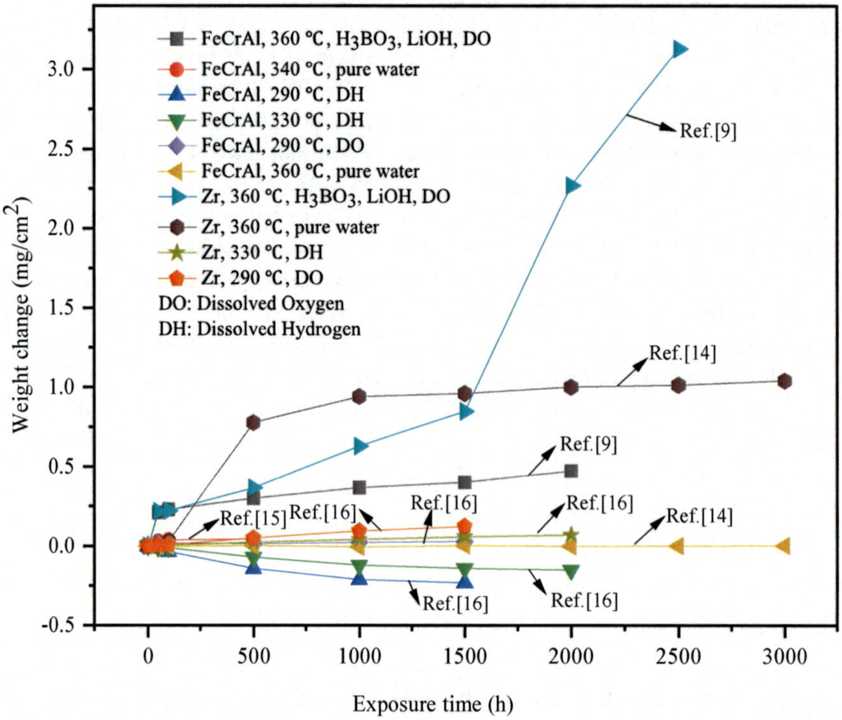

Fig. 3.9 Weight changes of FeCrAl and Zircaloy with exposure time in different corrosive environments

Due to differences in experimental conditions, there is no agreement on the corrosion kinetics of FeCrAl alloys at low temperature. However, studies with experimental conditions close to real reactors all point to the fact that the weight of FeCrAl alloys increases with the corrosion time in parabola. According to the research of Terrani et al. [65], the formation of oxides on the inner surface of FeCrAl alloys follows parabolic kinetics with exposure time in low temperature light water reactors environments. Due to the limitation of iron passing through the chromite layer to the surface, the mass loss of claddings caused by the dissolution of Fe-rich precipitates also displays parabolic kinetics. Pan et al. [63] found that FeCrAl had the same kinetic in terms of the mass loss under high temperature steam at 1200 °C. And, the weight gain rate of FeCrAl alloys is obviously lower than that of Zr-4 under the same state at 1000 °C. Besides, Lipkina et al. [66] studied the oxidation behavior of oxide dispersion strengthened FeCrAl in air and steam. Except Al_2O_3 was observed on the surface, white adhesive was also found. It seems to improve the transport of oxygen in oxides and plays an important role in oxidation kinetics. EDS-line scan results indicate that Zr increases, and O, Fe, Al as well as Cr decreases in the white adhesive in contrast to the surrounding alloy matrix. However, whether the adhesive is oxidation product or metallic should be further confirmed.

3.5.2 Corrosion Process and Mechanism

The growth of oxides on the surface of FeCrAl depends on the oxygen vacancies formed on the interface between the alloy matrix and oxide layers. The formation of oxygen vacancies consumes metal ions, which can be supplemented by matrix under the action of metal vacancies. The specific behavior of FeCrAl alloys in corrosive environment is determined by environment conditions and chemical reactions dominated by elements (e.g., Fe, Cr and Al). Park et al. [64] analyzed the oxidation behavior of FeCrAl alloys in pure water at 360 °C and 18.5 MPa and observed FeO, Fe_3O_4 and α-Fe_2O_3 on the outer surface of alloys after 500 days exposure. Raiman et al. [67] found that only Fe_2O_3 existed on the surface of FeCrAl in hydrogenated water chemistry. Through comparative experiments, it has demonstrated that FeCrAl is less seriously corroded in normal water chemistry without dissolved hydrogen. The enrichment of chromium in oxides around the metal matrix may be the reason for the enhanced protection of oxide layer in normal water chemistry.

As shown in Table 3.12, a single, duplex or triple oxide layer all may be formed on the surface of the FeCrAl alloy at typical working temperature (280–360 °C) of water chemistry. When oxygen is dissolved in the pure water, one protective film with Fe oxides or Fe-rich spinel will be produced while duplex oxides are easily observed in hydrogen water chemistry. The inner oxide is Cr-rich spinel and Fe-rich spinel or Fe oxides takes shape on the outer surface.In lithiated water chemistry, transition layer may exist except from inner and outer oxide layers. The diffusion of Ni from the autoclave material can explain the phenomenon that $NiFe_2O_4$ is reported in some studies. Different oxide compositions in the same environment may be attributed to the contents of balancing Fe in the alloy. In high-temperature oxidation atmosphere, single oxide which is mainly occupied by Al_2O_3 prevails in the alloy surface. α-Al_2O_3 seems to appear frequently as the central structure of Al_2O_3. As for FeCrAl corrosion at a medium temperature (e.g. 600 °C), relevant research all reveals the breakdown of primary corrosion protection and the formation of secondary protection, where an inward-growing spinel and an outward-growing Fe oxide (Fe_2O_3 or/and Fe_3O_4) are developed [68–72]. Corrosive anions will make primary protection easily lose efficacy and turn into multi-layered corrosion regime. Absolutely, different oxidation temperature and environments will lead to various oxide phases.

3.5.2.1 Low Temperature Oxidation

According to Eqs. 3.5–3.10 [74, 79], Fig. 3.10 depicts an oxidation process of FeCrAl alloy in low-temperature (<360 °C) water chemistry. As main contents of the matrix, Fe and Cr generally participate int the formation of oxide film. When water lies in the subcritical state, dissolved oxygen exerts a significant influence in the formation of protective oxides. Large percentage of Fe in the alloy guarantees its enough reaction with oxygen or Cr to form the Fe-rich layer. Among Fe-related oxides (e.g., FeO, Fe_2O_3, and Fe_3O_4), Fe_2O_3 is found to be the primary phase, and α-Fe_2O_3 is the common morphology instead [64]of γ-Fe_2O_3 [64].

$$2Fe + O_2 \rightarrow 2FeO \quad (3.5)$$

$$4Fe + 3O_2 \rightarrow 2Fe_2O_3 \quad (3.6)$$

Table 3.12 Corrosion research involving oxides compositions on the surface of FeCrAl

Fe (wt%)	Cr (wt%)	Al (wt%)	Other elements	Corrosion environments	Oxides	Ref.
77.5 81.8 79.4 78.8	10.0 13.1 13.0 13.0	6.0 5.3 6.0 7.2	Mo, Nb, Y, Si	290 °C, 9 MPa, 63 ppb dissolved hydrogen	$NiFe_2O_4$, Fe_3O_4, $FeCr_2O_4$	[67]
78.7	13.0	5.1	Mo, Nb, Y, Si	290 °C, 9 MPa, 63 ppb dissolved hydrogen	α-Fe_2O_3, $NiFe_2O_4$, Fe_3O_4, $FeCr_2O_4$	[67]
77.5 81.8 79.4 78.8	10.0 13.1 13.0 13.0	6.0 5.3 6.0 7.2	Mo, Nb, Y, Si	290 °C, 9 MPa 2000 ppb dissolved oxygen	Fe_2O_3, Fe_3O_4	[67]
85.2	10.0	4.8	Zr, Ni, Mo, Si, O, C, S,	290 °C, 7 MPa 0.3 ppm H_2, 5.6 pH	Outer: Fe-rich spinel Inner: Cr-rich spinel	[65]
85.2	10.0	4.8	Zr, Ni, Mo, Si, O, C, S	290 °C, 7 MPa 1.0 ppm O_2, 5.6 pH	α-Fe_2O_3	[65]
68.0 70.5 80.8 82.3 85.0	21.0 21.0 15.1 13.1 13.0	11.0 5.0 3.9 4.4 2.0	Mo, Y	320 °C, 14 MPa Proton and electron irradition	Outer: Fe_3O_4, Fe_2O_3, $NiFe_2O_4$ Inner: Cr/Al-rich oxides	[73]
85.2	10.0	4.8	Zr, Ni, Mo, Si, O, C, S,	330 °C, 15 MPa 3.57 ppm H_2, 7.2 pH	Outer: Fe-rich spinel Inner: Cr-rich spinel	[65]
72.4	14.4	8.9	Mo, Si, Ti, V, W, Ni, Y	340 °C, 15.16 MPa	Outer: α-Fe_2O_3 Inner: Cr-rich spinel	[74]
Balance	22.0	5.8	Si, Mn, C	360 °C, 18.5 MPa 7.0–7.5 pH	FeO, Fe_3O_4, α-Fe_2O_3	[64]

(continued)

Table 3.12 (continued)

Fe (wt%)	Cr (wt%)	Al (wt%)	Other elements	Corrosion environments	Oxides	Ref.
Balance Balance	12.9 12.9	5.0 5.0	Mo, Y, S, P, O, N, H	360 °C, 19 MPa 1200 ppm H_3BO_3 2.3 ppm LiOH <5 ppb dissolved oxygen	Outer: Fe_3O_4 Inner: $FeFe_{2-x-y-z}Cr_xAl_yMo_zO_4$ Transition: Cr-Al oxides	[75]
Balance	10.0	4.0	Si, Zr	600 °C, O_2 600 °C, O_2, H_2O 600 °C, O_2, H_2O, KCl	Outer: Fe_2O_3, Fe_3O_4 Inner: (Fe, Cr, Al)-spinel	[19]
72.0–92.0	5.0–25.0	1.0–6.0	Not mentioned	600 °C, O_2, K_2CO_3	Outer: Fe_2O_3, Fe_3O_4 Inner: $(Fe, Cr, M)_3O_4$	[62]
Balance	5.3	2.7	Not mentioned	1000 °C, O_2	FeO_x, $(Cr, Al)_2O_3$	[76]
Balance	13.1	4.3	Mo, Nb, C, N	1000 °C, H_2O (g)	FeO, α-Al_2O_3	[63]
66.0	20.8	11.1	O, Mo	1000–1200 °C, H_2O (g)	α-Al_2O_3, α-FeCr, χ-$Mo_5Cr_6Fe_{18}$	[77]
84.9 84.0	12.9 12.8	2.1 3.2	C	1200 °C, H_2O (g)	Fe_2O_3, Cr_2O_3, Al_2O_3, $FeCr_2O_4$	[78]
83.2 82.1 81.2	12.8 12.8 12.9	4.0 5.1 5.9	C	1200 °C, H_2O (g)	Al_2O_3	[78]
Balance	13.1	4.3	Mo, Nb, C, N	1200 °C, H_2O (g)	α-Al_2O_3, $FeAl_3$	[63]
Balance	22.0	5.8	Si, Mn, C	1200 °C, H_2O (g)	α-Al_2O_3	[64]

$$\frac{3}{2}Fe + O_2 = \frac{1}{2}Fe_3O_4 \tag{3.7}$$

$$\frac{4}{3}Cr + O_2 = \frac{2}{3}Cr_2O_3 \tag{3.8}$$

$$2Fe + 2Cr_2O_3 + O_2 = 2FeCr_2O_4 \tag{3.9}$$

$$2xCr + (3 - 2x)Fe + 2O_2(g) \rightarrow Fe(Fe_{1-x}Cr_x)_2O_4(s) \tag{3.10}$$

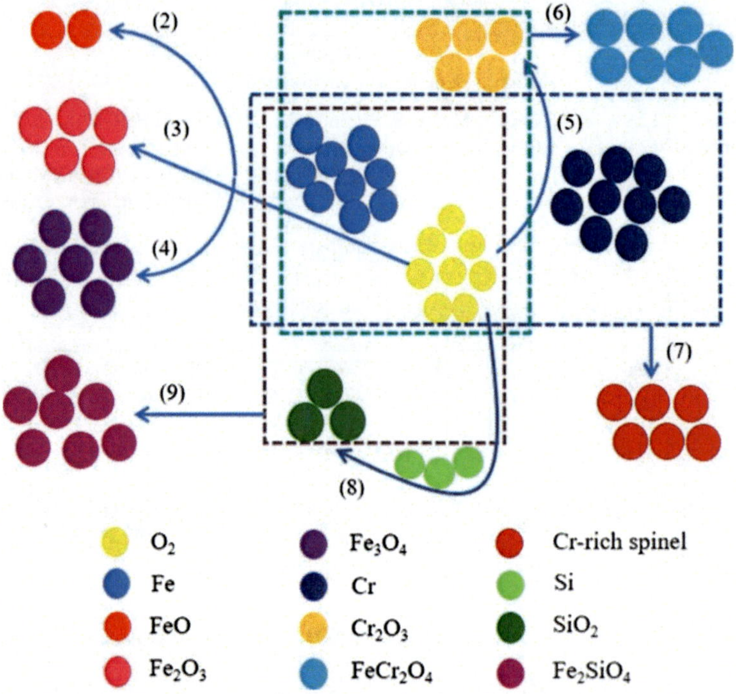

Fig. 3.10 Chemical reactions of FeCrAl alloys in low-temperature corrosive water chemistry

When there is a large amount of Si in the alloy, Fe will be oxidized into Fe_2SiO_4 with SiO_2 which follows Eqs. 3.11 and 3.12 [80]. As shown in Eq. 3.13, more vacancies will be formed on the surface of the matrix and oxidation of spinel to form gaseous $CrO_2(OH)_2$ will be accelerated under irradiation [74]. $CrO_2(OH)_2$ tends to deposit on the pipes' surface of reactors components with the flow of coolant, resulting in heat transfer deterioration and other accidents.

$$Si + O_2 = SiO_2 \tag{3.11}$$

$$2Fe + SiO_2 + O_2 = Fe_2SiO_4 \tag{3.12}$$

$$\frac{1}{2}(Cr_xFe_{1-x})_2O_3(s) + \frac{3}{4}xO_2(g) + xH_2O(l) \rightarrow xC_rO_2(OH)_2(g) + \frac{1}{2}(1-x)Fe_2O_3(g) \tag{3.13}$$

As shown in Fig. 3.11, Ning et al. [75] clarified the complete corrosion process of FeCrAl cladding in typical water chemistry. Corrosion products on the surface of the alloy are outer Fe_3O_4 and inner spinel. There is also a transition layer at the

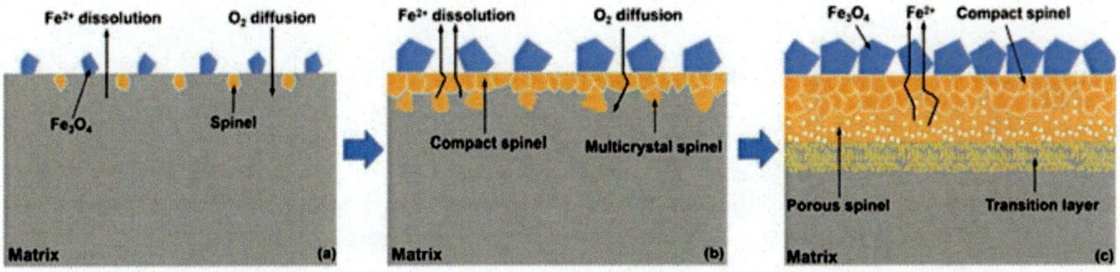

Fig. 3.11 Corrosion process of FeCrAl in water chemistry containing H_3BO_3 and LiOH [75]

interface between the matrix and the inner surface of oxide film. The experimental conditions are relatively close to water chemistry in actual reactors, so the result has great reference value.

In the process, low-concentration oxygen at the inner oxide/matrix interface leads to the polycrystalline structure of spinel oxide. Pores on the surface of polycrystalline spinel are attributed to that the outward diffusion rate is higher than the inward diffusion rate of oxygen [64]. The minimum oxygen concentration at the interface of oxide/matrix fails to provide enough driving force to make a large amount of Fe^{2+} released from the matrix. As a result, most Fe remains in the matrix as Fe^0. Al and Cr are preferentially oxidized to form Al/Cr-rich oxides because the critical oxygen partial pressure required for related oxides is less than that of Fe [80], and Al/Cr-rich oxides will constitute a transition layer with some Fe. The dissolution of a small amount of Fe^{2+} may also lead to the formation of pores in the transition layer [64]. The porosity on the surface of transition layer and polycrystalline spinel as well as the shedding of local oxides may be main reason for the weight loss of FeCrAl in water chemistry [81]. At high-temperature steam, the voids with different sizes are also observed on the alloy surface. This is because the aluminum element diffuses outward to form an alumina layer and leave the vacancies [78]. If temperature keeps rising, the diffusion will turn faster, bringing about thicker intermetallic layers and larger cavities [77].

3.5.2.2 High Temperature Oxidation

Ongoing nuclear fission and continuous irradiation increase temperature of claddings. Strong activation energy caused by high temperature makes elements in FeCrAl cladding participate in corrosive reactions more actively.

Figure 3.12 shows complex corrosive reaction network of FeCrAl under high temperature, which occur according Eqs. 3.14–3.32 [63, 71, 78]. Corrosion of the cladding is not caused by Fe, Cr and Al in isolation. Although Fe is involved in reactions, its products will react with Cr and Al as reactants Correspondingly, slowly formed Al/Cr oxides are closely connected with anti-corrosion protection of FeCrAl in the early stage of high temperature environment containing oxygen. Dense Al_2O_3 formed on the FeCrAl alloy surface in high temperature steam can prevent other elements being oxidized, so remarkably decreasing the oxidation rate

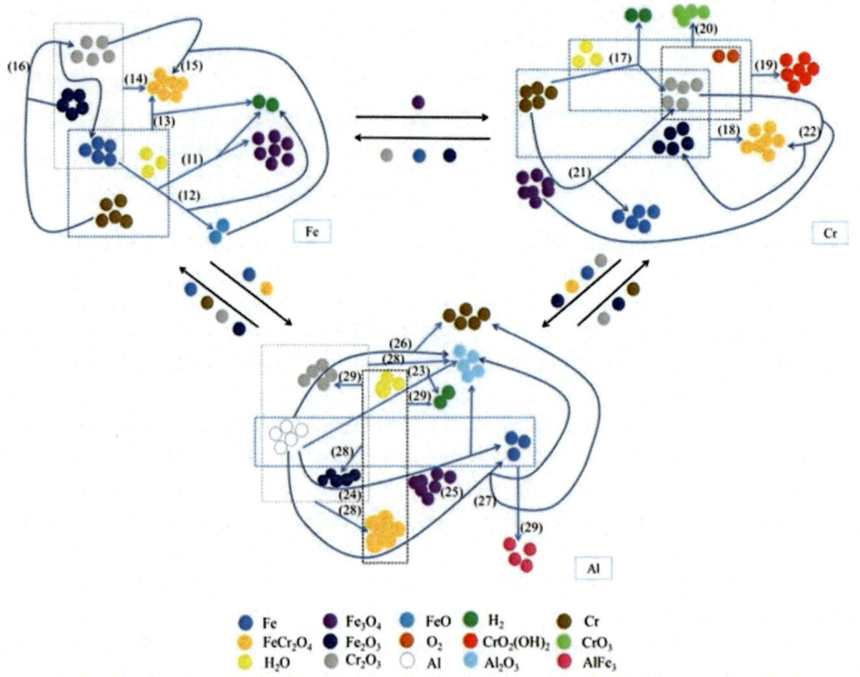

Fig. 3.12 Chemical reactions of FeCrAl alloys in high-temperature oxidation environment

$$x\mathrm{Fe}(s) + \mathrm{H_2O}(g) \rightarrow y\mathrm{Fe_3O_4}(s) + \mathrm{H_2}(g)\left[x = \frac{3}{4}\&y = \frac{1}{4}, x = \frac{2}{3}\&y = \frac{1}{3}\right] \tag{3.14}$$

$$\mathrm{Fe}(s) + \mathrm{H_2O}(g) \rightarrow \mathrm{FeO}(s) + \mathrm{H_2}(g) \tag{3.15}$$

$$\frac{1}{4}\mathrm{Fe}(s) + \frac{1}{2}\mathrm{Cr}(s) + \mathrm{H_2O}(g) \rightarrow \frac{1}{4}\mathrm{FeCr_2O_4}(s) + \mathrm{H_2}(g) \tag{3.16}$$

$$\frac{1}{3}\mathrm{Fe}(s) + \mathrm{Cr_2O_3}(s) + \frac{1}{3}\mathrm{Fe_2O_3}(s) \rightarrow \mathrm{FeCr_2O_4}(s) \tag{3.17}$$

$$\mathrm{FeO}(s) + \mathrm{Cr_2O_3}(s) \rightarrow \mathrm{FeCr_2O_4}(s) \tag{3.18}$$

$$\mathrm{Fe_2O_3}(s) + 2\mathrm{Cr}(s) \rightarrow \mathrm{Cr_2O_3}(s) + 2\mathrm{Fe}(s) \tag{3.19}$$

$$\frac{2}{3}\mathrm{Cr}(s) + \mathrm{H_2O}(g) \rightarrow \frac{1}{3}\mathrm{Cr_2O_3}(s) + \mathrm{H_2}(g) \tag{3.20}$$

$$\frac{1}{3}\mathrm{Cr}(s) + \frac{5}{6}\mathrm{Cr_2O_3}(s) + \frac{1}{2}\mathrm{Fe_2O_3}(s) \rightarrow \mathrm{FeCr_2O_4}(s) \tag{3.21}$$

$$\frac{1}{2}\mathrm{Cr_2O_3}(s) + \frac{3}{4}\mathrm{O_2}(g) + \mathrm{H_2O}(g) \rightleftharpoons \mathrm{CrO_2(OH)_2}(g) \tag{3.22}$$

$$2\mathrm{Cr_2O_3}(s) + 3\mathrm{O_2}(g) \rightarrow 4\mathrm{CrO_3}(g) \tag{3.23}$$

$$\frac{3}{4}\mathrm{Fe_3O_4}(s) + 2\mathrm{Cr}(s) \rightarrow \mathrm{Cr_2O_3}(s) + \frac{9}{4}\mathrm{Fe}(s) \tag{3.24}$$

$$\mathrm{Fe_3O_4}(s) + \mathrm{Cr_2O_3}(s) \rightarrow \mathrm{FeCr_2O_4}(s) + \mathrm{Fe_2O_3}(s) \tag{3.25}$$

$$\frac{2}{3}\mathrm{Al}(s) + \mathrm{H_2O}(g) \rightarrow \frac{1}{3}\mathrm{Al_2O_3}(s) + \mathrm{H_2}(g) \tag{3.26}$$

$$2\mathrm{Al}(s) + x\mathrm{Fe_2O_3}(s) \rightarrow y\mathrm{Fe}(s) + \mathrm{Al_2O_3}(s)\left[x = 1\&y = 2, x = \frac{3}{4}\&y = \frac{9}{4}\right] \tag{3.27}$$

$$2\mathrm{Al}(s) + \mathrm{Cr_2O_3}(s) \rightarrow 2\mathrm{Cr}(s) + \mathrm{Al_2O_3}(s) \tag{3.28}$$

$$2\mathrm{Al}(s) + \frac{3}{4}\mathrm{FeCr_2O_4}(s) \rightarrow \frac{3}{4}\mathrm{Fe}(s) + \frac{3}{2}\mathrm{Cr}(s) + \mathrm{Al_2O_3}(s) \tag{3.29}$$

$$\frac{1}{3}\mathrm{Al}(s) + \mathrm{Cr_2O_3}(s) + \frac{1}{2}\mathrm{Fe_2O_3}(s) \rightarrow \mathrm{FeCr_2O_4}(s) + \mathrm{Al_2O_3}(s) \tag{3.30}$$

$$\mathrm{Al}(s) + 3\mathrm{Fe}(s) \rightarrow \mathrm{AlFe_3}(s) \tag{3.31}$$

$$\mathrm{FeCr_2O_4}(s) + \mathrm{H_2O}(g) \rightarrow \mathrm{Fe_2O_3}(s) + 2\mathrm{Cr_2O_3}(s) + \mathrm{H_2}(g) \tag{3.32}$$

The morphology of Al_2O_3 shares nearly no similarity at different temperatures [82]. It is considered that alumina forming alloys commonly form protective α-Al_2O_3 on its surface at high temperatures, while metastable 'transient' Al_2O_3 (γ, δ, θ) can be found at medium temperatures (>500 °C) [83]. On the whole, the transformation of Al_2O_3 is in the order of γ-Al_2O_3 → θ-Al_2O_3 → α-Al_2O_3 or γ-Al_2O_3 → δ-Al_2O_3 → θ-Al_2O_3 → α-Al_2O_3. For FeCrAl alloys, the transformation of metastable Al_2O_3 into α-Al_2O_3 seems to occur within 900–1000 °C and α-Al_2O_3 is quickly formed at 1000 °C and higher temperatures [84]. However, the transition behavior is dependent on many variables such as water vapor level and heating rates. Besides, the formation of protective external Al_2O_3 scale from transient oxide follows different pathways as the contents of Cr and Al change. [85]. In addition to Al_2O_3, Fe_2O_3 and Cr_2O_3 are early oxidation products of the alloy in low temperature steam. At high temperature, iron and chromium oxides react with steam to generate volatile hydroxide, which peels off the alloy surface, and $FeCr_2O_4$ becomes main oxidation products. The formation of $AlFe_3$ is due to the external diffusion of Al atoms through exchanging with Fe atoms and the defects in ferrite [63].

3.5.2.3 Corrosion of FeCrAl Coating

When FeCrAl is coated on the surface of Zircaloy, it tends to be full of pores and cracks. The corrosion process has to take interaction between coating and matrix into consideration. Under high temperature steam, the diffusion between FeCrAl coating and Zr matrix will reduce the content of Cr and Al, and form loose iron oxide with poor protection on the surface of the sample. Zr-rich and Zr-poor bands are shaped due to selective oxidation of Zr caused by different standard Gibbs free energy [86]. It is obvious that the inter-diffusion between coating and substrate largely affects the corrosion of FeCrAl coating, which is the reason for different corrosion behaviors of FeCrAl tube and coating. Mo interlayer coating is often introduced to act as diffusion barriers, and thus high temperature oxidation resistance of specimen is improved [87]. Spraying process is also an influencing factor. The FeCrAl coating fabricated by conventional arc spraying has many surface defects like buried porosities and micro-cracks, which provide the chance for corrosive penetration and shorten the service life of coating. High current pulsed electron beam can be used as a surface modification of FeCrAl coating [88, 89]. The corrosion rate of arc-sprayed coating made by high current pulsed electron beam indicates a significant decline. Cold spray process is also reported [90]. The effects of different thermal and cold spraying techniques on the corrosion of FeCrAl coating need further comprehensive comparison. Optimal annealing temperature can reduce surface defects and increase coating adhesion, so producing FeCrAl coating with good properties [91]. Certainly, the corrosion resistance of FeCrAl coating has big connection with alloying elements contained. High Cr-containing coating tends to exhibit better oxidation resistance than does low Cr-containing coating [92].

3.5.3 Influence of Water Chemistry

The influence of water chemistry on the corrosion behavior of FeCrAl alloys is relatively obvious in low-temperature environment. In addition to boric acid and lithium hydroxide, typical primary water chemistry also contains dissolved hydrogen, dissolved oxygen, anions (Cl^-, F^-, SO_4^{2-}), cations (Mg^{2+}, Na^+, Si^{4+}, etc.), and added Zn or heavy metals [93]. These substances all impact the corrosion of alloys.

3.5.3.1 Dissolved Hydrogen and Oxygen

Dissolved hydrogen and oxygen will engender potential on oxides which grow on the surface of the alloy, thus promoting the hydrogen absorption [94]. Hydrides are further generated and deposited under the oxide film, so that hydrogen embrittlement appears. Compared with dissolved oxygen, dissolved hydrogen in water chemistry greatly determines the thickness of oxide layers formed.

According to the relationship between the thickness of oxide layers and the concentration of dissolved hydrogen in Ni-based alloys, it is inferred that similar effect will occur in FeCrAl alloys [95]. High dissolved hydrogen is beneficial to the stability of alloy elements, while low concentration of dissolved hydrogen can maintain a thick and dense oxide film and act as an ion diffusion barrier on the surface of the alloy, which reduces the sensitivity of primary water stress corrosion cracking [96]. Therefore, the concentration of dissolved hydrogen in water chemistry should be controlled below the critical concentration where the transition between metals in the matrix and their metal oxides happens [97]. However, when the concentration of dissolved hydrogen constantly decreases, Li and B in water chemistry will easily react to produce lithium metal ($LiBO_2$) following Eq. 3.33, consequently reducing the pH value of coolant and improving the anti-corrosion efficiency of alloys.

$$Li^{+} + H_3BO_3 \leftrightarrow LiBO_2 + H^{+} + H_2O \tag{3.33}$$

Based on previous studies, influencing mechanisms of dissolved hydrogen and oxygen on the structure of oxide layers formed in the corrosion can be concluded below. First, dissolved oxygen and hydrogen are responsible for the difference of diffusion speed of metal ions from alloys (e.g. $Fe^{2+} > Ni^{2+} \gg Cr^{3+}/Al^{+}$) and selective dissolution of Fe and Cr [98]. Second, dissolved hydrogen is absorbed by the matrix and extracts electrons from surrounding metal atoms when penetrating the metal, thus weakening metal bonds [96]. Third, H atoms reduce the adsorption and diffusion energy of oxygen in the matrix, making it easier for oxygen to penetrate the surface and facilitate the oxidation process [99].

3.5.3.2 Anions

The existence of anions can induce electrochemical corrosion, which is an omen of stress corrosion cracking of the alloys [100]. Corrosive power is exhibited by different ionic species [101]. The influence of different anions on intergranular stress corrosion cracking has the following reduction order of $SO_3{}^{2-}$, S^{2-}, $S_2O_3{}^{2-}$, $SO_4{}^{2-}$, $AsO_2{}^{-}$, OH^{-}, $SiO_3{}^{2-}$, $HPO_4{}^{2-}$, $PO_4{}^{3-}$, Cl^{-}, $CO_3{}^{2-}$, $B_4O_7{}^{2-}$, and $NO_3{}^{-}$[102]. Among them, chloride-induced stress corrosion cracking of alloys is commonly investigated [103–106]. Cl^{-} induces pitting corrosion in the early stage, which will combine with residual stress to cause stress corrosion cracking of cladding materials [107]. Cold spray deposition seems to have great potential in reducing chlorine induced stress corrosion cracking [108]. In addition, hydrogen fluoride can catalyze dissociation to produce fluorine on the alloy surface. The interaction between fluorine and alloy elements and their binding properties will lead to alloy corrosion [109]. Sulfate affects the behavior of the coolant at the inner interface of the alloy, thus affecting the formation of oxides.

3.5.3.3 Cations

Cations (e.g., Na^+, Mg^{2+}, Zn^{2+}, Al^{3+}, Zr^{4+}, and La^{3+}) in the coolant mainly come from makeup water and reloading during overhaul [110]. The solubility of these cationic salts and oxides is characterized by negative temperature coefficient. With the flow of coolant, they deposit on the surface of fuel elements, and inhibit the corrosion of alloys, especially Fe-based alloys. In a lab test, Mg^{2+} is proved to be the fastest to result in specimens failure, Ca^{2+} needs a longer time and Na^+ is the slowest [107]. The effect of Zn on corrosion behavior is often studied. Because of the highest site preference energy, Zn will replace Fe, Ni, Co and other elements in the lattice gap of spinel oxide, and form more stable Zn oxides [111]. On other hand, Zn will reconstruct the structure of spinel oxides on the alloy surface and reduce defects in the oxides, so significantly improving electrochemical corrosion behavior of the alloy in coolant [102]. Zn ions are also effective in suppressing the deposition of Co-60, which is the major radiation source in boiling water reactors [112].

3.5.4 Influence of Irradiation

The corrosion behavior of FeCrAl greatly depends on irradiation in real reactors. Irradiation promotes both overall corrosion and intergranular corrosion [113]. The behavior of FeCrAl under proton irradiation [73], electron irradiation [88] and ion irradiation [114, 115], all have been reported and many research in neutron irradiation can be found [116–118]. Proton, electron and ion irradiation incline to induce the transpassive dissolution of Cr-rich inner oxide, while the redistribution of alloying elements (especially Fe) due to fast neutron irradiation is considered as a primary contributor of enhancing corrosion [119]. In general, all these irradiation bands formed on FeCrAl surface aggravate the defects including dislocation loops and vacancies at grain boundaries to shorten the path of atomic diffusion, so accelerating the combination of metal with oxygen atoms, and increasing the corrosion rate [74, 120]. Obviously, the increasing rate of corrosion relies on the irradiation dose [121]. Radiolysis products caused by ionization and sputtering of water under high energy density irradiation all long influence corrosion process. Among all products including •e − aq, •HO, •H, $\bullet O_2^-$, $\bullet HO_2$, H_3O^+, OH^-, H_2O_2 and H_2. H_2O_2 is widely involved in the redox reaction of oxide surface with the flow of coolant, thus significantly increasing the electrochemical corrosion potential [73]. When the potential exceeds the transition potential between different phases of metal oxide on the surface, the type of oxides will change. By comparison, short-lived radicals such as •HO and •H may have less effect on the corrosion rate [122].

3.5.5 *Influence of Temperature*

In low temperature water chemistry, the effect of temperature on the corrosion behavior of the alloy is mainly reflected in increasing the activation energy and promoting the occurrence of chemical reaction to accelerate metal dissolution [123]. Temperature has larger influence on the corrosion rate in a gas environment compared with water vapor [124]. At present, the influence in high temperature steam on FeCrAl corrosion behavior has been reported. The growth mechanism of FeCrAl oxide film differs at diverse temperatures. At 900 °C, the main mode of oxidation is the diffusion of aluminum ions, while the oxidation at 1200 °C is determined by the diffusion of oxygen ions, which is accompanied by folding and peeling of the oxide film [63]. Therefore, reducing the heating rate during the temperature-raising period can improve the steam resistance of FeCrAl [125].

3.6 Corrosion Characteristics of Typical Zircaloy

Before discussing the corrosion of FeCrAl alloys, it is necessary to make a simple retrospect of the corrosion behavior of Zircaloy. Zircaloy has low thermal neutron capture cross section, excellent corrosion resistance and adequate mechanical properties. Zircaloy–1 is the first alloyed Zr, which commonly contains 2.5 wt% Sn to mitigate the negative effects of N, and then Zircaloy–2 (1.5 wt% Sn, 0.12 wt% Fe, 0.1 wt% Cr, 0.05 wt% Ni) with better corrosion resistance is developed. Zircaloy–3 has less Sn (0.25–0.5 wt%) and Fe (0.2–0.4 wt%) than does Zircaloy–2. Typical compositions of Zircaloy–4 are 1.2–1.7 wt% Sn, 0.18–0.24 wt% Fe and 0.07–0.13 wt% Cr, while Ni is not contained [126]. Compared with Zircaloy–2 and Zircaloy–3, Zircaloy–4 has a weaker hydrogen absorption capacity under the same mechanical properties, so the possibility of hydrogen corrosion is reduced. Thus, Zircaloy–4 is widely applied as fuel cladding material in commercial light water reactors [127].

3.6.1 *Corrosion Process*

A complete corrosion process of Zircaloy is shown in Fig. 3.13. The corrosion mechanism is combined with dissociation, adsorption, oxidation and reduction. Transport of corrosive species such as hydrogen, electron and oxygen is the decisive step of corrosion rate. Zircaloy mainly occurs uniform corrosion as Eq. 3.34 [128] to form a protective oxide film in coolant or high temperature steam, and part of the hydrogen

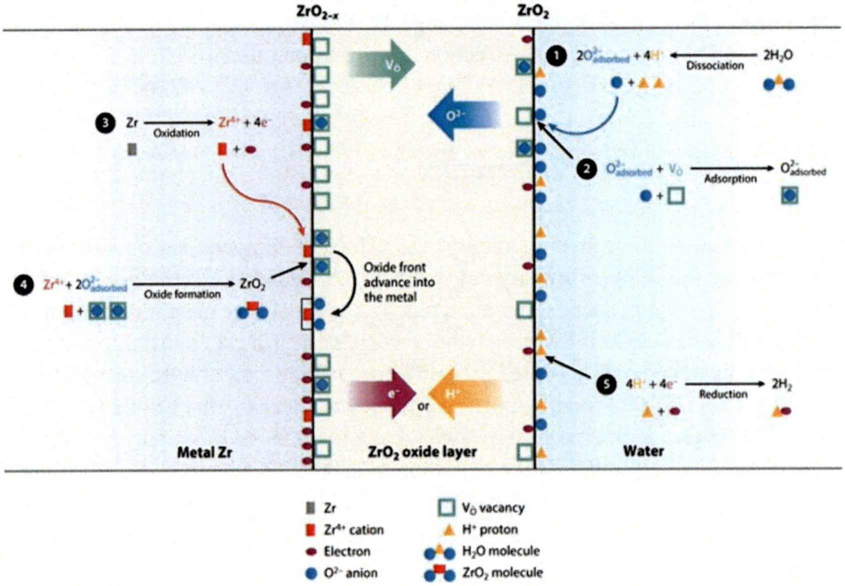

Fig. 3.13 Corrosion process of Zircaloy [131]

generated will be absorbed by the matrix. The thickness of the oxide film increases with corrosion time. After reaching a certain thickness, stress will cause the oxide film to break or fall off, leading to turning point [129]. Moya et al. [130] discovered that a ZrC thin film (~1 μm) was formed on the surface of ZrO_2. As exhibited in Eq. 3.35, ZrC is a result of carbon reduction of the oxide layer top surface. Herein, C needed for the reaction can be derived from stainless steel component (e.g., grids, and walls) in the autoclave.

$$Zr + 2H_2O \rightarrow ZrO_2 + 2H_2 \quad (3.34)$$

$$ZrO_2 + C \rightarrow ZrC + O_2 \quad (3.35)$$

The oxidation of Zircaloy generally follows parabolic or sub-parabolic kinetics until a turning point occurs after which the kinetics become linear [132]. Before the turning point, the surface of alloys is black and viscous monoclinic $ZrO_{2\text{-}n}$ (n < 0.05). After the turning point, the oxide film transforms into grayish white and loose non-protective ZrO_2. The transition time rests with temperature and the type of Zircaloy [133]. It is noted that dissolved oxygen usually exists in actual water chemistry. Its high concentration will lead to nodular corrosion with the formation of grey and lenticular oxides growing fast on the local surface of zirconia [134]. Therefore, it is highly important to control water chemistry to prevent non-uniform corrosion of Zircaloy. There are two common corrosion types of Zircaloy, that is, uniform corrosion tends to occur in pressurized water reactors and nodular corrosion mainly occurs in boiling water reactors [135].

Furthermore, shadow corrosion is usually observed as small patches of thicker oxide at connection points between fuel cladding and grids, and it is also a nonnegligible issue when it comes to the widespread application of Zircaloy in boiling water reactors [136]. At present, there is no unified understanding on the enhanced corrosion mechanism of Zircaloy. Cox [119]. demonstrated four feasible statements from existing assumptions. The first is displacement damage arouse from fast neutron and increasing diffusion rate in the protective oxide. The second is crevice corrosion effects caused by LiOH. The third is enhanced electronic conductivity of zirconia film resulted from β and γ radiation. The last one is residual potential differences between metallic phases. Undoubtedly, numerous publications make it difficult to determine whether it is a single mechanism or multiple mechanisms working together. Therefore, the mechanisms of Zircaloy in low temperature water chemistry and in high temperature oxidation environment need more in-depth study with interdisciplinary collaboration among physics, material science and chemistry [137].

3.6.2 *Influence of Alloying Elements*

Elements have always exerted considerable influence in corrosion of alloys. Harding [138]. calculated true oxidation states of various alloying elements in zirconia and vindicated that the production of an aliovalent charge was a significant factor in the anti-corrosion protection of Zircaloy. Gong et al. [139] systematically studied the effect of elements on the formation of nano pores in the oxide film of Zircaloy, and they thought that these pores were caused by the enrichment of alloying elements at grain boundaries and the subsequent uneven diffusion of cations. The experimental results from Bell et al. [129] showed that Sc, Sb and Nb would form substitutional Zr defects in t–ZrO_2. Sc will reduce the corrosion resistance. The addition of Sb makes Sb–Zr binary alloys exhibit significant corrosion. Zr–Nb–Sb ternary alloys are the slowest to be corroded and absorb the least hydrogen. This indicates the beneficial effect of Nb in improving the corrosion resistance, which is expounded by Chen et al. [140]. Low oxidation state of Nb^{2+}/Nb^{3+} can balance the accumulation of positive space-charge in the oxide layer to increase oxygen vacancy and electron mobility, so causing the occurrence of near-parabolic corrosion kinetics [141]. The Nb content and annealing temperature largely influence corrosion characteristics of Zr–Nb alloys [142]. Sn can improve the corrosion resistance of Zircaloy by primarily suppressing the deleterious effects of nitrogen [143]. But high Sn will have a negative influence, especially when Cl^- exists in the surrounding environment. The active reaction of Sn with Cl^- will form chlorination complexes (e.g. $SnCl^+$), which create favorable conditions for the adsorption and penetration of corrosive chloride ions [144].

Similarly, the addition of slight Si helps improve the oxidation resistance and limit the sensitivity of pitting corrosion, while high silicon content will lead to lower anti-oxidant activity and impact the structural integrity when used as the element of coating [145]. Fe, Cr and Ni are favorable in increasing the corrosion resistance of Zircaloy under high temperature. Besides, the addition of Cr can also advance the creep resistance and mechanical strength and the presence of Fe gives rise to low hydrogen absorption [143]. Notably, the impact of elements on the corrosion behavior of Zircaloy often changes significantly after thermal treatment, which is attributed to the distribution of alloying elements and their oxides in different phases [146]. Duan et al. [127] summarized the optimum values of main components in Zircaloy as below. O is within the limits of 0.11–0.16 wt%, and N is no more than 0.07 wt%.The specific content of Sn should take negative impact of N on corrosion resistance into account. The content of Fe ought to be considered together with Cr and Ni, and their total amount may be controlled between 0.18–0.38 wt%. Mo should not exceed 0.1 wt%. Si falls in the range of 0.008–0.014 wt%, which is relatively consistent with the results obtained by Hong et al. [147]. As for Nb, its addition is better in the range of 0.2–0.3 wt% so that the annealing temperature effect on corrosion resistance can be neglected [148]. Influences of other subordinate elements, such as Al, Cu, Ge, need to be further determined and their optimal amounts must be combined with the above main components to determine in order to improve the performance of Zircaloy in harsh environment.

3.6.3 *Influence of pH*

Typical pH of coolant in nuclear reactors is usually controlled by H_3BO_3 and LiOH. Severe corrosion behavior of Zircaloy in LiOH has been studied by most researchers, but it is hard to reach a conclusion on corrosion mechanism.

Table 3.13 lists some research about the influencing mechanisms of LiOH and H_3BO_3 on the corrosion of Zircaloy. Although experimental conditions are different, a consensus has been reached that LiOH exerts a negative influence on the corrosion resistance while H_3BO_3 plays a converse role. The increment of LiOH requirs more H_3BO_3 to offset its stimulation on the corrosion rate. Generally speaking, LiOH can increase pores in oxides and thus aggravates the corrosion rate, but the corresponding mechanism is still unclear. The substitution of Li^+ for Zr^{4+} in the oxide layer owing to similar ion size increases anion vacancies, thus the corrosion is accelerated under the diffusion of vacancies [149]. The acceleration is also probably related to the change of ZrO_2 morphology [150]. Characteristics of LiOH solution lead to the dissolution of cubic or tetragonal ZrO_2 prior to monoclinic ZrO_2, because under the same conditions, the rapid corrosion acceleration of Zircaloy can be observed in LiOH, but only slightly increasing corrosion is observed in NaOH, and almost no corrosion occurs in KOH and NH_4OH [151]. It has also been attributed to the increase in hydrogen concentration, resulting in the oxidation of metals containing hydrides faster than those without hydrides. Kim et al. [152] confirmed this statement, but in their experiments, the normal samples showed a faster corrosion rate over time. This can be explained that the zirconium hydride reaches a certain thickness so that the hydrogen absorption in the sample is inhibited. It is noteworthy that LiOH may

Table 3.13 Influencing mechanisms of LiOH and H_3BO_3 on the corrosion of Zircaloy

Temperat-ure (°C)	LiOH (ppm)	H_3BO_3 (ppm)	Exposure time (h)	Major findings	Ref.
360	2.4–12,000	2000, 4000, 8000	96	H_3BO_3 can inhibit the reactions between LiOH and anion vacancies to prevent the formation of porosity	[153]
300	240, 2400, 24,000	6666.7–53,333.3	696–1632	H_3BO_3 can prevent the deep penetration of pores in oxides	[154]
–	–	–	–	LiOH largely influences ZrO_2 morphology	[150]
350	102.9, 754.3	–	13,200	Li^+ incorporated into ZrO_2 determines the microstructure and growth mechanism of oxide at the metal/oxide interface	[149]
–	6.9–240	666.7–66,666.7	5760	Boron functions as a blocking agent in the form of $H_4BO_4^-$	[155]
350	960	–	48	The concentration of LiOH intensely affects the corrosion rate by transforming t–ZrO_2 into m–ZrO_2	[156]
Room temperature	342.9	66.7–26,666.7	–	The ameliorating influence of H_3BO_3 is combined with pH effect and blocking mechanism	[157]
Room temperature	7.5–24,000	333.3–26,666.7	–	The inhibition of B is attributed to pH neutralizing and its participation in the corrosion process LiOH leads to accelerated corrosion by pH effect and porosity	[158]

(continued)

Table 3.13 (continued)

Temperat-ure (°C)	LiOH (ppm)	H_3BO_3 (ppm)	Exposure time (h)	Major findings	Ref.
Room temperature	7.9	10,000	0.5	Slight Cl^- enrichment exists in the pit after H_3BO_3 addition	[159]
230	0–9.3	–	11.8–112	The concentration of Li is highly connected with corrosion deposition	[160]

– Means that the information is not mentioned in the reference

have synergistic effect with F^- [151]. The solution first penetrates oxides through the pores formed during kinetic rate transformation, so increasing the solubility of ZrO_2. The dissolved zirconia is recrystallized in porous oxides to recover F^- and returns to the solution. Therefore, the dissolution process will be self-catalyzed, leading to oxidation out of control.

Definitely, the concentration LiOH is highly connected with the change mechanism of corrosion rate. Jeong et al. [149]revealed that the critical concentration of LiOH to quicken corrosion is in the range from 7.5 to 103 ppm. However, Waheed et al. [161] found that when the concentration of LiOH was 7.5 ppm, the corrosion rate was greatly reduced, while the corrosion rate got fast when the concentration of LiOH reached 12 ppm. The critical concentration of LiOH within 12,000–24,000 ppm was reported by Kandil et al. [157]. Experimental temperature and Zircaloy type should be responsible for the different results above. The alkalinisation of the solution caused by the elevated concentration of LiOH has been proved to reduce typical oxides formed on the matrix under simulated pressurized water reactor conditions [160]. Most existing studies mainly focus on the role of Li in LiOH but ignore that of OH^-.Different from above research, Liu et al. [156] reported that the corrosion of Zircaloy was more controlled by OH^- than Li^+, and LiOH advanced the growth rate of oxide by OH^- diffusion to the interface and reacting with Zr to form ZrO_2 and hydrogen. Thus, theoretical support for the opposite effect of the concentration change and specific action of solutes in corrosion behavior needs to be further explored.

The existence of H_3BO_3 can reduce the permeability of pores, so alleviating the accelerated effect of LiOH on the corrosion of Zircaloy. The behavior of H_3BO_3 on the corrosion alleviation is a result of combined effect. In summary, the addition of H_3BO_3 will neutralize and dilute LiOH to hinder the reaction between LiOH and anion vacancies. The ionized form of boric acid and its complexes can also be used as a sealing agent to prevent the oxide from entering the metal/oxides interface. However, the dosage of H_3BO_3 is restricted and its role in decreasing corrosion depends on the concentration of Li, exposure time and heat transfer conditions [154, 155, 158]. In addition, the corrosion of Zircaloy in oxidation environment is also related to irradiation in reactors [162], thermodynamic parameters (e.g., the temperature of corrosive environment, cooling temperature and cooling rate) [163, 164], the texture of the alloy[165], hydride effect [158, 166], and impurities in coolant [167, 168], etc. Therefore, the influence of single factor on the corrosion behavior of Zircaloy and the coupling mechanism of multi elements require to be further clarified.

3.6.4 Influence of Secondary Phase Precipitates

When alloying elements (e.g., Fe, Cr, Ni, Nb, and Sn) are higher than their solubility limits in zirconium, intermetallic phases which are called secondary phase precipitates or particles will be formed [169]. In Zircaloy-2, main secondary phase precipitate is Zr_2(Ni, Fe), while that of Zircaloy-4 is Zr(Fe, $Cr)_2$ with Laves structure at room temperature [170]. Table 3.14 lists secondary phase precipitates of some alloys in relevant studies. It is clear that various precipitates are found in Zr-based alloys. Different elements in Zircaloy produce unsimilar precipitates, which exhibit distinct corrosion behaviors due to structural differences. For example, Zr(Fe, $Cr)_2$ forms a more elongated shape upon oxidation with a strongly banded elemental segregation, while Zr_2Fe shows uniform corrosion and the produced oxide does not match t-ZrO_2 [171]. Related studies have revealed that Cu and Sn decrease the average size of secondary phase precipitates without strongly changing their crystal structures, while Cr may markedly refine precipitates and trigger new precipitates with Laves phases [170, 172, 173]. When suffering from irradiation, Fe in Zr-Fe based precipitates will deplete and give rise to precipitates amorphization which deteriorates irradiation resistance of alloys [174]. Moreover, radiation induces the

Table 3.14 Secondary phase precipitates of alloys reported in relevant studies

Precipitate	Type	Alloy	Ref.
δ-ZrH_x	FCC	Zr–Sn–Fe–Cr (Zircaloy-4)	[177]
Zr_2(Fe, Ni)	BCT	Zr–Sn–Nb–Fe–Cr–Ni	[178]
Zr(Fe, Cr, $Nb)_2$ (Zr, $Nb)_2$Fe	HCP FCC	Zr–Sn–Nb–Fe–O/Zr–Sn–Nb–Fe–Cr–O	[172]
Zr_3Fe Zr_2Fe	Orthorhombic Tetragonal	Zr–Sn–Nb–Fe–Cu–O Zr–Sn–Nb–Fe–Cr–Cu–O	
Zr(Fe, $Cr)_2$	HCP	Zr–Sn–Fe–Cr (Zircaloy-4)	[171]
β-Nb	BCC	Zr–Nb–Fe	[179]
Zr-(Fe, Cu)	Hexagonal	Zr–Sn–Nb–Fe–Cu	[176]
$ZrCr_2$	FCC	Zr–Cr–Fe–Mo–Bi	[180]
Zr(Fe, $Nb)_2$	HCP	Zr–Nb–Sn–Fe	[181]
Zr_9S_2	Tetragonal	Zr–Sn–Nb–Fe–Cr-S	[182]

Herein, FCC, HCP, BCT and BCC represent Face-centered Cubic, Hexagonal Close-packed, Body-centered Tetragonal and Body-centered Cubic, respectively redistribution of Nb, thus influencing corrosion performance of Zr–Nb alloys [175]. The addition of Ni seems to have a positive effect on anti-irradiation performance of Zircaloy, but it requires more discussion [176].

Based on existing information, precipitates effects on Zircaloy corrosion are mainly embodied in four aspects. First of all, the oxidation of precipitates produces amorphous ZrO_2 which acts as a barrier to suppress the inner-diffusion of oxygen ion into oxide film [179]. Secondly, additional stress caused by oxidation products of secondary phase precipitates makes defects on the oxide/matrix interface develop into pores in the inner surface of oxide film. The oxide layer becomes loose and thus corrosion accelerates [179, 182, 183]. The detrimental influence of oxidation of precipitates on the corrosion behavior of Zircaloy will be magnified if the surrounding environment contains corrosive Li^+ and/or OH^- ions [183]. Thirdly, the hybridizations and pseudogap effects among atoms such as Mo, Bi and Fe stir up different distribution trends and site preferences of alloying elements, and then cracks in the oxide are produced, so affecting the integrity of protective oxide films [180–184]. Fourthly, elements in precipitates can migrate and change chemical compositions of Zircaloy exposed to irradiation, so reducing corrosion resistance [185].

3.6.5 *Steam Oxidation in Loss of Coolant Accident*

When the loss of primary coolant accident takes place in nuclear reactors, the Zircaloy fuel cladding will be oxidized in high temperature air and steam mixtures with the phase changes of the superficial ZrO_2 layer from monoclinic to tetragonal and cubic [186]. Different from parabolic or sub-parabolic then linear kinetics under operating conditions, the high temperature oxidation kinetics of Zircaloy (especially Zircaloy-4) is sub-parabolic or fully parabolic at accident conditions although total reaction still follows Eq. 3.34 [187–190]. Meanwhile, once the loss of primary coolant accident occurs, ZrO_2 will be formed on the cladding surface while a fraction of oxygen diffuses in the metal, causing the formation of an oxygen stabilized α-Zr(O) metallic phase [191]. At around 1000 °C, the appearance of breakaway oxidation will increase oxidation rate and hydrogen pickup [192]. This negative effect on the integrity of cladding is linked to cracks in the pre-oxide layer which shapes in low temperature steam [193]. Le Saux et al. [191] put forward a complete mechanism to account for the occurrence of breakaway oxidation. In their proposal, the local destabilization of t-ZrO_2 is an incentive to breakaway oxidation. As for pre- and post-breakaway regimes, the diffusion of oxygen vacancies in the oxide is the rate-determining step which dominates the reaction kinetics. [194].

Apparently, the steam oxidation behavior of Zircaloy is closely connected with temperature, which influences not only the existence of breakaway oxidation but also the transport of oxygen and hydrogen. Its oxidation rate will decrease due to low activation energy if heating rate is increased [195]. The effects of pre-oxide can not be ignored. A pre-existing oxide layer helps to retard the oxidation rate at high temperatures [196]. As a corrosive medium, steam is the dominant factor for the Zircaloy-4 oxidation at high temperatures. On the one hand, the reactions between H_2O and Zircaloy-4 suppress the presence of SnO, which is responsible for the inhibition of ZrO_2 grain growth and t-ZrO_2 transformation into m-ZrO_2. On other hand, the transport of H_2O can modify the anionic defect ZrO_2 state because of oxygen vacancy annihilation [197, 198]. Higher Nb content makes alloys have superior high temperature oxidation resistance as a result of the stabilization of β-Zr suppressing metal matrix oxidation [187, 199]. Lower Sn content achieves similar oxidation resistance effect, but the segregation of Sn at the oxide/matrix can induce the nodular corrosion of Zircaloy.[134]. The influence of pressure is also reported. High pressure reduces the surface energy for transformation, so t-ZrO_2 is converted into poor-protection m-ZrO_2 with many cracks [200].

References

1. P. Xu, L.Y. Zhao, K. Sridharan, T.R. Allen, Oxidation behavior of grain boundary engineered alloy 690 in supercritical water environment. J. Nucl. Mater. **422**(1–3), 143–151 (2012)
2. X. Zhong, E.H. Han, X. Wu, Corrosion behavior of alloy 690 in aerated supercritical water. Corros. Sci. **66**, 369–379 (2013)
3. X. Zhong, X. Wu, E.H. Han, Characteristics of oxidation and oxygen penetration of alloy 690 in 600 °C aerated supercritical water. J. Mater. Sci. Technol. **34**, 561–569 (2018)
4. M. Payet, L. Marchetti, M. Tabarant, J.P. Chevalier, Corrosion mechanism of a Ni-based alloy in supercritical water: impact of surface plastic deformation. Corros. Sci. **100**, 47–56 (2015)
5. J. Gao, F.R. Wan, G.P. Cao, K. Sridharan, T. Allen, Analysis of oxides formed on the surface of the alloy 690 in hydrogenated supercritical water. Acta Metall. Sin-Engl. **29**(8), 774–781 (2016)
6. W. Wang, X. Cao, Y. Lu, X. Ding, T. Shoji, TEM study on oxide mechanism of the dissimilar welds between 316 stainless steels/Inconel 182 in high temperature and high pressure water. Mater. Charact. **131**, 339–347 (2017)
7. J. Yang, S. Wang, Y. Li, Z. Jiang, J. Zhang, D. Xu, K. Wang, Oxidation-sulfidation attacks on alloy 600 in supercritical water containing organic sulfides. Mater. Lett. **263**, 127218 (2020)
8. Y. Li, D.D. Macdonald, J. Yang, J. Qiu, S. Wang, Point defect model for the corrosion of steels in supercritical water: part I, film growth kinetics. Corros. Sci. **163**, 108280 (2020)
9. M. Payet, L. Marchetti, M. Tabarant, F. Jomard, J.P. Chevalier, Corrosion mechanisms of 316L stainless steel in supercritical water: the significant effect of work hardening induced by surface finishes. Corros. Sci. **157**, 157–166 (2019)
10. Y. Li, S. Wang, J. Yang, D. Xu, Y. Guo, L. Qian, W. Song, Corrosion characteristics of a nickel-base alloy C-276 in harsh environments. Int. J. Hydrogen Energy **42**(31), 19829–19835 (2017)
11. H. Shi, Z. Gao, Z. Fan, Y. Ding, Y. Qiao, Z. Zhu, Corrosion behavior of alloy C-276 in supercritical water. Adv. Mater. Sci. Eng. **2018**, 1–6 (2018)
12. Q. Zhang, R. Tang, K. Yin, X. Luo, L. Zhang, Corrosion behavior of hastelloy C-276 in supercritical water. Corros. Sci. **51**(9), 2092–2097 (2009)
13. Q. Zhang, R. Tang, C. Li, X. Luo, C. Long, K. Yin, Corrosion behavior of Ni base alloys in supercritical water. Nucl. Eng. Technol. **41**(1), 107–112 (2009)
14. M. Sun, X. Wu, Z. Zhang, E.H. Han, Analyses of oxide films grown on alloy 625 in oxidizing supercritical water. J. Supercrit. Fluids **47**(2), 309–317 (2008)
15. K.H. Chang, S.M. Chen, T.K. Yeh, J.J. Kai, Effect of dissolved oxygen content on the oxide structure of Alloy 625 in supercritical water environments at 700 °C. Corros. Sci. **81**, 21–26 (2014)

16. Y. Watanabe, T. Adschiri, K. Sue, K. Arai, Estimation of metal oxide solubility and understanding corrosion in supercritical water. Corrosion **107**, 702–705 (2001)
17. D. Rodriguez, A. Merwin, Z. Karmiol, D. Chidambaram, Surface chemistry and corrosion behavior of Inconel 625 and 718 in subcritical, supercritical, and ultrasupercritical water. Appl. Surf. Sci. **404**, 443–451 (2017)
18. J. Yang, S. Wang, X. Tang, Y. Wang, Y. Li, Effect of low oxygen concentration on the oxidation behavior of Ni-based alloys 625 and 825 in supercritical water. J. Supercrit. Fluids **131**, 1–10 (2018)
19. E. Asselin, A. Alfantazi, S. Rogak, Corrosion of nickel–chromium alloys, stainless steel and niobium at supercritical water oxidation conditions. Corros. Sci. **52**(1), 118–124 (2010)
20. C. Schroer, J. Konys, J. Novotny, J. Hausselt, Aspects of the corrosion kinetics for binary nickel-chromium alloys under conditions found in supercritical water oxidation plants processing chlorinated substances. Corrosion **62**(5), 444–459 (2005)
21. K. Chen, J. Wang, Z. Shen, D. Du, X. Guo, B. Gong, J. Liu, L. Zhang, Comparison of the stress corrosion cracking growth behavior of cold worked Alloy 690 in subcritical and supercritical water. J. Nucl. Mater. **520**, 235–244 (2019)
22. K. Chen, J. Wang, Z. Shen, D. Du, X. Guo, L. Zhang, P.L. Andresen, Effect of intergranular carbides on the cracking behavior of cold worked alloy 690 in subcritical and supercritical water. Corros. Sci. **164**, 108313 (2020)
23. N. Hara, S. Tanaka, S. Soma, K. Sugimoto, Corrosion resistance of Fe-Cr and Ni-Cr alloys in oxidizing supercritical HCl solution. NACE Int. (2002)
24. X. Tang, S. Wang, D. Xu, Y. Gong, J. Zhang, Y. Wang, Corrosion behavior of Ni-based alloys in supercritical water containing high concentrations of salt and oxygen. Ind. Eng. Chem. Res. **52**(51), 18241–18250 (2013)
25. S. Lin, H. Li, L. Xu, Y. Zhang, C. Cui, A novel experimental device for electrochemical measurements in supercritical fluids up to 700 °C/1000 bar and its application in the corrosion study of superalloy inconel 740H. RSC Adv. **7**(54), 33914–33920 (2017)
26. T. Fujii, K. Sue, S.I. Kawasaki, Effect of pressure on corrosion of inconel 625 in supercritical water up to 100MPa with acids or oxygen. J. Supercrit. Fluids **95**, 285–291 (2014)
27. L. Tan, X. Ren, K. Sridharan, T.R. Allen, Corrosion behavior of Ni-base alloys for advanced high temperature water-cooled nuclear plants. Corros. Sci. **50**(11), 3056–3062 (2008)
28. P. Shi, R. Hu, T. Zhang, L. Yuan, J. Li, Grain boundary character distribution and its effect on corrosion of Ni–23Cr–16Mo superalloy. Mater. Sci. Technol. **33**(1), 84–91 (2016)
29. A. Selvig, X. Huang, D.J. Kim, D. Guzonas, Surface oxide formation on IN625 and plasma sprayed NiCrAlY after high density and low density supercritical water testing. Mater. Corros. **65**(8), 768–777 (2014)
30. Y. Daigo, Y. Watanabe, K. Sue, Effect of chromium ion from autoclave material on corrosion behavior of Nickel-based alloys in supercritical water. Corrosion **63**(2), 277–284 (2007)
31. S.H. Son, J.H. Lee, C.H. Lee, Corrosion phenomena of alloys by subcritical and supercritical water oxidation of 2-chlorophenol. J. Supercrit. Fluids **44**(3), 370–378 (2008)
32. Y. Li, T. Xu, S. Wang, J. Yang, J. Li, T. Xu, D. Xu, Characterization of oxide scales formed on heating equipment in supercritical water gasification process for producing hydrogen. Int. J. Hydrogen Energy **44**(56), 29508–29515 (2019)
33. J. Konys, S. Fodi, J.H. Hausselt, C.V. Schmidt, Corrosion of high-temperature alloys in chloride-containing supercritical water oxidation systems. Corrosion **55**(1), 45–51 (1999)
34. N. Boukis, W. Habicht, G. Franz, E. Dinjus, Behavior of Ni-base alloy 625 in methanol-supercritical water systems. Mater. Corros. **54**, 326–330 (2003)
35. R. Fujisawa, K. Nishimura, T. Nishida, M. Sakaihara, Y. Kurata, Y. Watanabe, Corrosion behavior of nickel-based alloys and type 316 stainless steel in slightly oxidizing or reducing supercritical water. Corrosion **62**(3), 270–274 (2006)
36. N. Saito, Y. Tsuchiya, Y. Akai, H. Omura, T. Takada, N. Hara, Corrosion performance of metals for supercritical water, oxidation-utilized organic waste-processing reactors. Corrosion **62**(5), 383–394 (2006)

37. H. Kim, D.B. Mitton, R.M. Latanision, Effect of pH and temperature on corrosion of nickel-base alloys in high temperature and pressure aqueous solutions. J. Electrochem. Soc. **157**(5), C194–C199 (2010)
38. H. Kim, D.B. Mitton, R.M. Latanision, Corrosion behavior of Ni-base alloys in aqueous HCl solution of pH 2 at high temperature and pressure. Corros. Sci. **52**(3), 801–809 (2010)
39. H. Kim, D.B. Mitton, R.M. Latanision, Stress corrosion cracking of alloy 625 in ph 2 aqueous solution at high temperature and pressure. Corrosion **67**, 035002 (2011)
40. B. Hertweck, T.G. Steigerwald, N.S.A. Alt, E. Schluecker, Different corrosion behaviour of autoclaves made of nickel base alloy 718 in ammonobasic and ammonoacidic environments. J. Supercrit. Fluids **95**, 158–166 (2014)
41. L. Tan, T.R. Allen, Y. Yang, Corrosion behavior of alloy 800H (Fe–21Cr–32Ni) in supercritical water. Corros. Sci. **53**, 703–711 (2011)
42. M. Fulger, M. Mihalache, D. Ohai, S. Fulger, S.C. Valeca, Analyses of oxide films grown on AISI 304L stainless steel and Incoloy 800HT exposed to supercritical water environment. J. Nucl. Mater. **415**, 147–157 (2011)
43. M. Fulger, D. Ohai, M. Mihalache, M. Pantiru, V. Malinovschi, Oxidation behavior of Incoloy 800 under simulated supercritical water conditions. J. Nucl. Mater. **385**, 288–293 (2009)
44. S.H. Nie, Y. Chen, X. Ren, K. Sridharan, T.R. Allen, Corrosion of alumina-forming austenitic steel Fe–20Ni–14Cr–3Al–0.6Nb–0.1Ti in supercritical water. J. Nucl. Mater. **399**, 231–235 (2010)
45. L. Tan, X. Ren, K. Sridharan, T.R. Allen, Effect of shot-peening on the oxidation of alloy 800H exposed to supercritical water and cyclic oxidation. Corros. Sci. **50**, 2040–2046 (2008)
46. L. Tan, K. Sridharan, T.R. Allen, The effect of grain boundary engineering on the oxidation behavior of INCOLOY alloy 800H in supercritical water. J. Nucl. Mater. **348**, 263–271 (2006)
47. H. Akhiani, M. Nezakat, S. Penttilä, J. Szpunar, The oxidation resistance of thermo-mechanically processed Incoloy 800HT in supercritical water. J. Supercrit. Fluids **101**, 150–160 (2015)
48. K.I. Choudhry, S. Mahboubi, G.A. Botton, J.R. Kish, I.M. Svishchev, Corrosion of engineering materials in a supercritical water cooled reactor: characterization of oxide scales on Alloy 800H and stainless steel 316. Corros. Sci. **100**, 222–230 (2015)
49. K.I. Choudhry, D.A. Guzonas, D.T. Kallikragas, I.M. Svishchev, On-line monitoring of oxide formation and dissolution on alloy 800H in supercritical water. Corros. Sci. **111**, 574–582 (2016)
50. S. Mahboubi, G.A. Botton, J.R. Kish, On the oxidation resistance of alloy 800HT exposed in supercritical water (SCW). NACE Int. **7**, 992–1002 (2015)
51. S. Bsat, X. Huang, S. Penttila, Corrosion behaviour of bare and NiCrAlY coated alloy 214 in supercritical water at 700 °C. J. Nucl. Eng. Radia Sci. **4**, 011017 (2017)
52. L. Tan, K. Sridharan, T.R. Allen, Altering corrosion response via grain boundary engineering. Mater. Sci. Forum **595–598**, 409–418 (2008)
53. L. Tan, K. Sridharan, T.R. Allen, R.K. Nanstad, D.A. McClintock, Microstructure tailoring for property improvements by grain boundary engineering. J. Nucl. Mater. **374**, 270–280 (2008)
54. K.I. Choudhry, D.T. Kallikragas, I.M. Svishchev, in Hydrogen Release Rate and Corrosion Measurements on Alloy 800H and Stainless Steel 316 in Supercritical Water Conditions. The 7th International Symposium on Supercritical Water-Cooled Reactors (2015)
55. K.I. Choudhry, I.M. Svishchev, L. Orlovskaya, Alloy 800H under supercritical water conditions: a flow reactor study of corrosion and hydrogen evolution. Mater. Corros. **66**, 1430–1434 (2015)
56. Y. Li, S. Wang, X. Tang, D. Xu, Y. Guo, J. Zhang, L. Qian, Effects of sulfides on the corrosion behavior of Inconel 600 and Incoloy 825 in supercritical water. Oxid. Met. **84**, 509–526 (2015)
57. J. Yang, S. Wang, D. Xu, Y. Guo, C. Yang, Y. Li, Effect of ammonium chloride on corrosion behavior of Ni-based alloys and stainless steel in supercritical water gasification process. Int. J. Hydrogen Energy **42**, 19788–19797 (2017)
58. Y. Behnamian, A. Mostafaei, A. Kohandehghan, B.S. Amirkhiz, D. Serate, Y. Sun, S. Liu, E. Aghaie, Y. Zeng, M. Chmielus, W. Zheng, D. Guzonas, W. Chen, J.L. Luo, A comparative

study of oxide scales grown on stainless steel and nickel-based superalloys in ultra-high temperature supercritical water at 800 °C. Corros. Sci. **106**, 188–207 (2016)
59. S. Mahboubi, Y. Jiao, W. Cook, W. Zheng, D.A. Guzonas, G.A. Botton, J.R. Kish, Stability of chromia (Cr2O3)-based scales formed during corrosion of austenitic Fe-Cr-Ni alloys in flowing oxygenated supercritical water. NACE Int. **72**, 1170–1180 (2016)
60. X. Tang, S. Wang, L. Qian, Y. Li, Z. Lin, D. Xu, Y. Zhang, Corrosion behavior of nickel base alloys, stainless steel and titanium alloy in supercritical water containing chloride, phosphate and oxygen. Chem. Eng. Res. Des. **100**, 530–541 (2015)
61. Y. Daigo, Y. Watanabe, K. Sugahara, T. Isobe, Compatibility of nickel-based alloys with supercritical water applications: aging effects on corrosion resistance and mechanical properties. NACE Int. **62**, 174–181 (2006)
62. B. Qiu, J. Wang, Y. Deng, M. Wang, S.Z. Qiu, A review on thermohydraulic and mechanical-physical properties of SiC, FeCrAl and Ti3SiC2 for ATF cladding. Nucl. Eng. Technol. **52**, 1–13 (2020)
63. D. Pan, R. Zhang, H. Wang, C. Lu, Y. Liu, Formation and stability of oxide layer in FeCrAl fuel cladding material under high-temperature steam. J. Alloys Compd. **684**, 549–555 (2016)
64. D.J. Park, H.G. Kim, J.Y. Park, Y. Jung, J.H. Park, Y.H. Koo, A study of the oxidation of FeCrAl alloy in pressurized water and high-temperature steam environment. Corros. Sci. **94**, 459–465 (2015)
65. K.A. Terrani, B.A. Pint, Y.J. Kim, K.A. Unocic, Y. Yang, C.M. Silva, H.M. Meyer, R.B. Rebek, Uniform corrosion of FeCrAl alloys in LWR coolant environments. J. Nucl. Mater. **479**, 36–47 (2016)
66. K. Lipkina, D. Hallatt, E. Geiger, B.W.N. Fitzpatrick, M.H.A. Piro, A study of the oxidation behaviour of FeCrAl-ODS in air and steam environments up to 1400 °C. J. Nucl. Mater. **541**, 152305 (2020)
67. S.S. Raiman, K.G. Field, R.B. Rebak, Y. Yamamoto, K.A. Terrani, Hydrothermal corrosion of 2nd generation FeCrAl alloys for accident tolerant fuel cladding. J. Nucl. Mater. **536**, 152221 (2020)
68. B. Pujilaksono, T. Jonsson, H. Heidari, M. Halvarsson, J.E. Svensson, L.G. Johansson, Oxidation of binary FeCr alloys (Fe-2.25Cr, Fe-10Cr, Fe-18Cr and Fe-25Cr) in O_2 and in O_2 + H_2O environment at 600 degrees C. Oxid. Met. **75**, 183–207 (2011)
69. T. Jonsson, H. Larsson, S. Karlsson, H. Hooshyar, M. Sattari, J. Liske, J.E. Svensson, L.G. Johansson, High-temperature oxidation of FeCr(Ni) alloys: the behaviour after breakaway. Oxid. Met. **87**, 333–341 (2017)
70. T. Jonsson, B. Pujilaksono, H. Heidari, F. Liu, J.E. Svensson, M. Halvarsson, L.G. Johansson, Oxidation of Fe–10Cr in O_2 and in O_2+H_2O environment at 600°C: a microstructural investigation. Corros. Sci. **75**, 326–336 (2013)
71. J. Eklund, B. Jönsson, A. Persdotter, J. Liske, J. Svensson, T. Jonsson, The influence of silicon on the corrosion properties of FeCrAl model alloys in oxidizing environments at 600 °C. Corros. Sci. **144**, 266–276 (2018)
72. A. Persdotter, J. Eklund, J. Liske, T. Jonsson, Beyond breakaway corrosion—influence of chromium, nickel and aluminum on corrosion of iron-based alloys at 600 °C. Corros. Sci. **177**, 108961 (2020)
73. P. Wang, S. Grdanovska, D.M. Bartels, G.S. Was, Effect of radiation damage and water radiolysis on corrosion of FeCrAl alloys in hydrogenated water. J. Nucl. Mater. **533**, 152108 (2020)
74. W. Wu, G. Ran, Y. Li, S. Cong, Y. Sun, Early corrosion behaviour of irradiated FeCrAl alloy in a simulated pressurized water reactor environment. Corros. Sci. **174**, 108824 (2020)
75. F. Ning, X. Wang, Y. Yang, J. Tan, Z. Zhang, D. Jia, X. Wu, E. Han, Uniform corrosion behavior of FeCrAl alloys in borated and lithiated high temperature water. J. Mater. Sci. Technol. **70**, 136–144 (2021)
76. Y. Niu, S. Wang, F. Gao, Z.G. Zhang, F. Gesmundo, The nature of the third-element effect in the oxidation of Fe–xCr–3 at.% Al alloys in 1atm O_2 at 1000 °C. Corros. Sci. **50**, 345–356 (2008)

77. X. Han, Y. Wang, S. Peng, H. Zhang, Oxidation behavior of FeCrAl coated Zry-4 under high temperature steam environment. Corros. Sci. **149**, 45–53 (2019)
78. Y. Qiao, P. Wang, W. Qi, S. Du, Z. Liu, F. Meng, X. Zhang, K. Wang, Q. Liu, Z. Yao, C. Bai, X. Wang, Mechanism of Al on FeCrAl steam oxidation behavior and molecular dynamics simulations. J. Alloys Compd. **828**, 154310 (2020)
79. R. Qiang, A. Leong, J. Zhang, M.P. Short, Corrosion behavior of Fe–Cr–Si alloys in simulated PWR primary water environment. J. Nucl. Mater. **526**, 151735 (2019)
80. F. Ning, X. Wu, J. Tan, Crevice corrosion behavior of Alloy 690 in high-temperature water. J. Nucl. Mater. **515**, 326–337 (2019)
81. K. Hong, J.R. Barber, M.D. Thouless, W. Lu, Cracking and spalling of the oxide layer developed in high-burnup Zircaloy-4 cladding under normal operating conditions in a PWR. J. Nucl. Mater. **512**, 46–55 (2018)
82. X. Huang, X.Y. Li, X.D. Fang, Research progress in FeCrAl alloys for accident-tolerant fuel cladding. Cailiao Gongcheng-J. Mater. Eng. **48**, 19–33 (2020)
83. H. Josefsson, F. Liu, J.E. Svensson, M. Halvarsson, L.G. Johansson, Oxidation of FeCrAl alloys at 500–900 °C in Dry O_2. Materials and corrosion-werkstoffe Und Korrosion. Mater. Corros. **56**, 801–805 (2005)
84. D.J. Young, in *Chapter 5—Oxidation of Alloys I: Single Phase Scales.* High Temperature Oxidation and Corrosion of Metals (Elsevier, 2016), pp. 193–260
85. S. Yoneda, S. Hayashi, S. Ukai, The transition from transient oxide to protective Al_2O_3 scale on Fe–Cr–Al alloys during heating to 1000 °C. Oxid. Met. **89**, 81–97 (2018)
86. Y. Wang, W. Zhou, Q. Wen, X.C. Ruan, F. Luo, G.H. Bai, Y.C. Qing, D.M. Zhu, Z.B. Huang, Y.W. Zhang, Behavior of plasma sprayed Cr coatings and FeCrAl coatings on Zr fuel cladding under loss-of-coolant accident conditions. Surf. Coat. Technol. **344**, 141–148 (2018)
87. H. Yeom, B. Maier, G. Johnson, T. Dabney, J. Walters, K. Sridharan, Development of cold spray process for oxidation-resistant FeCrAl and Mo diffusion barrier coatings on optimized ZIRLO^{TM}. J. Nucl. Mater. **507**, 306–315 (2018)
88. S. Hao, L. Zhao, D. He, Surface microstructure and high temperature corrosion resistance of arc-sprayed FeCrAl coating irradiated by high current pulsed electron beam. Nucl. Instrum. Meth. **312**, 97–103 (2013)
89. S. Hao, D. He, L. Zhao, Microstructure and corrosion resistance of FeCrAl coating after high current pulsed electron beam surface modification. Procedia Eng. **27**, 1700–1706 (2012)
90. D.J. Park, H.G. Kim, Y.I. Jung, J.H. Park, J.H. Yang, Y.H. Koo, Microstructure and mechanical behavior of Zr substrates coated with FeCrAl and Mo by cold-spraying. J. Nucl. Mater. **504**, 261–266 (2018)
91. J. Chen, G. Li, G. Lin, P. Du, R. Zhang, Effect of annealing on microstructure, shear and tensile strength of diffusion bonded FeCrAl/Zy4 clad plate. J. Nucl. Mater. **535**, 152174 (2020)
92. T. Dabney, G. Johnson, H. Yeom, B. Maier, J. Walters, K. Sridharan, Experimental evaluation of cold spray FeCrAl alloys coated zirconium-alloy for potential accident tolerant fuel cladding. Nucl. Mater. Energy **21**, 100715 (2019)
93. C. Harrington, A. Baron-Wiechec, R. Burrows, R. Holmes, R. Clark, S. Walters, T.L. Martin, R. Springell, J. Oijerholm, R. Becker, Chemistry and corrosion research and development for the water cooling circuits of European DEMO. Fusion Eng. Des. **146**, 478–481 (2019)
94. J. Wang, T. Zhu, K. Chen, D. Du, L. Zhang, Effect of dissolved oxygen and hydrogen on the stress corrosion cracking behavior of alloy 600 in high temperature water. J. Nucl. Mater. **543**, 152603 (2020)
95. J. Xu, T. Shoji, C. Jang, The effects of dissolved hydrogen on the corrosion behavior of alloy 182 in simulated primary water. Corros. Sci. **97**, 115–125 (2015)
96. T. Kim, K.J. Choi, S.C. Yoo, J.H. Kim, Effects of dissolved hydrogen on the crack-initiation and oxidation behavior of nickel-based alloys in high-temperature water. Corros. Sci. **106**, 260–270 (2016)
97. J. Xu, T. Shoji, The corrosion behavior of Alloy 182 in a cyclic hydrogenated and oxygenated water chemistry in high temperature aqueous environment. Corros. Sci. **104**, 248–259 (2016)

98. S.-H. Jeon, E.-H. Lee, D.H. Hur, Effects of dissolved hydrogen on general corrosion behavior and oxide films of alloy 690TT in PWR primary water. J. Nucl. Mater. **485**, 113–121 (2017)
99. J. Nong, Q. Guo, J. Xu, T. Shoji, Effects of alternating dissolved oxygen and dissolved hydrogen on the corrosion behavior of alloy 52 in high temperature high pressure water. J. Nucl. Mater. **540**, 152396 (2020)
100. Y. Reda, A.M. El-Shamy, K.M. Zohdy, A.K. Eessaa, Instrument of chloride ions on the pitting corrosion of electroplated steel alloy 4130. Ain Shams Eng. J. **11**, 191–199 (2020)
101. J.L. Trompette, L. Arurault, S. Fontorbes, L. Massot, Influence of the anion specificity on the electrochemical corrosion of anodized aluminum substrates. Electrochim. Acta **55**, 2901–2910 (2010)
102. F. Arjmand, J. Wang, L. Zhang, Zinc addition and its effect on the corrosion behavior of a 30% cold forged alloy 690 in simulated primary coolant of pressurized water reactors. J. Alloys Compd. **791**, 1176–1192 (2019)
103. P.S. Kumar, S.G. Acharyya, Controlling chloride induced stress corrosion cracking of AISI 316L stainless steel by application of buffing. Mater. Today Proc. **15**, 138–144 (2019)
104. B. Krawczyk, P. Cook, J. Hobbs, D.L. Engelberg, Atmospheric chloride-induced stress corrosion cracking of laser engraved type 316L stainless steel. Corros. Sci. **142**, 93–101 (2018)
105. W.N.L.W. Zakaria, K.E. Kee, M.C. Ismail, The effect of sensitization treatment on chloride induced stress corrosion cracking of 304L stainless steel using U-bend test. Mater. Today Proc. **29**, 75–81 (2020)
106. P.M. Scott, P. Combrade, General corrosion and stress corrosion cracking of alloy 600 in light water reactor primary coolants. J. Nucl. Mater. **524**, 340–375 (2019)
107. Y. Xie, J. Zhang, Chloride-induced stress corrosion cracking of used nuclear fuel welded stainless steel canisters: a review. J. Nucl. Mater. **466**, 85–93 (2015)
108. H. Yeom, T. Dabney, N. Pocquette, K. Ross, F.E. Pfefferkorn, K. Sridharan, Cold spray deposition of 304L stainless steel to mitigate chloride-induced stress corrosion cracking in canisters for used nuclear fuel storage. J. Nucl. Mater. **538**, 152254 (2020)
109. C.-L. Ren, H. Han, W.B. Gong, C.B. Wang, W. Zhang, C. Cheng, P. Huai, Z.Y. Zhu, Adsorption and diffusion of fluorine on Cr-doped Ni(111) surface: fluorine-induced initial corrosion of non-passivated Ni-based alloy. J. Nucl. Mater. **478**, 295–302 (2016)
110. L. Xinfu, The Effect of Primary Water Chemistry on the Integrity of Nuclear Fuel Cladding Tube. Shanghai Jiao Tong University (2007)
111. Z. Hai, C. Xin, H. Wang, Corrosion behavior of nuclear structural materials in Zn-added coolant. Rare Met. Mater. Eng. **48**, 683–687 (2019)
112. C.C. Lin, A review of corrosion product transport and radiation field buildup in boiling water reactors. Prog. Nucl. Energy **51**, 207–224 (2009)
113. P. Deng, Q. Peng, E.H. Han, W. Ke, C. Sun, Z. Jiao, Effect of irradiation on corrosion of 304 nuclear grade stainless steel in simulated PWR primary water. Corros. Sci. **127**, 91–100 (2017)
114. J.C. Haley, S.A. Briggs, P.D. Edmondson, K. Sridharan, S.G. Roberts, S. Lozano-Perez, K.G. Field, Dislocation loop evolution during in-situ ion irradiation of model FeCrAl alloys. Acta Mater. **136**, 390–401 (2017)
115. C.L. Chen, A. Richter, R. Kögler, G. Talut, Dual beam irradiation of nanostructured FeCrAl oxide dispersion strengthened steel. J. Nucl. Mater. **412**, 350–358 (2011)
116. D. Zhang, S.A. Briggs, P.D. Edmondson, M.N. Gussev, K.G. Field, Influence of welding and neutron irradiation on dislocation loop formation and α' precipitation in a FeCrAl alloy. J. Nucl. Mater. **527**, 151784 (2019)
117. K.G. Field, S.A. Briggs, K. Sridharan, R.H. Howard, Y. Yamamotoo, Mechanical properties of neutron-irradiated model and commercial FeCrAl alloys. J. Nucl. Mater. **489**, 118–128 (2017)
118. K.G. Field, S.A. Briggs, X. Hu, Y. Yamamotoo, R.H. Howard, K. Sridharan, Heterogeneous dislocation loop formation near grain boundaries in a neutron-irradiated commercial FeCrAl alloy. J. Nucl. Mater. **483**, 54–61 (2017)

119. B. Cox, Some thoughts on the mechanisms of in-reactor corrosion of zirconium alloys. J. Nucl. Mater. **336**, 331–368 (2005)
120. I. Ioka, Y. Ishijima, K. Usami, N. Sakuraba, Y. Kato, K. Kiuchi, Radiation hardening and IASCC susceptibility of extra high purity austenitic stainless steel. J. Nucl. Mater. **417**, 887–891 (2011)
121. S.A. Briggs, P.D. Edmondson, K.C. Littrell, Y. Yamamotoo, R.H. Howard, C.R. Daily, K.A. Terrani, K. Sridharan, K.G. Field, A combined APT and SANS investigation of α' phase precipitation in neutron-irradiated model FeCrAl alloys. Acta Mater. **129**, 217–228 (2017)
122. S.S. Raiman, A. Flick, O. Toader, P. Wang, N.A. Samad, Z. Jiao, G.S. Was, A facility for studying irradiation accelerated corrosion in high temperature water. J. Nucl. Mater. **451**, 40–47 (2014)
123. V. Konovalova, The effect of temperature on the corrosion rate of iron-carbon alloys. Mater. Today Proc. (2020)
124. Z. Liang, T. Guo, S. Deng, Q. Zhao, High-temperature corrosion of an Fe–Ni-based alloy HR6W under various conditions at 750 °C and 810 °C: Effect of the temperature, water vapor, simulated ash and SO_2. Mater. Chem. Phys. **256**, 123670 (2020)
125. C. Tang, A. Jianu, M. Steinbrueck, M. Grosse, A. Weisenburger, H.J. Seifert, Influence of composition and heating schedules on compatibility of FeCrAl alloys with high-temperature steam. J. Nucl. Mater. **511**, 496–507 (2018)
126. S. Suman, M.K. Khan, M. Pathak, R.N. Singh, J.K. Chakravartty, Hydrogen in zircaloy: Mechanism and its impacts. Int. J. Hudrogen Energy **40**, 5976–5994 (2015)
127. Z. Duan, H. Yang, Y. Satoh, K. Murakami, S. Kano, Z. Zhao, J. Shen, H. Abe, Current status of materials development of nuclear fuel cladding tubes for light water reactors. Nucl. Eng. Des. **316**, 131–150 (2017)
128. C. Duriez, D. Drouan, G. Pouzadoux, Reaction in air and in nitrogen of pre-oxidised Zircaloy-4 and M5^{TM} claddings. J. Nucl. Mater. **441**, 84–95 (2013)
129. B.D.C. Bell, S.T. Murphy, P.A. Burr, R.J. Comstock, J.M. Partezana, R.W. Grimes, M.R. Wenman, The influence of alloying elements on the corrosion of Zr alloys. Corros. Sci. **105**, 36–43 (2016)
130. J.S. Moya, M. Diaz, J.F. Bartolomé, E. Roman, J.L. Sacedon, J. Lzquierdo, Zirconium oxide film formation on zircaloy by water corrosion. Acta Mater. **48**, 4749–4754 (2000)
131. A.T. Motta, A. Couet, R.J. Comstock, Corrosion of zirconium alloys used for nuclear fuel cladding. Ann. Rev. Mater. Res. **45**, 311–343 (2015)
132. J.L. Vandegrift, P.M. Price, J.P. Stroud, C.J. Parga, I.J. Royen, B.J. Jaques, D.P. Butt, Oxidation behavior of zirconium, zircaloy-3, zircaloy-4, zr-1Nb, and Zr-2.5Nb in air and oxygen. Nucl. Mater. Energy **20**, 100692 (2019)
133. M. Steinbrück, M. Böttcher, Air oxidation of Zircaloy-4, M5® and ZIRLO^{TM} cladding alloys at high temperatures. J. Nucl. Mater. **414**, 276–285 (2011)
134. S. Xie, B. Zhou, X. Liang, W. Liu, H. Li, Q. Li, M. Yao, J. Zhang, A novel mechanism for nodular corrosion of zircaloy-4 corroded in 773K superheated steam. Corros. Sci. **126**, 44–54 (2017)
135. V.V. Likhanskii, I.A. Evdokimov, Effect of additives on the susceptibility of zirconium alloys to nodular corrosion. J. Nucl. Mater. **392**, 447–452 (2009)
136. P. Barberis, M. Skocic, D. Kaczorowski, D. Perche, Y. Wouters, Shadow corrosion: experiments and modeling. J. Nucl. Mater. **523**, 310–319 (2019)
137. E. Chajduk, A. Bojanowska-Czajka, Corrosion mitigation in coolant systems in nuclear power plants. Prog. Nucl. Energy **88**, 1–9 (2016)
138. J.H. Harding, The effect of alloying elements on zircaloy corrosion. J. Nucl. Mater. **202**, 216–221 (1993)
139. W. Gong, H. Zhang, C. Wu, H. Tian, X. Wang, The role of alloying elements in the initiation of nanoscale porosity in oxide films formed on zirconium alloys. Corros. Sci. **77**, 391–396 (2013)
140. X. Chen, Z. Li, J. Zhou, F. Tian, Summarizing for effect of alloying elements on corrosion resistance of zirconium alloy. Hot Working Technol. **44**(14–16), 19 (2015)

141. B.D.C. Bell, S.T. Murphy, R.W. Grimes, M.R. Wenman, The effect of Nb on the corrosion and hydrogen pick-up of Zr alloys. Acta Mater. **132**, 425–431 (2017)
142. H.G. Kim, S.Y. Park, M.H. Lee, H.J. Yong, D.K. Sun, Corrosion and microstructural characteristics of Zr–Nb alloys with different Nb contents. J. Nucl. Mater. **373**, 429–432 (2008)
143. Y.R. Than, R.W. Grimes, B.D.C. Bell, M.R. Wenman, Understanding the role of Fe, Cr and Ni in zircaloy-2 with special focus on the role of Ni on hydrogen pickup. J. Nucl. Mater. **530**, 151956 (2020)
144. A. Liens, B. Ter-Ovanessian, N. Courtois, D. Fabregue, J. Chevalier, Effect of alloying elements on the microstructure and corrosion behavior of TiZr-based bulk metallic glasses. Corros. Sci. **177**, 108854 (2020)
145. J. Liu, Z. Cui, D. Ma,J. Lu, Y. Cui, C. Li, W. Liu, Z. Hao, P. Hu, M. Yao, P. Huang, G. Bai, D. Yun, Investigation of oxidation behaviors of coated zircaloy as accident-tolerant fuel with CrAlN and CrAlSiN coatings in high-temperature steam. Corros. Sci. **175**, 108896 (2020)
146. M. Geetha, U. Kamachi Mudali, A.K. Gogia, R. Asokamani, B. Raj, Influence of microstructure and alloying elements on corrosion behavior of Ti–13Nb–13Zr alloy. Corros. Sci. **46**, 877–892 (2004)
147. H.S. Hong, S.J. Kim, K.S. Lee, Influence of dilute silicon addition on the oxidation resistance and tensile properties of modified zircaloy-4. J. Nucl. Mater. **304**, 8–14 (2002)
148. Y.H. Jeong, H.G. Kim, T.H. Kim, Effect of β phase, precipitate and Nb-concentration in matrix on corrosion and oxide characteristics of Zr–xNb alloys. J. Nucl. Mater. **317**, 1–12 (2003)
149. Y.H. Jeong, J.H. Baek, S.J. Kim, H.J. Kim, H. Ruhmann, Corrosion characteristics and oxide microstructures of zircaloy-4 in aqueous alkali hydroxide solutions. J. Nucl. Mater. **270**, 322–333 (1999)
150. B. Cox, Is zirconium oxide morphology on fuel cladding largely determined by lithium hydroxide concentration effects. J. Nucl. Mater. **249**, 87–90 (1997)
151. Y.M. Wong, B. Cox, N. Ramasubramanian, V.C. Ling, Synergistic effects of LiOH and F-in accelerating the corrosion of Zircaloy-4. J. Nucl. Mater. **265**, 178–202 (1999)
152. S.J. Kim, K. Ho Kim, J. Hyuk Baek, B.K. Choi, Y.H. Jeong, Y.H. Jung, The effect of hydride on the corrosion of zircaloy-4 in aqueous LiOH solution. J. Nucl. Mater. **256**, 114–123 (1998)
153. N. Ramasubramanian, P.V. Balakrishnan, Aqueous chemistry of lithium hydroxide and boric acid and corrosion of Zircaloy-4 and Zr-2.5Nb alloys (United States). ASTM Int. **378** (2009)
154. B. Cox, C. Wu, Transient effects of lithium hydroxide and boric-acid on zircaloy corrosion. J. Nucl. Mater. **224**, 169–178 (1995)
155. P. Billot, S. Yagnik, N. Ramasubramanian, J. Peybernes, D. Pecheur, in The Role of Lithium and Boron on the Corrosion of Zircaloy-4 Under Demanding PWR-Type Conditions. Zirconium in the Nuclear Industry: Thirteenth International Symposium, vol. 1423 (2002), pp. 169–188
156. W. Liu, B. Zhou, Q. Li, M. Yao, Detrimental role of LiOH on the oxide film formed on zircaloy-4. Corros. Sci. **47**, 1855–1860 (2005)
157. A.-A.F.W. Abdel-Hakim Kandil, T.M.H. Hani, Water chemistry effect on corrosion of nuclear fuel cladding material, zircaloy-4 (Zr-4). Int. J. Adv. Res. **2**, 149–162 (2014)
158. A.-A.F. Waheed, A.-H.T. Kandil, H.M. Hamed, Electrochemical corrosion of zircaloy-2 under PWR water chemistry but at room temperature. Ann. Nucl. Energy **94**, 168–174 (2016)
159. S. Wang, H. Ming, J. Ding, Z. Zhang, J. Wang, E. Han, A. Atrens, Effect of H_3BO_3 on corrosion in 0.01M NaCl solution of the interface between low alloy steel A508 and alloy 52M. Corros. Sci. **102**, 469–483 (2016)
160. S. Cassineri, A. Cioncolini, M. Curioni, J. Morrison, A. Banks, N. Stevens, F. Scenini, Understanding the effect of Li and flow velocity on corrosion deposition at 230 °C hydrogenated water. Corros. Sci. **168**, 108588 (2020)
161. A.A.F. Waheed, A.H. Kandil, Y.M. Khawassek, A.H. Ali, H.M. Hamed, The impact of elevated lithium concentration 3.5 ppm on the corrosion rate of zirconium. J. Nucl. Mater. **495**, 27–32 (2017)

162. M. Tupin, B. Venague, G. Gutierrez, Influence of light ion irradiation of the oxide layer on the oxidation rate of zircaloy-4: irradiation of the post-transition oxide layer. Corros. Sci. **161**, 108190 (2019)
163. M. Tupin, Chapter 7—understanding the corrosion processes of fuel cladding in pressurized water reactors. Nucl. Corros. 251–299 (2020)
164. R. Birch, S. Wang, V.S. Tong, B. Britton, The effect of cooling rate and grain size on hydride microstructure in Zircaloy-4. J. Nucl. Mater. **513**, 221–225 (2019)
165. K. Vaibhaw, S.V.R. Rao, S.K. Jha, N. Saibaba, R.N. Jayaraj, Texture and hydride orientation relationship of zircaloy-4 fuel clad tube during its fabrication for pressurized heavy water reactors. J. Nucl. Mater. **383**, 71–77 (2008)
166. C.C. Tseng, M.H. Sun, C.K. Chao, Hydride effect on crack instability of Zircaloy cladding. Nucl. Eng. Des. **270**, 427–435 (2014)
167. V.I. Belozerov, E.R. Sitdikov, Y.V. Varseev, Effects of impurities on heat transfer in lead coolants. Nucl. Energy Technol. **2**, 132–135 (2016)
168. P. Fejes, L. Ljungberg, L. Unneberg, Impurity effects in BWR water chemistry. J. Press. Vessels Piping **34**, 47–57 (1988)
169. C. Degueldre, J. Raabe, G. Kuri, S. Abolhassani, Zircaloy-2 secondary phase precipitate analysis by X-ray microspectroscopy. Talanta **75**, 402–406 (2008)
170. G. Cao, Y. Yun, L. Yang, G. Yuan, Q. Yue, G. Shao, J. Hu, The formation and stacking faults of Fe and Cr containing laves phase in zircaloy-4 alloy. Mater. Lett. **191**, 203–205 (2017)
171. K. Annand, M. Nord, I. MacLaren, M. Gass, The corrosion of $Zr(Fe, Cr)_2$ and Zr_2Fe secondary phase particles in zircaloy-4 under 350 °C pressurised water conditions. Corros. Sci. **128**, 213–223 (2017)
172. B. Tao, R. Qiu, Y. Zhao, Y. Liu, X. Tan, B. Luan, Q. Liu, Effects of alloying elements (Sn, Cr and Cu) on second phase particles in Zr-Sn-Nb-Fe-(Cr, Cu) alloys. J. Alloys Compd. **748**, 745–757 (2018)
173. Y. Liu, R. Qiu, X. Tan, B. Tao, Y. Zhao, A new hexagonal Zr-(Fe, Cu) precipitate in Zr-1.0Sn-0.3Nb-0.3Fe-0.1Cu alloy. J. Nucl. Mater. **527**, 151792 (2019)
174. H.H. Shen, S.M. Peng, X. Xiang, F.N. Naab, K. Sun, X.T. Zu, Proton irradiation effects on the precipitate in a Zr–1.6Sn–0.6Nb–0.2Fe–0.1Cr alloy. J. Nucl. Mater. **452**, 335–342 (2014)
175. G. He, J. Liu, K. Li, C. Grovenor, Investigating the stability of second phase particles in Zr-Nb alloys under irradiation. J. Nucl. Mater. **526**, 151738 (2019)
176. J. Faldowski, A. Motta, L.M. Howe, P.R. Okamoto, Influence of stacking faults and alloy composition on irradiation induced amorphization of Zrcr2, Zrfe2 And Zr3(Fei. X,Nix). MRS Proc. **398** (2011)
177. G. Yuan, G. Cao, Q. Yue, L. Yang, J. Hu, G. Shao, Formation of nanocrystalline δ-ZrHx in zircoloy-4: orientation Relationship and twinning. J. Alloys Compd. **658**, 494–499 (2016)
178. C. Li, R. Zuo, Z. Li, B. Shen, Transmission electron microscopy investigation of Zr2(Fe, Ni) particles incorporated in the oxide film formed on a zirconium alloy. Thin Solid Films **461**, 272–276 (2004)
179. C. Ramos, C. Saragovi, M.S. Granovsky, Some New experimental results on the Zr–Nb–Fe system. J. Nucl. Mater. **366**, 198–205 (2007)
180. B.F. Luan, J.M. Wang, R.S. Qiu, B.R. Tao, Q. Liu, The distribution trends and site preferences of alloying elements in precipitates within a Zr alloy: a combined first-principles and experimental study. J. Alloys Compd. **678**, 456–462 (2016)
181. G. Yuan, G. Cao, Q. Yue, L. Yang, Y. Yun, G. Shao, J. Hu, Formation and fine-structures of nano-precipitates in ZIRLO. J. Alloys Compd. **687**, 451–457 (2016)
182. J. Huang, M. Yao, B. Chen, Y. Mao, X. Liang, Oxidation behavior of Zr9S2 precipitates in Zr-0.8Sn-1.0Nb-0.3Fe-0.1Cr-xS alloys. Corros. Sci. **120**, 82–89 (2017)
183. M.Y. Yao, C.Y. Gao, J. Huang, J.C. Peng, X. Liang, J.L. Zhang, B.X. Zhou, Q. Li, Oxidation behavior of β-Nb precipitates in Zr–1Nb–0.2Bi alloy corroded in lithiated water at 360 °C. Corros. Sci. **100,** 169–176 (2015)
184. J. Huang, M.-Y. Yao, C.Y. Gao, X. Liang, J.C. Peng, J.L. Zhang, B.X. Zhou, The influence of second phase particles on the crack formation in oxide films formed on zirconium alloys. Corros. Sci. **99**, 172–177 (2015)

185. S. Kanazawa, Y. Kaneno, H. Inoue, W.Y. Kim, T. Takasugi, Microstructures and defect structures in ZrCr2 laves phase based intermetallic compounds. Intermetallics **10**, 783–792 (2002)
186. M. Steinbrück, J. Birchley, A.V. Boldyrev, A.V. Goryachev, M.S. Veshchunov, High-temperature oxidation and quench behaviour of zircaloy-4 and E110 cladding alloys. Prog. Nucl. Energy **52**, 19–36 (2010)
187. H.H. Kim, J.H. Kim, J.Y. Moon, H.S. Lee, J.J. Kim, Y.S. Chai, High-temperature oxidation behavior of zircaloy-4 and zirlo in steam ambient. J. Mater. Sci. Technol. **26**, 827–832 (2010)
188. T.K. Sawarn, S. Banerjee, A. Samanta, B.N. Rath, S. Kumar, Study of oxide and α-Zr(O) growth kinetics from high temperature steam oxidation of zircaloy-4 cladding. J. Nucl. Mater. **467**, 820–831 (2015)
189. M. Steinbrueck, F.O. da Silva, M. Grosse, Oxidation of zircaloy-4 in steam-nitrogen mixtures at 600–1200 °C. J. Nucl. Mater. **490**, 226–237 (2017)
190. M. Steinbrück, N. Vér, M. Große, Oxidation of advanced zirconium cladding alloys in steam at temperatures in the range of 600–1200 °C. Oxid. Met. **76**, 215–232 (2011)
191. R. Zino, R. Chosson, M. Ollivier, E. Serris, Breakaway characterization of zircaloy-4 oxidized in steam and in oxygen at high temperatures using HT- XRD analysis. Corros. Sci. **176**, 109028 (2020)
192. M. Le Saux, J.-C. Brachet, V. Vandenberghe, A. Ambard, R. Chosson. Breakaway oxidation of zirconium alloys exposed to steam around 1000 °C. Corros. Sci. **176**, 108936 (2020)
193. M. Gestin, M. Mermoux, O. Coindreau, C. Duriez, M. Pijolat, V. Peres, L. Favergeon, Experimental study of oxidation in oxygen, nitrogen and steam mixtures at 850°C of pre-oxidized zircaloy-4. J. Nucl. Mater. **519**, 302–314 (2019)
194. R. Zino, R. Chosson, M. Ollivier, E. Serris, L. Favergeon, Parallel mechanism of growth of the oxide and α-Zr(O) layers on zircaloy-4 oxidized in steam at high temperatures. Corros. Sci. **179**, 109178 (2021)
195. M. Amaya, F. Nagase, The relationship between the amount of oxidation and activation energy on the steam oxidation reaction of zircaloy-4 cladding. J. Nucl. Mater. **440**, 457–466 (2013)
196. M.L. Saux, J.C. Brachet, V. Vandenberghe, E. Rouesne, S. Urvoy, A. Ambard, R. Chosson, Effect of a pre-oxide on the high temperature steam oxidation of zircaloy-4 and m5framatome alloys. J. Nucl. Mater. **518**, 386–399 (2019)
197. H. Buscail, R. Rolland, C. Issartel, S. Perrier, L. Latu-Romain, Stress determination by in situ X-ray diffraction—influence of water vapour on the zircaloy-4 oxidation at high temperature. Corros. Sci. **134**, 38–48 (2018)
198. R. Rolland, H. Buscail, C. Issartel, Influence of water vapour on the zircaloy-4 oxidation at 700 °C: stress determination by in situ X-ray diffraction. Oxid. Met. **87**, 491–500 (2017)
199. J.H. Baek, Y.H. Jeong, Steam oxidation of Zr–1.5Nb–0.4Sn–0.2Fe–0.1Cr and zircaloy-4 at 900–1200 °C. J. Nucl. Mater. **361**, 30–40 (2007)
200. K. Park, S. Yang, K. Ho, The effect of high pressure steam on the oxidation of Low-Sn zircaloy-4 at temperatures between 700 and 900 °C. J. Nucl. Mater. **420**, 39–48 (2012)

Chapter 4
Corrosion Mechanisms of Alloys in Supercritical Water

4.1 Oxidation Processes and Involved Chemical Reactions

Owing to unique properties, supercritical water (SCW) can be used as a working fluid in fossil-fuel power plants and in supercritical water oxidation systems. Supercritical water-cooled reactor has a more simplified core and a higher thermal efficiency than does light water reactor. It is a significant option in the six generation nuclear power plant design due to many advantages, such as fuel cycle sustainability, waste minimization, good economics, safety, reliability and enhanced proliferation resistance [1–3]. However, its efficiency improvement requires higher temperature and pressure of steam in high-temperature pipelines and facilities including superheater and reheater [4]. The dramatical change of water properties from subcritical to supercritical state affects significantly the interaction between water and structural material [1, 5]. SCW is considered to be more corrosive to metal material, compared with water or steam in traditional nuclear reactors (i.e., light water reactor and pressurized water reactor) [6]. High temperature and oxidative nature of SCW increases the corrosion rate of alloy and the exfoliation rate of corresponding oxide scale [7]. So reducing the mechanical stability of the used material. Similarly, chemical and electrochemical corrosion of structural material has also become a key problem in developing a supercritical water oxidation reactor for organic waste treatment due to high temperature, pressure and oxygen content [8, 9].

Overall, corrosion potential is dependent on particular chemical environment [10, 11] and oxidation is regarded as a typical corrosion mechanism for most alloy materials [12]. The rate-determining step of aqueous corrosion generally is the dissolution of a protective surface salt (mostly an oxide) [13]. The corrosion resistance of an alloy is primarily attributed to the formation of a stable and protective oxide film, which can suppress the further oxidation of the material in a harsh environment for a long period of time. Hence, it is of significance to explore the formation and development mechanisms of oxide films and involved chemical reactions of corrosion-resistant alloys in SCW [14]. However, to the best of our knowledge, there is no a review

D. Xu and S. Guo, *Corrosion Characteristics, Mechanisms and Control Methods of Candidate Alloys in Sub- and Supercritical Water*,
https://doi.org/10.1007/978-981-16-5525-8_4

article now on this topic. Therefore, this review work systematically summarizes and analyzes current research on oxidation processes and involved chemical reactions of corrosion-resistant alloys in SCW. It pays more attention to the formation mechanisms of Cr/Ni/Fe oxides and Ni/Fe hydroxides in SCW. This information is valuable for corrosion control and material selection in developing supercritical water-cooled reactors and supercritical water oxidation reactors.

4.1.1 Reactions of Typical Elements

Oxidation occurs after a corrosion-resistant alloy is exposed to SCW, which is a heterogeneous chemical reaction process at the initial stage. The first stage of oxidation can be shown by Reaction 4.1: metal elements lose electrons and are released as ions, which diffuse from the alloy substrate to the oxide/water interface. They then react with oxygen (see Reaction 4.2), high-temperature water and anions to form oxides and hydroxides, and finally precipitate on the alloy surface. As for Cr, Ni and Fe, Reaction 4.1 can further be expressed by Reactions 4.3–4.6. The second stage takes place some reactions among the oxides and hydroxides above. The two reaction stages can be explained by the solid-state growth mechanism and the metal dissolution/oxide precipitation mechanism [15]. In general, diffusion phenomenon plays an important role in the oxidation process, and the growth rate of oxide film is determined by the diffusion rates of metal ions and O^{2-} [16].

$$M \rightarrow M^{n+} + ne^- \tag{4.1}$$

$$xM_{(s)} + \frac{y}{2}O_{2(g)} \leftrightarrow M_xO_{y(g)} \quad [17] \tag{4.2}$$

$$Cr \rightarrow Cr^{3+} + 3e^- \quad [18] \tag{4.3}$$

$$Ni \rightarrow Ni^{2+} + 2e^- \quad [16, 19] \tag{4.4}$$

$$Fe \rightarrow Fe^{3+} + 3e^- \quad [18] \tag{4.5}$$

$$Fe \rightarrow Fe^{2+} + 2e^- \quad [18] \tag{4.6}$$

4.1.1.1 Reactions of Cr

At the initial stage of oxidation, dissolved oxygen diffuses to metal interior through short circuit paths (e.g., micro-pores and grain boundaries), and then preferentially reacts with Cr to form Cr-rich oxides (mainly Cr_2O_3) at the initial metal/H_2O interface through Reaction 4.7. This can be attributed to the higher oxygen affinity of Cr than that of Fe and Ni [20–22]. As a result, a continuous compact inner layer is formed due to the high lattice diffusion coefficient of Cr ion in Cr_2O_3 [23] and the high diffusion rate of O along grain boundaries [21]. In addition, Cr_2O_3, CrO_3 and $Cr_2O_7{}^{2-}$ may also be produced via Reactions 4.8–4.11 in the processes above.

The corrosion resistance of an alloy is mainly dependent on the properties of oxide film formed on its surface. A continuous layer of Cr_2O_3 can stop or remarkably slow down the diffusion of metal ions and oxygen and thus exhibits a good protective ability, because it contributes to the low oxygen partial pressure on the Cr_2O_3/substrate interface [24]. Apparently, a continuous protective oxide scale is significantly affected by the Cr content. Although the alloy with a high Cr content tends to form a dense, continuous and protective Cr_2O_3 layer than do the alloy with a low Cr content, the optimum Cr content is approximately 20 wt% commonly [25, 26].

$$2Cr_{(s)} + \frac{3}{2}O_{2(g)} \leftrightarrow Cr_2O_{3(g)} \quad [18, 25, 27-29] \tag{4.7}$$

$$Cr_{(s)} + \frac{3}{2}O_2 \leftrightarrow Cr_2O_{3(g)} \quad [18, 27] \tag{4.8}$$

$$Cr_2O_{3(s)} + \frac{3}{2}O_{2(g)} \leftrightarrow 2CrO_{3(g)} \quad [18, 27] \tag{4.9}$$

$$2Cr + 3H_2O \rightarrow Cr_2O_3 + 3H_2 \quad [24, 30-32] \tag{4.10}$$

$$4Cr^{3+} + 3O_2 + 8H_2O \rightarrow 2Cr_2O_7^{2-} + 16H^+ \quad [26] \tag{4.11}$$

However, Cr_2O_3 can evaporate in the forms of CrO_3, $CrO_2(OH)$ and especially $CrO_2(OH)_2$ in SCW with a high oxygen content [33]. More specifically, Cr_2O_3 and CrO_3 can react with water and oxygen to form an oxide hydroxide (i.e., $CrO_2(OH)_2$ (chromium (VI)) by Reactions 4.12 and 4.13. $CrO_2(OH)_2$ is an active volatile species and probably formed in the metal wastage process in high-temperature and high-pressure water on the oxide/substrate interface [34]. Chromium (VI) species (likely $CrO_2(OH)_2$) is generated on the surface of austenitic TP347H exposed to SCW only in the very early phase and then gradually disappears as exposure time increases [35]. The formation and evaporation of $CrO_2(OH)_2$ results in severe Cr depletion in the protective layer and in the vicinity of the alloy substrate. This facilitates Fe diffusion into the base scale to fill in the Cr vacant lattice sites to form a less-protective Fe-rich Cr_2O_3-based scale or Fe-rich oxide nodules [36], so ultimately causing the damage of the protective oxide scale [33]. Moreover, formed pores in the outer layer may be ascribed to the vaporization of $CrO_2(OH)_2$ [34]. In other words, high content of dissolved oxygen may cause the formation of chromium (VI) (i.e., $CrO_2(OH)_2$), because its formation in SCW requires the lower oxygen partial pressure than that of CrO_3 [37]. However, as the reaction proceeds, dissolved oxygen is gradually consumed, and the lower oxygen partial pressure cannot guarantee the formation of $CrO_2(OH)_2$, so resulting in the evaporation of $CrO_2(OH)_2$. Generally, $CrO_2(OH)_2$ may exist in high-temperature steam (>700 °C) and in strong oxidizing SCW with at least ppm level of oxygen [37, 38]. Besides, high pH-value favors the transformation from chromium (III) to chromium (VI) [13].

Cr diffusing to the alloy/oxide interface is significant to maintain its protective effect due to Cr depletion. Hence, Cr diffuses outwards through grain boundaries and increases the Cr content on the alloy/oxide interface, and thereby Cr-rich spinel grows again. This can be regarded as cyclic reactions and finally results in an alternate band of Cr-rich oxides and Fe-rich oxides [31]. However, if Cr supply is insufficient to compensate for Cr loss by evaporation, an increasing amount of tiny pores will be formed in the inner and outer layers [33].

$$2Cr_2O_{3(s)} + 3O_{2(g)} + 4H_2O_{(g)} \rightarrow 4CrO_2(OH)_{2(g)} \quad [18, 20, 27, 28, 31, 33, 36, 39\text{–}41] \tag{4.12}$$

$$CrO_{3(g)} + H_2O \rightarrow CrO_2(OH)_{2(g)} \quad [42] \tag{4.13}$$

Besides, Reactions 4.7–4.9 and 4.12 can reach equilibrium if external conditions do not change in some aspects such as temperature and pressure [27]. $Cr(OH)_3$ is formed on alloy surface and then is converted into Cr_2O_3 via the following reactions (see Reactions 4.14–4.16). Therefore, Cr oxides eventually exist in the form of Cr_2O_3.

$$Cr^{3+} + 3OH^- \rightarrow Cr(OH)_3 \quad [43] \tag{4.14}$$

$$Cr(OH)_3 + Cr + 3OH^- \rightarrow Cr_2O_3 + 3H_2O + 3e^- \quad [43] \tag{4.15}$$

$$Cr(OH)_3 + Cr \rightarrow Cr_2O_3 + 3H^+ + 3e^- \quad [44] \tag{4.16}$$

The critical Cr content in an alloy is the minimum content to form a continuous Cr_2O_3 layer to hinder internal oxidation. The critical Cr content for the Fe–Cr alloy is expressed as follows [40, 45, 46]:

$$N_{Cr}^{O} > \left(\frac{\pi g^*}{3} N_O^S \frac{D_O V_m}{\tilde{D}_{Fe-Cr} V_{CrO_{1.5}}} \right)^{0.5} \tag{4.17}$$

where N_{Cr}^{O} is the criterion content of Cr, N_O^S is the oxygen solubility in the alloy, D_O is the oxygen diffusivity in the alloy, V_m is the molar volume of the alloy, $\tilde{D}_{Fe-Cr}$ is the interdiffusion coefficient of the alloy, $V_{CrO_{1.5}}$ is the molar volume of Cr oxides, and the value of g^* is assumed to be 0.3.

The diffusion coefficients of Fe and Cr in volume and in grain boundary can be calculated by following equations [39, 47]:

$$D_{Fe}^{V} = 1.6 \times 10^{-3} \exp(-\frac{213298}{RT}) \tag{4.18}$$

$$\delta D_{Fe}^{GB} = 2 \times 10^{-11} \exp(-\frac{16986}{RT}) \tag{4.19}$$

$$D_{Cr}^{V} = 0.61 \exp(-\frac{263708}{RT}) \tag{4.20}$$

$$\delta D_{Cr}^{GB} = 1.3 \times 10^{-7} \exp(-\frac{150691}{RT}) \tag{4.21}$$

where D_{Fe}^{V} and D_{Cr}^{V} are separately the diffusion coefficients of Fe and Cr in the volume, D_{Fe}^{GB} and D_{Cr}^{GB} are the diffusion coefficients of Fe and Cr in the grain boundary, and δ is the width of the grain boundary.

Moreover, the critical concentration of a specific element can be determined below [21]:

$$N_M \geq \left(\frac{\pi g^*}{2\gamma} N_O \frac{D_O V_M}{D_M V_{OX}}\right)^{\frac{1}{2}} \tag{4.22}$$

where N_M is the critical concentration of a specific element, N_O is the oxygen partial pressure, D_O is the diffusion rate of oxygen, and D_M is the diffusion rate of cation. When a stable oxide scale is formed, the inward diffusion of oxygen is inhibited, that is, the value of D_O reduces. This results in a low oxygen partial pressure on the interface between alloy substrate and fluid, so decreasing the value of N_M.

The relations between standard Gibbs free energy change and temperature are expressed as follows [48]:

$$\Delta G_{Cr_2O_3} = -1139.7(kJ/mol) - 81.2(J/mol)T \tag{4.23}$$

$$\Delta G_{FeCr_2O_3} = -345.3(kJ/mol) - 34.9(J/mol)T \tag{4.24}$$

$$\Delta G_{Ni(OH)_2} = -126.6(kJ/mol) - 21(J/mol)T \tag{4.25}$$

where ΔG_T is the standard Gibbs free energy change, and T is the temperature. Thus, $\Delta G_{Cr_2O_3}$ is always more negative than $\Delta G_{FeCr_2O_4}$ and $\Delta G_{Ni(OH)_2}$ at the same temperature, showing the greater affinity of O to Cr according to the classical thermodynamic theory [49]. Therefore, the formation of Cr_2O_3 is more thermodynamically preferred than that of $Ni(OH)_2$ and $FeCr_2O_4$.

4.1.1.2 Reactions of Ni

Ni dealloying and Cr oxidizing of are dominant reactions in high-temperature subcritical water [29]. Almost at the same time of Cr_2O_3 formation, Ni^{2+} derived from Reaction 4.26 diffuses outwards to the interface of oxides and SCW along oxide grain boundaries and reacts with oxygen (see Reaction 4.27), so leading to the formation of NiO. Besides, NiO can also be formed by the reactions of Ni and H_2O/oxygen based on Reactions 4.28–4.30.

$$Ni_{(s)} + 2H^{+}_{(aq)} \rightarrow Ni^{2+}_{(aq)} + H_{2(g)} \quad [50] \tag{4.26}$$

$$Ni^{2+} + \frac{1}{2}O_2 + 2e^- \rightarrow NiO \quad [16] \tag{4.27}$$

$$Ni_{(s)} + H_2O \leftrightarrow NiO_{(s)} + 2H^+ + 2e^- \quad [51] \tag{4.28}$$

$$Ni_{(s)} + H_2O \rightarrow NiO_{(s)} + H_{2(g)} \quad [50, 52] \tag{4.29}$$

$$Ni_{(s)} + \frac{1}{2}O_{2(g)} + 2H^{+}_{(aq)} \rightarrow Ni^{2+}_{(aq)} + H_2O_{(l)} \quad [29] \tag{4.30}$$

NiO is stable in mildly basic oxidizing SCW and acts as a barrier to inhibit ion diffusion, whereas it is unstable in a neutral oxidizing SCW. Due to the reduction of the oxygen partial pressure on the interface of oxide film and alloy substrate, the decomposition of NiO can occur via Reactions 4.31 and 4.32 [16].

$$NiO + H_2O \rightarrow Ni(OH)_2 \quad [10] \tag{4.31}$$

$$NiO + H_2 \rightarrow Ni + H_2O \quad [10] \tag{4.32}$$

Ni^{2+} can also react with high temperature water and OH^- derived from water ionization to form Ni hydroxide through Reactions 4.33–4.35, Herein, the ionization of water is realized via Reactions 4.36 and 4.37. However, $Ni(OH)_2$ can be decomposed and reduced into NiO and Ni in SCW by Reactions 4.38 and 4.39, respectively.

$$Ni^{2+}_{(aq)} + nH_2O \rightarrow Ni(OH)^{2-n}_{n(aq)} + nH^{+}_{(aq)} \quad [50], \text{ where } n = 1, 2 \tag{4.33}$$

$$Ni^{2+} + 2H_2O \rightarrow Ni(OH)_2 + 2H^+ \quad [19] \tag{4.34}$$

$$Ni^{2+} + 2OH^- \rightarrow Ni(OH)_2 \quad [15] \tag{4.35}$$

$$H_2O \rightarrow H^+ + OH^- \quad [15] \tag{4.36}$$

$$4H_2O + 8e^- \rightarrow 4H_2 + 4O^{2-} \quad [18] \tag{4.37}$$

$$Ni(OH)_2 \rightarrow NiO + H_2O \quad [43] \tag{4.38}$$

$$Ni(OH)_2 + H_2 \rightarrow Ni + 2H_2O \quad [10] \tag{4.39}$$

Surface treatment process affects chemical reactions of elements in alloys. The exposure experiments of mild steel (1015) specimen with Ni-W-P coating in supercritical water oxidation show that W and P atoms release from the coating and the P atoms react with Ni after the break of Ni cells. This results in the formation of Ni_5P_4, which may be partially gasified at high temperatures [53]. The reactions between Ni and P are expressed as follows:

$$2Ni + P \rightarrow Ni_2P \tag{4.40}$$

$$5Ni + 4P \rightarrow Ni_5P_4 \tag{4.41}$$

$$5Ni_2P + 3P \rightarrow 2Ni_5P_4 \tag{4.42}$$

4.1.1.3 Reactions of Fe

As shown by Reactions 4.5 and 4.6, Fe can diffuse to oxide surface and exists in the forms of Fe^{2+} and Fe^{3+}, and then reacts with oxygen, H_2O and O^{2-} to form a Fe-rich outer oxide layer. Thus, the outward diffusion of iron cation through the inner and outer layers [51] is an overall oxidation mechanism of the outer oxide layer. Of course, this process also involves the inward diffusion of oxygen along short-circuit paths (e.g., defects and grain boundaries).

At the initial stage of exposure in SCW, especially when the protectiveness of oxide scale becomes weak, Fe diffuses outwards along oxide grain boundaries, short-circuit paths and pores to the inner oxide layer, so leading to the formation of magnetite (Fe_3O_4) and the release of H_2 (see Reaction 4.43)). Magnetite contains high porosity and thus enhancs effective diffusivity, and is very stable in the temperature range of 400–600 °C based on thermodynamics [54]. Since the formation of Cr_2O_3 induces Cr depletion in alloy substrate [24], Fe can diffuse through Cr-rich oxides and reacts with oxygen to form Fe_3O_4 (see Reaction 4.44). Nevertheless, most Fe cations diffuse through Fe–Cr spinel because Fe diffusion in a Fe–Cr spinel layer is about three times faster than is Cr diffusion [24]. Besides, iron cations (i.e., Fe^{2+} and Fe^{3+}) react with oxygen to generate Fe_3O_4 on the outer layer/SCW interface via Reaction 4.45, so leading to the formation of the outer oxide layer rich in Fe_3O_4. However, formed Fe_3O_4 is lightly soluble in SCW, which can be regarded as a volatilization process, as shown by Reactions 4.46–4.50.

$$3Fe + 4H_2O \rightarrow Fe_3O_4 + 4H_2 \quad [21, 24, 30, 31, 55] \tag{4.43}$$

$$3Fe + 2O_2 \rightarrow Fe_3O_4 \quad [21, 25] \tag{4.44}$$

$$Fe^{2+} + 2Fe^{3+} + 4O^{2-} \rightarrow Fe_3O_4 \quad [18] \tag{4.45}$$

$$2Fe_3O_4 + H_2O \rightarrow 3Fe_2O_3 + H_2 \quad [30, 31] \tag{4.46}$$

$$Fe_3O_4 + 3H_2O \rightarrow 3Fe(OH)_{2(g)} + \frac{1}{2}O_2 \quad [24, 26, 31] \tag{4.47}$$

$$Fe_3O_4 + 4H_2 \rightarrow 3Fe + 4H_2O \quad [31] \tag{4.48}$$

$$\frac{1}{3}Fe_3O_4 + 2H^+_{(aq)} + \frac{1}{3}H_{2(g)} \leftrightarrow Fe^{2+}_{(aq)} + \frac{4}{3}H_2O_{(l)} \quad [56] \tag{4.49}$$

$$2Fe_3O_4 + \frac{1}{2}O_2 \rightarrow 3Fe_2O_3 \quad [57] \tag{4.50}$$

Studies on reaction dynamics simulations show that initially reacted O in O_2 will activate surface Fe atoms to react with O dissolved in SCW and absorb generated protons after SCW dissociation. An appropriate oxygen content can promote the formation of a protective oxide layer, while an excessive oxygen content may cause the destruction of internal Fe oxides [58].

After Fe_3O_4 formation, hematite (i.e., Fe_2O_3) is produced on the Fe_3O_4/H_2O interface and at the network of the porous outer oxide layer through Fe_3O_4 grain boundaries and grows up. Fe_2O_3 is easily formed in high-temperature environment with a high oxygen partial pressure [30], and this is also a dominant reaction under subcritical water conditions [55] (see Reaction 4.51). Similar to Fe_3O_4, formed Fe_2O_3 is also partially dissolved in SCW via Reactions 4.52–4.54.

$$2Fe + 3H_2O \rightarrow Fe_2O_3 + 3H_2 \quad [55] \tag{4.51}$$

$$2Fe_2O_3 + 4H_2O \rightarrow 4Fe(OH)_2 + O_2 \quad [26, 31] \tag{4.52}$$

$$3Fe_2O_3 + 3H_2 + 3H_2O \rightarrow 6Fe(OH)_2 \quad [55] \tag{4.53}$$

$$\frac{1}{2}Fe_2O_3 + 3H^+_{(aq)} \leftrightarrow Fe^{3+}_{(aq)} + \frac{3}{2}H_2O_{(l)} \quad [56] \tag{4.54}$$

Surface treatment process affects oxide (e.g., Fe_2O_3 and Fe_3O_4) contents on alloy surface. After exposure to SCW at 500 °C (see Fig. 4.1a), GBE (grain boundary engineering)-treated Incoloy 800H has an outer layer with a high hematite content and a low magnetite content than does as-received Incoloy 800H. Meanwhile, Fe oxides also undergo following phase transitions with temperature increasing based on Fe oxidation thermodynamics [26]:

$$Fe_3O_4 \xrightarrow{200°C} \gamma{-}Fe_2O_3 \xrightarrow{400°C} \alpha{-}Fe_2O_3 \tag{4.55}$$

In addition to the formation of Fe oxides, Fe ions also dissolve and precipitate as hydroxide (i.e., $Fe(OH)_2$) through Reactions 4.56 and 4.57. $Fe(OH)_2$ is an unstable and volatile intermediate product and will rapidly react with other oxides or hydroxides via the reactions reported in the following Sect. 4.3. Furthermore, $Fe(OH)_2$ is related to the formation of pores in the outer oxide layer and to the metal wastage process when Fe-based alloys are exposed to high-temperature water [24]. Notably, Reactions 4.58 and 4.59 are proposed to explain the conversion and decomposition of $Fe(OH)_2$.

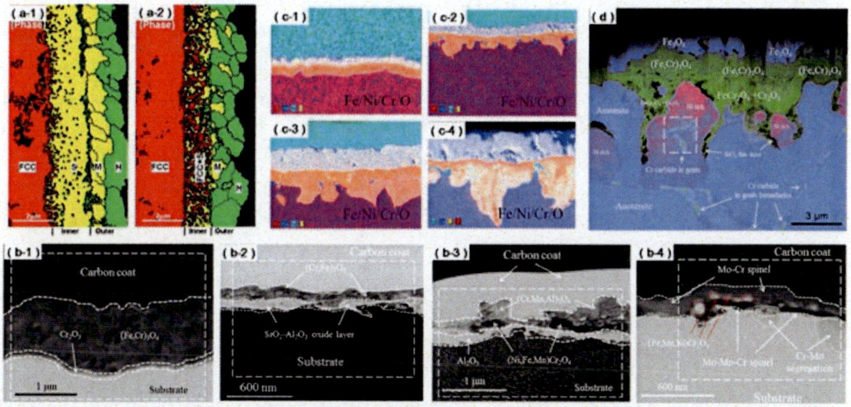

Fig. 4.1 Cross-sectional images of different corrosion-resistant alloys exposed to SCW. **a1-2** Cross-sectional EBSD (electron back-scatter diffraction) maps of phase distribution after as-received Incoloy 800H exposed to SCW at 500 °C, 25 MPa for 3 weeks and after GBE (grain boundary engineering)-treated Incoloy 800H exposed to SCW at 500 °C, 25 MPa for 4 weeks, respectively.[60] The labels FCC, S, M, and H denote the phases of substrate austenite, spinel ($(Fe,Cr)_3O_4$), magnetite (Fe_3O_4), and hematite (Fe_2O_3), respectively; **b1-4** Cross-sectional TEM (transmission electron microscope) images of 316 SS, Incoloy 800H, Alloy 214, and Hastelloy C-276 after exposed to SCW at 800 °C, 25 MPa for 12 h, respectively;[61] **c1-4** Cross-sectional EDS (energy dispersive spectrometer) maps of oxide layers formed on 316L SS after exposed to SCW at 500 °C, 25 MPa for 500 h, 5000 h, 10,000 h and 20,000 h, respectively;[62] **d** Cross-sectional EDS (energy dispersive spectrometer) overlay map of 310S SS after exposed to SCW at 500 °C, 25 MPa for 20,000 h.[20]

$$Fe + 2O_2 \rightarrow Fe(OH)_2 + H_2 \quad [34] \tag{4.56}$$

$$2Fe + 2H_2O + O_2 \rightarrow 2Fe(OH)_2 \quad [34] \tag{4.57}$$

$$4Fe(OH)_2 + 2H_2O + O_2 \rightarrow 4Fe(OH)_3 \quad [34] \tag{4.58}$$

$$6Fe(OH)_2 \rightarrow 2Fe_3O_4 + 2H_2 + 4H_2O \quad [55] \tag{4.59}$$

4.1.1.4 Reactions of Mo, Ti, Al and Si

Although Mo, Ti, Al, Si and other elements have relatively low contents in corrosion-resistant alloys, they also undergo a series of chemical reactions in SCW. For example, Mo is oxidized at 600 °C (see Reaction 4.60) while it is absent on alloy surface at 650 and 700 °C [15]. Mo is also dissolved with the formation of hexavalent acids (i.e., H_2MoO_4) in oxidizing SCW (see Reaction 4.61), thereby losing its protective ability [26]. MoO_3 is produced through Reaction 4.60 [59] or the reactions between Mo and H_2O when the ferritic-martensitic steel T91 sample is exposed to SCW at 540 °C for 10 h. Due to the relatively high equilibrium oxygen partial pressure of MoO_3, the protective Cr-rich layer cannot completely prevent the outward diffusion of Mo from the alloy substrate [59].

$$Mo + O_2 \rightarrow MoO_2 \tag{4.60}$$

$$Mo + H_2O + \frac{3}{2}O_2 \rightarrow H_2MoO_4 \tag{4.61}$$

$$\frac{2}{3}Mo + O_{2(g)} = \frac{2}{3}MoO_3 \tag{4.62}$$

During SCW exposure, Ti departs from alloy surface into fluid and transforms into ionic states (i.e., Ti^{2+}, Ti^{3+} and Ti^{4+}), which further react with SCW and dissolved oxygen to form TiO, Ti_2O_3 and TiO_2, respectively. Similarly, a triplex oxide layer structure (consisting of TiO_2, Ti_2O_3 and TiO from outer to inner layer) is formed on TA10 exposed to SCW (400 °C, 25 MPa) containing sodium chloride, sodium triphosphate and hydrogen peroxide [60]. These titanium oxides contribute greatly to good corrosion resistance of titanium alloys. Relevant reactions are listed as follows:

$$Ti + O_2 \rightarrow TiO_2 \tag{4.63}$$

$$2Ti + O_2 \rightarrow 2TiO \tag{4.64}$$

$$4Ti + 3O_2 \rightarrow 2Ti_2O_3 \quad (4.65)$$

In addition, there are some studies reporting the solubility of titanium dioxide in SCW. For instance, the solubility of titanium dioxide in high temperature water can be estimated by the following equation [61]:

$$[TiO_2]_{totaldissolved} = \left[Ti(OH)_3^+\right] + [Ti(OH)_4] + \left[Ti(OH)_5^-\right] \quad (4.66)$$

Al can react with oxygen to form Al_2O_3 via Reaction 4.67. Similar to the protective effect of Cr_2O_3 and the Cr-rich spinel layer, Al_2O_3 is the foundation of high corrosion resistance and can hinder the inward diffusion of oxygen and the outward diffusion of Fe. When the corrosion resistance of an alloy is mainly attributed to the Cr_2O_3 oxide layer, the protection effect of Al_2O_3 should not be ignored. An inner layer of tightly adherent Al_2O_3 is formed on alloy 214 surface after exposure to SCW at 800 °C for 12 h (see Fig. 4.1b-1), and herein the higher Al content is one of the reasons for its good corrosion resistance in high temperature service. Moreover, Si is selectively oxidized into SiO_2 via Reaction 4.68.

$$2Al + \frac{3}{2}O_2 \rightarrow Al_2O_3 \quad [21] \quad (4.67)$$

$$Si + O_2 \rightarrow SiO_2 \quad (4.68)$$

Although the formation of Al_2O_3 and SiO_2 acts as a protective diffusion barrier to slow down further oxidation, some reactions consuming these oxides occur to form volatile hydroxides and thus weaken the protective ability. A series of reactions (see Reactions 4.69–4.72) are proposed to explain the volatility of Al_2O_3 and SiO_2, and the volatility of some other oxides can be expressed by Reaction 4.73.

$$\frac{1}{2}Al_2O_3 + \frac{3}{2}H_2O_{(g)} \rightarrow Al(OH)_{3(g)} \quad [42, 62] \quad (4.69)$$

$$SiO_2 + 2H_2O_{(g)} \rightarrow Si(OH)_{4(g)} \quad [42, 62] \quad (4.70)$$

$$SiO_2 + H_2O_{(g)} \rightarrow SiO(OH)_{2(g)} \quad [42, 62] \quad (4.71)$$

$$SiO_2 + \frac{1}{2}H_2O_{(g)} \rightarrow \frac{1}{4}O_2 + SiO(OH)_{(g)} \quad [42, 62] \quad (4.72)$$

$$MO_x + nH_2O_{(g)} + mO_2 \rightarrow MO_{(x+n+2m)}H_{(2n)(g)} \quad [62] \quad (4.73)$$

Figure 4.2 elucidates the schematic diagram of corrosion-resistant alloy oxidation. Herein, the numbers in parentheses correspondingly represent the numbers of chemical reactions mentioned in this work. Notably, these reactions do not occur simultaneously on an alloy, but some reactions therein will occur in a specific environment. As illustrated in Fig. 4.2, the migration and transformation of metal elements (especially Ni, Cr and Fe) plays an important role in alloy oxidation processes in SCW. Stabilizing substances derived from the three elements above consist of NiO, Cr_2O_3, Fe_2O_3, Fe_3O_4, as well as some hydroxides including $Ni(OH)_2$, $Cr(OH)_3$, $Fe(OH)_2$ and $Fe(OH)_3$. However, Fig. 4.2 only covers closely relevant chemical reactions that have been reported so far, and other products may be generated via potential reaction pathways. Subsequent research should pay more attention to formation mechanisms of various components in the oxide layer.

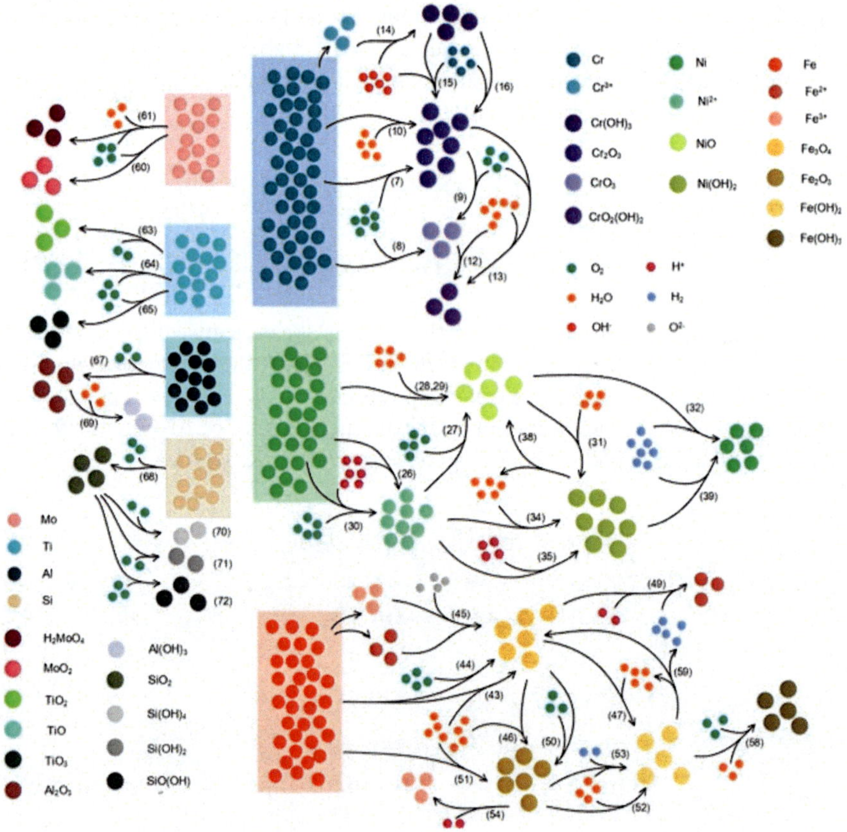

Fig. 4.2 Schematic diagram for the oxidation processes of corrosion-resistant alloys in SCW

4.1.2 Reactions of Ni/Cr/Fe Oxides and Ni/Fe Hydroxides

With the development of oxidation, partial NiO further reacts with Cr_2O_3 to form spinel $NiCr_2O_4$, which is more stable than NiO in SCW according to Reaction 4.74. Some other NiO reacts with Fe^{3+} and OH^- to form $NiFe_2O_4$ via Reaction 4.75 in the outer oxide layer. Moreover, spinel can also be formed by Reaction 4.76.

$$NiO + Cr_2O_3 \rightarrow NiCr_2O_4 \quad [15, 16, 23] \tag{4.74}$$

$$NiO + 2Fe^{3+} + 6OH^- \rightarrow NiFe_2O_4 + 3H_2O \quad [23] \tag{4.75}$$

$$Ni^{2+} + 2OH^- + Cr_2O_3 \rightarrow NiCr_2O_4 + H_2O \quad [15] \tag{4.76}$$

Spinel is likely derived from hydroxide reactions. Formed $Ni(OH)_2$ and $Fe(OH)_2$ (see Sects. 4.1.2 and 4.1.3) precipitate on alloy surface and then form the outer layer of nickel ferrite via Reactions 4.77 and 4.78. The instability of $Ni(OH)_2$ results in its reaction with Fe^{3+} and OH^- to form spinel, as expressed by Reaction 4.79.

$$Ni(OH)_2 + 2Fe(OH)_2 \rightarrow NiFe_2O_4 + 2H_2O + H_2 \quad [23] \tag{4.77}$$

$$\begin{aligned}(2 + z)Fe(OH)_2 + (1 - z)Ni(OH)_2 \leftrightarrow Ni_{(1-z)}Fe_{(2+z)}O_4 \\ + 2H_2O + H_{2(aq)} \quad [52]\end{aligned} \tag{4.78}$$

$$Ni(OH)_2 + 2Fe^{3+} + 6OH^- \rightarrow NiFe_2O_4 + 4H_2O \quad [15] \tag{4.79}$$

At the same time, the Cr_2O_3 inner layer can further react with Fe and/or Fe oxide to form Fe–Cr spinel $FeCr_2O_4$ based on Reactions 4.80–4.83, so resulting in the decrease of the Cr_2O_3 amount. A Cr content of more than 25 wt% is imperative for the formation of an uniform Cr_2O_3 layer [63], and continuous Cr_2O_3 or Cr-rich oxides cannot be formed when the Cr content is lower than 20 wt% [21].Of course, the formation of Fe–Cr spinel can also be achieved by Reaction 4.84.

$$Cr_2O_3 + H_2O + Fe \rightarrow FeCr_2O_4 + H_2 \quad [24, 30, 31] \tag{4.80}$$

$$2Fe + 2Cr_2O_3 + O_2 \rightarrow 2FeCr_2O_4 \quad [21, 25, 40, 63] \tag{4.81}$$

$$4Cr_2O_3 + Fe_3O_4 + Fe \rightarrow 4FeCr_2O_4 \quad [32] \tag{4.82}$$

$$FeO_{(s)} + Cr_2O_{3(s)} \rightarrow FeCr_2O_{4(s)} \quad [57] \tag{4.83}$$

$$Fe^{2+} + Fe^{3+} + Cr^{3+} + 4O^{2-} \rightarrow FeCr_2O_4 \quad [18] \tag{4.84}$$

Notably, $(Cr_xFe_{1\text{-}x})_2O_3$ is easy to vaporize in SCW containing O_2, and thus resulting in some pores formed on oxide film surface [34]. The following reaction is likely to occur,

$$\frac{1}{2}(Cr_xFe_{1-x})O_{3(s)} + \frac{3}{4}xO_{2(g)} + xH_2O \rightarrow xCrO_2(OH)_{2(g)} + \frac{1}{2}(1-x)Fe_2O_{3(g)} \quad [22, 31, 36] \tag{4.85}$$

In addition, the formations of $FeCr_2O_4$, $NiCr_2O_4$ and $NiFe_2O_4$ can be attributed to the reactions among Fe, Cr, Ni and H_2O via the following Reactions 4.86–4.88. Nevertheless, with the transport of ions along oxide grain boundaries and short-circuit paths, both the Fe_3O_4 layer and the Fe–Cr spinel layer become thicker and some pores nucleate by the trapping or ejection of H_2, so eventually causing the exfoliation of oxide layer [31].

$$Fe + 2Cr + 4H_2O \rightarrow FeCr_2O_4 + 4H_{2(aq)} \quad [52] \tag{4.86}$$

$$Ni + 2Cr + 4H_2O \rightarrow NiCr_2O_4 + 4H_{2(aq)} \quad [52] \tag{4.87}$$

$$Ni + 2Fe + 4H_2O \rightarrow NiFe_2O_4 + 4H_{2(aq)} \quad [52] \tag{4.88}$$

4.1.3 Reactions in Complicated SCW Environment

Alloys may be severely attacked in the presence of aggressive ionic species in reactants and products, such as acids, salts, and gases in supercritical water oxidation [10, 12]. It is necessary to study alloy corrosion processes in complex SCW environment containing aggressive substances so as to develop candidate materials for supercritical water oxidation reactors. For example, H_2S is formed via the reaction between Na_2S and H_2O in SCW containing sodium sulfide. Thus, the exposure environment can be regarded as a H_2S-SCW atmosphere with a higher steam partial pressure, in which H_2S reacts with metal elements to form sulfides and then a portion of Ni_3S_2 may react with FeS to produce $NiFe_2S_4$ [32]. The following reactions (see Reactions 4.89–4.92) take place during the whole oxidation processes.

$$Na_2S + 2H_2O \rightarrow H_2S_{(g)} + NaOH \quad [32] \tag{4.89}$$

$$3Ni + 2H_2S \rightarrow Ni_3S_2 + 2H_2 \quad [32] \tag{4.90}$$

$$Cr + H_2S \rightarrow CrS + H_2 \quad [32] \tag{4.91}$$

$$Ni_3S_2 + 6FeS + 2S_2 \rightarrow 3NiFe_2S_4 \quad [32] \tag{4.92}$$

Aggressive species (e.g., SO_2, HCl, Cl_2, O_2, NaCl and Na_2SO_4) can directly react with the alloy substrate and its protective oxide layer, so leading to a significant weight loss after a long exposure time. [68]Reactions 4.93–4.96 are likely to occur on alloys in the environment containing HCl, Cl_2 and NH_3. Corrosion behaviors of some typical corrosion-resistant alloys have been explored in SCW with phosphate, sulfate, chloride and/or oxygen [65]. In addition to Ni/Fe/Cr oxides, phosphates (i.e., $FePO_4$, $Ni_3(PO_4)_2$ and $CrPO_4$) are detected on alloy surface, and possible chemical reactions are inferred to include Reactions 4.97–4.99. These water-insoluble phosphates formed on alloy surface contribute to the improvement of alloy corrosion resistance.

$$Ni_{(s)} + 2HCl_{(g)} \leftrightarrow NiCl_{(g)} + H_{2(g)} \quad [64] \tag{4.93}$$

$$NiO_{(s)} + Cl_{2(g)} \leftrightarrow NiCl_{2(g)} + \frac{1}{2}O_{2(g)} \quad [64] \tag{4.94}$$

$$NiO_{(s)} + 2HCl_{(g)} \leftrightarrow NiCl_{2(g)} + H_2O_{(g)} \quad [64] \tag{4.95}$$

$$Ni(OH)^{2-n}_{n(aq)} + NH_{3(aq)} \rightarrow Ni(OH)_n(NH_3)^{2-n}_{(aq)} \quad \text{where } n = 0, 1, 2 \ [50] \tag{4.96}$$

$$Fe^{3+} + PO_4^{3-} \rightarrow FePO_4 \tag{4.97}$$

$$Cr^{3+} + PO_4^{3-} \rightarrow CrPO_4 \tag{4.98}$$

$$3Ni^{2+} + 2PO_4^{3-} \rightarrow Ni_3(PO_4)_2 \tag{4.99}$$

4.1.4 Overall Oxidation Process Evaluation

In summary, the occurrence of chemical reactions depends on the diffusion rates of metal ions and the stabilities of reactants in SCW. The diffusion rates of typical metal ions are in the order of Fe > Ni > Cr, so the speeds of the three atoms in oxides follow the order of Fe > Ni > > Cr [26] and Fe oxidation occurs preferentially. Diffusion rate can be affected by the steady-state oxygen partial pressure on the interface between oxide and fluid, which is expressed by the following equation [26]:

$$j = -D\frac{dC}{dx} \tag{4.100}$$

where j is the ion flux, D is the diffusion coefficient, C is the ion concentration, and x is the oxide scale thickness.

Fe can diffuse through alloy interior substrate and grain boundaries to the alloy surface [24], and Fe first diffuses to the oxide-metal interface. However, due to the higher Mn content of Nitronic 50 that originates from Mn incorporated spinel compounds, the diffusion of Fe through the inner oxide layer can be mitigated. This results in the formation of the inner layer of Nitronic 50 containing $MnCr_2O_4$, Mn_2O_3 and Cr_2O_3, which is different from the oxide structure on austenitic steel exposed to SCW [66]. The thermodynamic stabilities of oxides formed in SCW are in the orders of $Al_2O_3 > TiO_2 > SiO_2 > MnO > Cr_2O_3 > FeCr_2O_4 > NiCr_2O_4 > NiFe_2O_4 > NiO$, and $Cr_2O_3 > Fe_3O_4 > Fe_2O_3$ [23, 34, 63]. Obviously, Al/Ti/Si/Mn/Cr oxides are more stable and NiO is easily decomposed in high temperature water. Exposure time significantly affects the diffusion and distribution of elements in the oxide layer. As exposure time increases (see Fig. 4.1c), oxide layer is no longer flat and more oxides penetrate into alloy interior, and Ni is rich in the inner oxide layer.

The following equation can be used to determine whether a metal can form protective oxides [27].

$$R_{PB} = \frac{V_{oxide}}{V_{metal}} = \frac{M_{oxide}\rho_{metal}}{nM_{metal}\rho_{oxide}} \tag{4.101}$$

where R_{PB} is the Pilling–Bedworth ratio, V (L) is the molar volume, M (mol/L) is the atomic or molecular mass, ρ (kg/m^3) is the density, n (mol) is the number of metal atoms in per molecule of oxides. If $R_{PB} < 1$, oxide layer is too thin to provide good protection for the metal; if $1 < R_{PB} < 2$, the oxide layer takes place passivation and can prevent the metal surface from being further oxidized; and if $R_{PB} > 2$, the oxide layer exfoliates and loses its protection effect on the metal.

Oxidation reactions in SCW commonly lead to the formation of oxide layers on alloy surface, which is an important basis for the corrosion resistance of the alloy. Corrosion-resistant alloys (e.g., Inconel 690, Hastelloy C-276, Inconel 625, 316L stainless steel, Incoloy 800HT, Incoloy 800, AISI 304L stainless steel, and AFA steel) exposed to SCW have multiple layers of oxide structures, in which the compact inner layer is rich in Cr (e.g., Cr oxides) while the loose outer layer is enriched in Ni and Fe (e.g., hematite, magnetite, and Fe–Ni/Fe–Cr spinel oxides). A three-layer oxide scale is formed on Incoloy 800H after exposure to SCW at 800 °C for 12 h (see Fig. 4.1b-2). It consists of an outer layer rich in Cr_3O_4 and Fe_3O_4, a discontinuous intermediate layer mainly containing SiO_2, and a discontinuous inner layer with abundant Al_2O_3. The order (from inside to outside) of these formed oxides is consistent with their stability order (i.e., $Al_2O_3 > SiO_2 > Cr_2O_3$).

Various phases are detected on 310S SS surface after exposure to SCW at 500 °C for 20,000 h (see Fig. 4.1d) and its inner and outer oxide layers have obvious boundaries. Nonetheless, there is commonly no obvious interface between the inner and outer oxide layers in most cases. Notably, a very thin Cr_2O_3 inner layer can exist in the outer Fe_3O_4 layer on 316 SS after exposure to SCW at 800 °C for 12 h (see Fig. Fig. 4.1b-3). As mentioned in Sect. 4.1.1, a low Cr content in 316 SS is insufficient to produce continuous protective Cr_2O_3, due to Cr depletion around grain boundaries. A discontinuous Ni oxide layer is also observed around the interface of inner and outer layers on Hastelloy C-2000 after exposure to SCW at 800 °C for 12 h, as depicted in Fig. 4.1b-4.

Figure 4.3 summarizes the studies on oxide film structures and compositions of corrosion-resistant alloys in SCW in different reports. It can be seen that the inner oxide layers of most alloys contain chromium oxides (e.g., Cr_2O_3), while an excessive oxygen content may cause the decomposition of iron oxides (e.g., hematite and magnetite). This proves that the stability of iron oxides in SCW is lower than that of chromium oxides. Besides, some ternary metal oxides (e.g., $NiFe_2O_4$, $NiCr_2O_4$ and $FeCr_2O_4$) are likely formed through the reactions mentioned in Sect. 4.3 and thus can be detected on alloy surface. It suggests that some binary metal oxides formed at the first stage continue to react with each other in the presence of oxygen and water.

Fig. 4.3 Summary of structures and compositions of oxide films on typical corrosion-resistant alloys tested in SCW

In addition to iron oxides, the outermost oxide layer sometimes contains some nickel oxides such as NiO and $NiFe_2O_4$. This has good agreement with the fact that the stability of these nickel oxides is higher than that of iron oxides in SCW.

Obviously, corrosion mechanism is significant for studying corrosion characteristics and corrosion control methods under all conditions envisioned in SCW technologies. In fact, in the investigations of alloy corrosion characteristics in SCW, researchers have obtained several corrosion mechanisms which primarily focus on general corrosion. There is still a lack of comprehensive study on the chemical reactions of metal elements, the properties of oxides formed on alloy surface and their protective mechanisms. Even in the literatures cited in this work, the researchers did not focus on discussing chemical reaction processes on alloy surface. It should be stated that tests need to consider the growth, development and exfoliation of oxide layer, which rely on material determination and theoretical examination after experiments. Numerous previous corrosion tests were performed with surface analysis but do not take into account chemical reactions on the sample surface, even if oxidation processes is the most important part of corrosion mechanism.

4.2 Total Corrosion Processes and Mechanisms

The corrosion resistance of one material in SCW depends on chemical stability, tightness, arrangement, strength and thermal expansion coefficient of various corrosion products. Corrosion mechanism in supercritical water is more complex than that in high-temperature water due to the changes in the properties of water. Many of studies have been carried out on the corrosion mechanisms of alloys in supercritical water environments. According to the investigations of Macdonald [67, 68], two corrosion mechanisms can be postulated depending on the water density. On the one hand, at a lower density, the dominant mechanism seems to be a direct chemical corrosion based on the molecular interactions between metal and oxygen and transport processes involving metal cations within oxide films. On other hand, when water density is high, the mechanism changes to an electrochemical mechanism involving charge transfer via local anodes and cathodes [6, 37, 69].

As for the formation of multi-layer oxide films on various alloys in SCW, presently accepted theories mainly include the solid-stated growth mechanism [70], the metal dissolution/oxide precipitation mechanism [71–73], and the mixed mechanism [74]. The solid-state growth mechanism and the mixed mechanism (i.e., the formation of the inner layer by solid state growth process and the formation of the outer layer by a metal dissolution-oxide precipitation mechanism [23, 74–76]) depending on the water density below or above 100–200 kg/m^3, have been widely employed to describe corrosion in subcritical- and supercritical water [35, 59, 66, 77–82]. That is, the formation and thickening of oxide scales on alloys in low-density supercritical water environments is basically a solid-state transport processes of metal cations and oxide ions within the oxide scales without the "flickering" separation of local anodic and cathodic processes on the surface, while it follows that the barrier (inner) oxide layer forms via charge transfer processes (as described in the Point Defect Model) in subcritical and high-density supercritical aqueous systems [68, 83]. The solid-state growth mechanism has been used extensively in describing oxide growth on various metals and alloys in high temperature gaseous environments such as a single gas (e.g., O_2, N_2, CO_2), air, high temperature steam, and series of mixture gases. However, it is necessary to point out that many are based upon Fickian diffusion alone and ignore the role of migration in the transfer of ions across the film. The latest extension of PDM, noted as SCW_PDM, to describe oxide growth in supercritical water is just available [84]. Noting that the transport rate of metal ions in oxides depends upon the elements substitutionally present in the lattice, and also recognizing that segregation of minor alloying elements into the barrier oxide layer occurs via point defect generation/annihilation reactions at interfaces [85, 86].

It is expected that the alloying elements present in the alloy substrate will have a profound impact on the oxide growth rate. Thus, it is generally recognized that the barrier layer grows via the inward transport of oxygen ions (O^{2-}) or the outward transport of oxygen vacancies (V_O..), and the outer layer develops from the outward transport of cations (either via the inward transport of cation vacancies ($V_M{}^{\chi'}$) or as metal interstitials ($M_i{}^{\chi+}$)). Hence, anything that inhibits the transport of these species will inhibit the growth of the two layers. One possibility is that alloying element cations substituted on the cation sublattice of the barrier layer interact electrostatically with moving defects and reduce their diffusivity, which inhibits both diffusion and migration, as described in the Solute-Vacancy Interaction Model (SVIM) that is an extension of PDM [87]. Thus, some alloying elements with relative slow removal rates are retained and enriched in the inner layer, while the outer layer grows by the outward diffusion of metal ions, especially iron and nickel in Ni/Fe-based alloys, through the inner layer and then reacting with gas species at scales/environment interface [82]. However, the mixed corrosion mechanism, obtaining wide application in condensed aqueous systems, emphasizes that the outer layer is formed by a metal passive dissolution-oxide precipitation mechanism, and precipitated metal ions are released from the barrier layer. Dissolved metal cations combine with anions such as OH^- in an aqueous environment to form oxides or hydroxides, which then precipitate onto the specimens surface to form and/or thicken the outer layer [88]. For the alloys exposed to SCW at 450–500 °C, in which the water density is close to the critical value of 100–200 kg/m^3, the formation and growth of the outer layer of the formed oxide film generally not only is a dissolution–precipitation process, but also involves the solid-state transport of metal cations in the barrier layer and possibly also in the outer layer.

Inconel 690 is a good representative of investigating corrosion behavior of the Ni–Cr alloys. Thus, the corrosion of Inconel 690 in SCW at 400–500 °C is taken as an example for analyzing all involved corrosion mechanisms of Ni-based alloys in SCW, and corresponding conclusions can be used as a reference for other Ni-based alloys. Based on the summarization of existing studies on corrosion mechanism, two basic corrosion processes of Inconel 690 in SCW are proposed in this work. Figure 4.4 illustrates schematically all mechanisms that are probably involved in the corrosion of Inconel 690 in SCW.

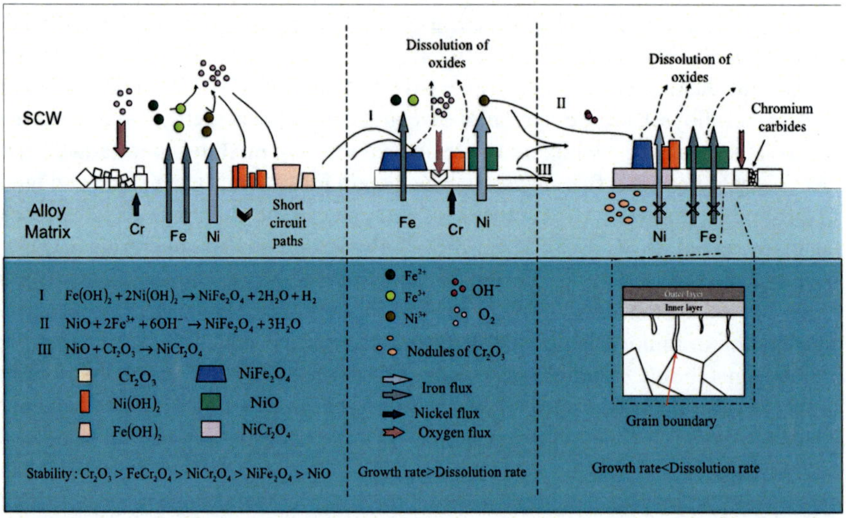

Fig. 4.4 Schematic diagram of corrosion mechanism of Inconel 690 exposed to supercritical water with dissolved oxygen

Overall, corrosion processes primarily include the formation, growth, dissolution, and exfoliation of oxides. As shown in Fig. 4.4, alloy elements first lose electrons to react with oxygen diffusing into the metal matrix surface to form oxides. In this process, chemical reactions and electrochemical heterogeneous reactions between the alloy and the corrosion medium cause metal elements being transformed into their oxidized states, that is, from the original zero valence to a positive valence in corrosion products. For example, metal cations (e.g., Fe^{3+}) detach from the alloy surface and react with dissolved oxygen in SCW to form Fe_2O_3 [26, 49]. Cr preferentially combines with dissolved oxygen to form Cr_2O_3, while Fe and Ni are selectively dissolved in high-density SCW in the active region [23]. Cr is more easily oxidized compared with other common alloying elements, and its transport rate in the oxide layer is much smaller than that of Fe and Ni. Therefore, Cr generally is restricted in the inner oxide layer. Finally, a continuous, protective inner layer consisting of Cr oxides is formed when the Cr content in the Ni-based alloy is higher than 14 wt.% [70]. The inward supply of oxide anions (O^{2-}) to the metal substrate/barrier layer (m/bl) interface leads to the formation of a dense inner layer of Cr_2O_3 at the m/bl interface via the generation of oxygen vacancies [68, 89]. Mo has a relatively low diffusion rate in the oxide layer, and thus is also restricted in the inner oxide layer.

As noted above, the inward growth of the Cr-rich inner layer occurs via the inward transport of oxygen ions (outward transport of oxygen vacancies) and the outward transport of metal cations either via a cation vacancy mechanism or as interstitials through the inner layer, so leading to the growth of the porous outer layer. Point defect model (PDM) is developed at the atomistic level to describe the growth of multi-layer anodic oxide films in electrolyte solutions at subcritical temperatures. It has just been extended to describe the growth of scales at supercritical temperatures [68]. The transport rate of Ni in oxides is higher than that of Cr [194]. Ni^{2+} generated at the metal/film interface migrates outward to the interface between the barrier layer and environment, and then reacts with dissolved oxygen or some other species [26], thereby resulting in the formation and growth of the outer oxide layer with NiO being commonly predominant component [89].

Predominant Cr-rich oxides in the inner layer usually have a dense structure and thus possess a good ability to inhibit further corrosion [91], and they even remain stable in pH = 2 SCW [92]. However, the spinel ($FeCr_2O_4$)-based oxide in the external layer owns almost negligible corrosion resistance, especially in subcritical- and high-density supercritical aqueous systems [93, 94]. As the amount of dissolved metal increases, metal cations may form hydroxides (e.g., $Ni(OH)_2$ and $Fe(OH)_2$), which deposit on the alloy surface and react with each other to form the outer layer of $NiFe_2O_4$ spinel (see Eq. 4.102). As the oxidation progresses, NiO reacts with the ions (i.e., Fe^{3+} and OH^-) in the solution to form $NiFe_2O_4$, and part of NiO reacts with Cr_2O_3 to generate $NiCr_2O_4$ spinel (see Eqs. 4.103 and 4.104). Fe oxides also slightly dissolve and react in SCW to form $Fe(OH)_2$ [90], as shown in Eqs. 4.105 and 4.106.

$$Ni(OH)_2 + 2Fe(OH)_2 = NiFe_2O_4 + 2H_2O + H_2 \tag{4.102}$$

$$NiO + 2Fe^{3+} + 6OH^- = NiFe_2O_4 + 3H_2O \tag{4.103}$$

$$NiO + Cr_2O_3 = NiCr_2O_4 \tag{4.104}$$

$$2Fe_3O_4 + 6H_2O = 6Fe(OH)_2 + O_2 \tag{4.105}$$

$$2Fe_2O_3 + 4H_2O = 4Fe(OH)_2 + O_2 \tag{4.106}$$

It is worth noticing that the formation and dissolution of outer oxides occur simultaneously at different or similar rates. Initially, the growth rate is higher than the dissolution rate, and the outer layer develops gradually. As corrosion proceeds, the increase in the thickness of the Cr-rich inner layer inhibits the outward diffusion of cations, so resulting in a decrease in the growth rate of outer oxides until being close to its dissolution rate if no spalling occurs [95]. As Fe and Ni diffuse outward through the inner layer of the formed oxide film, oxide deposition and solid-state reactions together contribute the formation of the outer layer in SCW at 450–500 °C, while this process is likely only solid-state growth in a low-density SCW with a density of less than 100 kg/m^3. Payet et al. [95] reported that the oxides produced on Inconel 690 in SCW had the following characteristics, that is, the outer oxide layer was mainly composed of Ni and Fe oxides such as $Ni_xFe_{3-x}O_4$, while the inner oxide layer consisted of α-chromia Cr_2O_3 and chromites, such as (Mn, Fe, Ni)Cr_2O_4.

Moreover, the growth of oxides depends on alloy microstructure, and grain boundaries favor the formation of Cr oxides while Fe and Cr oxides are usually formed above intragranular regions.

The presence of oxygen and some aggressive ions affects the corrosion behavior of Ni-based alloys in SCW. Different metal oxides have various standard Gibbs free formation energies and thus own a variety of oxygen affinities according to the classical thermodynamic theory. Cr_2O_3 generally forms preferentially and can prevent further corrosion, because the standard Gibbs free energy of Cr_2O_3 is more negative than those of $Ni(OH)_2$ and $FeCr_2O_3$ at the same conditions. However, when the oxygen concentration is extremely high, Cr undergoes transpassive dissolution [96]. to form Cr hexavalent compounds [109], such as soluble chromate [64, 97, 98]. Aggressive ions is an important factor of affecting corrosion behavior [99]. Passivity breakdown occurs locally due to anion-catalyzed cation vacancy generation at the barrier layer/solution interface, so resulting in pitting corrosion [83]. Cation vacancy condensation at the metal/film interface is considered to be the cause of passivity breakdown, and leads to decohesion of the barrier layer from the metal and ultimately to film rupture because of the presence of tensile stresses in the film coupled with film dissolution [100].

Some researchers have also conducted experimental studies on corrosion behavior of Ni-based alloys in complex SCW containing oxygen and different salts or corrosive substances. For example, Kritzer et al. [101] studied the corrosion behavior of Inconel 625 in oxygen-containing, phosphate aqueous systems under high temperature and high pressure conditions. The corrosion properties of Inconel 625, Inconel 686 and pure chromium were examined at 500 °C and 46.5 MPa in the model fluid consisting of $HCl + H_2O + NaCl +$ methanol with H_2O_2 as oxidant [102]. Kritzer [13] stated that chloride could penetrate and destroy the protective oxide layer, so resulting in severe local corrosion, pitting corrosion, and stress corrosion cracking. However, the evidence for chloride penetration is scarce and highly controversial [83]. Tang et al. [98]explored the corrosion characteristics of Incoloy 800, Incoloy 825, Inconel 625, and Hastelloy C-276 in chloride-containing SCW, and found that the synergistic effect between salt and oxygen led to severe corrosion. They also studied the corrosion behavior of SS 316, Inconel 600, Inconel 625, Hastelloy C-276, and TA10 in SCW containing sodium chloride, sodium triphosphate, and hydrogen peroxide, and confirmed that phosphate had a corrosion inhibitory effect on alloys in SCWO environment [60]. Ma et al. [65] reported the corrosion properties of SS 316, SS 316L, Inconel 625, Incoloy 825, and Hastelloy C-276 in SCW containing phosphate, sulfate, chloride, and oxygen, and found that water-insoluble phosphate products that formed on the alloy surface helped to inhibit further corrosion. It should be mentioned that even without oxygen, supercritical water is very corrosive for Ni-based alloys if for instance there is some chloride.

This work mainly focuses the corrosion behavior of Ni-based alloys in various supercritical aqueous systems and also synoptically introduces corrosion mechanisms of typical Ni-based alloys in supercritical water. More review on fundamental issues related to the properties of supercritical aqueous systems and their impacts on the corrosion of metals and alloys, in-site measurement of corrosion rate and passivity and passivity breakdown, and on development of mechanistic determination especially at the atomic level, is required in subsequent work.

References

1. J. Macák, R. Novotný, P. Sajdl, V. Bystrianský, L. Tůma, M. Novák, In-situ electrochemical impedance measurements of corroding stainless steel in high subcritical and supercritical water. Corros. Sci. **150**, 9–16 (2019)
2. N. Tepylo, X. Huang, S. Jiang, S. Penttilä, Performance of aluminide and cr-modified aluminide pack cementation-coated stainless steel 304 in supercritical water at 700 °C. J. Nucl. Eng. Radiat. Sci. **5**, 0110141 (2019)
3. Y. Li, T. Xu, S. Wang, J. Yang, B. Fekete, J. Yang, A. Wu, J. Qiu, Y. Xu, D.D. Macdonald, Predictions and analyses on the growth behavior of oxide scales formed on ferritic–martensitic in supercritical water. Oxid. Met. **92**, 27–48 (2019)
4. Z. Zhu, H.I. Khan, Q. Cao, N. Zhang, Oxidation behavior of ferritic steel T22 exposed to supercritical water. High Temp. Mater. Proc. **38**, 476–484 (2019)
5. M. Payet, L. Marchetti, M. Tabarant, F. Jomard, J.P. Chevalier, Corrosion mechanisms of 316L stainless steel in supercritical water: the significant effect of work hardening induced by surface finishes. Corros. Sci. **157**, 157–166 (2019)
6. Y. Li, T. Xu, S. Wang, B. Fekete, J. Yang, J. Yang, J. Qiu, A. Xu, J. Wang, Y. Xu, D.D. Macdonald, Modelling and analysis of the corrosion characteristics of ferritic-martensitic steels in supercritical water. Materials **12**, 409 (2019)
7. G. Jákli, A.R. Imre, Á. Horváth, Investigation of corrosion resistance of alloys with potential application in supercritical water-cooled nuclear reactors. Period. Polytech.-Chem. **63**, 328–332 (2019)
8. J. Yang, S. Wang, Y. Li, Y. Zhang, D. Xu, Novel design concept for a commercial-scale plant for supercritical water oxidation of industrial and sewage sludge. J. Environ. Manage. **233**, 131–140 (2019)
9. V. Vadillo, J. Sánchez-Oneto, J.R. Portela, E.J.M. de la Ossa, Problems in supercritical water oxidation process and proposed solutions. Ind. Eng. Chem. Res. **52**, 7617–7629 (2013)
10. P.A. Marrone, G.T. Hong, Corrosion control methods in supercritical water oxidation and gasification processes. J. Supercrit. Fluids **51**, 83–103 (2009)
11. M.D. Bermejo, M.J. Cocero, Supercritical water oxidation: a technical review. AlChE J. **52**, 3933–3951 (2006)
12. C. Sun, R. Hui, W. Qu, S. Yick, Progress in corrosion resistant materials for supercritical water reactors. Corros. Sci. **51**, 2508–2523 (2009)
13. P. Kritzer, Corrosion in high-temperature and supercritical water and aqueous solutions: a review. J. Supercrit. Fluids **29**, 1–29 (2004)
14. D. Rodriguez, A. Merwin, Z. Karmiol, D. Chidambaram, Surface chemistry and corrosion behavior of Inconel 625 and 718 in subcritical, supercritical, and ultrasupercritical water. Appl. Surf. Sci. **404**, 443–451 (2017)
15. Z. Zhu, H. Xu, H.I. Khan, N. Zhang, Oxidation behaviour of Nimonic 263 in high-temperature supercritical water. Corros. Eng. Sci. Techn. **53**, 617–624 (2018)
16. H. Shi, Z. Gao, Z. Fan, Y. Ding, Y. Qiao, Z. Zhu, Corrosion behavior of alloy C-276 in supercritical water. Adv. Mater. Sci. Eng. **2018**, 1027640 (2018)
17. G. Wright, R.B. Dooley, A review of the oxidation behaviour of structural alloys in steam. Int. Mater. Rev. **55**, 129–167 (2013)
18. X. Huang, Y. Shen, Oxidation behavior of an 11Cr ferritic–martensitic steel after Ar-ions irradiation in supercritical water. J. Nucl. Mater. **461**, 1–9 (2015)
19. H. Kim, D.B. Mitton, R.M. Latanision, Stress corrosion cracking of alloy 625 in pH 2 aqueous solution at high temperature and pressure. Corrosion **67**, 0350021 (2011)

20. Y. Behnamian, A. Mostafaei, A. Kohandehghan, B.S. Amirkhiz, J. Li, R. Zahiri, E. Aghaie, W. Zheng, D. Guzonas, M. Chmielus, W. Chen, J.L. Luo, Internal oxidation and crack susceptibility of alloy 310S stainless steel after long term exposure to supercritical water at 500 °C. J. Supercrit. Fluids **120**, 161–172 (2017)
21. X. Guo, K. Chen, W. Gao, Z. Shen, L. Zhang, Corrosion behavior of alumina-forming and oxide dispersion strengthened austenitic 316 stainless steel in supercritical water. Corros. Sci. **138**, 297–306 (2018)
22. Z. Shen, K. Chen, X. Guo, L. Zhang, A study on the corrosion and stress corrosion cracking susceptibility of 310-ODS steel in supercritical water. J. Nucl. Mater. **514**, 56–65 (2019)
23. X. Zhong, E.H. Han, X. Wu, Corrosion behavior of alloy 690 in aerated supercritical water. Corros. Sci. **6**, 369–379 (2013)
24. X. Guo, K. Chen, W. Gao, Z. Shen, P. Lai, L. Zhang, A research on the corrosion and stress corrosion cracking susceptibility of 316L stainless steel exposed to supercritical water. Corros. Sci. **127**, 157–167 (2017)
25. Y. Behnamian, A. Mostafaei, A. Kohandehghan, B.S. Amirkhiz, R. Zahiri, W. Zheng, D. Guzonas, M. Chmielus, W. Chen, J.L. Luo, A comparative study on the oxidation of austenitic alloys 304 and 304-oxide dispersion strengthened steel in supercritical water at 650 °C. J. Supercrit. Fluids **119**, 245–260 (2017)
26. J. Yang, S. Wang, X. Tang, Y. Wang, Y. Li, Effect of low oxygen concentration on the oxidation behavior of Ni-based alloys 625 and 825 in supercritical water. J. Supercrit. Fluids **131**, 1–10 (2018)
27. X. Huang, Y. Shen, J. Zhu, Influence of Ar-ions irradiation on the oxidation behavior of ferritic–martensitic steel P92 in supercritical water. J. Nucl. Mater. **457**, 18–28 (2015)
28. X. Zhang, Q. Yan, Y. Yang, Z. Hong, L. Zhang, C. Ge, Corrosion resistance of chromium-coated ferritic-martensitic steel in supercritical water. Corrosion **71**, 50–56 (2015)
29. H. Kim, D.B. Mitton, R.M. Latanision, Effect of pH and temperature on corrosion of nickel-base alloys in high temperature and pressure aqueous solutions. J. Electrochem. Soc. **157**, C194–C199 (2010)
30. Z. Zhang, Z.F. Hu, L.F. Zhang, K. Chen, P.M. Singh, Effect of temperature and dissolved oxygen on stress corrosion cracking behavior of P92 ferritic-martensitic steel in supercritical water environment. J. Nucl. Mater. **498**, 89–102 (2018)
31. X. Zhong, X. Wu, E.H. Han, Effects of exposure temperature and time on corrosion behavior of a ferritic–martensitic steel P92 in aerated supercritical water. Corros. Sci. **90**, 511–521 (2015)
32. Y. Li, S. Wang, X. Tang, D. Xu, Y. Guo, J. Zhang, L. Qian, Effects of sulfides on the corrosion behavior of Inconel 600 and Incoloy 825 in supercritical water. Oxid. Met. **84**, 509–526 (2015)
33. N. Zhang, Q. Cao, J. Gui, M. Li, Y. Ni, H. Xu, Oxidation and chromia evaporation of austenitic steel TP347HFG in supercritical water. Mater. High Temp. **35**, 461–468 (2017)
34. K.K. Choudhry, D.A. Guzonas, D.T. Kallikragas, I.M. Svishchev, On-line monitoring of oxide formation and dissolution on alloy 800H in supercritical water. Corros. Sci. **111**, 574–582 (2016)
35. Y. Li, S. Wang, P. Sun, D. Xu, M. Ren, Y. Guo, G. Lin, Early oxidation mechanism of austenitic stainless steel TP347H in supercritical water. Corros. Sci. **128**, 241–252 (2017)
36. S. Mahboubi, Y. Jiao, W. Cook, W. Zheng, D.A. Guzonas, G.A. Botton, J.R. Kish, Stability of chromia (Cr_2O_3)-based scales formed during corrosion of austenitic Fe–Cr–Ni alloys in flowing oxygenated supercritical water. Corrosion **72**, 1170–1180 (2016)
37. D.G. Briceno, F. Blazquez, A.S. Maderuelo, Oxidation of austenitic and ferritic/martensitic alloys in supercritical water. J. Supercrit. Fluids **78**, 103–113 (2013)
38. R. Gordon, Holcomb, Steam oxidation and chromia evaporation in ultrasupercritical steam boilers and turbines. J. Electrochem. Soc. **156**, C292–C297 (2009)
39. Z. Zhu, H. Li, Q. Cao, Oxidation behavior of austenitic steels in supercritical water containing dissolved oxygen. J. Mater. Eng. Perform. **28**, 44–52 (2018)
40. H. Li, Q. Cao, Z. Zhu, Oxidation behaviour of Super 304H stainless steel in supercritical water. Corros. Eng. Sci. Technol. **53**, 293–301 (2018)

41. X. Guo, Y. Fan, W. Gao, R. Tang, K. Chen, Z. Shen, L. Zhang, Corrosion resistance of candidate cladding materials for supercritical water reactor. Ann. Nucl. Energy **127**, 351–363 (2019)
42. S.R.J. Saunders, M. Monteiro, F. Rizzo, The oxidation behaviour of metals and alloys at high temperatures in atmospheres containing water vapour: a review. Prog. Mater. Sci. **53**, 775–837 (2008)
43. J. Huang, X. Wu, E.H. Han, Electrochemical properties and growth mechanism of passive films on Alloy 690 in high-temperature alkaline environments. Corros. Sci. **52**, 3444–3452 (2010)
44. A. Machet, A. Galtayries, S. Zanna, L. Klein, V. Maurice, P. Jolivet, M. Foucault, P. Combrade, P. Scott, P. Marcus, XPS and STM study of the growth and structure of passive films in high temperature water on a nickel-base alloy. Electrochim. Acta. **49**, 3957–3964 (2004)
45. Z. Zhu, D. Jiang, Q. Cao, H. Xu, H.I. Khan, N. Zhang, Oxidation behavior of austenitic steel Sanicro25 and TP347HFG in supercritical water. Mater. Corros. **70**, 1087–1098 (2019)
46. C. Wagner, Theoretical analysis of the diffusion processes determining the oxidation rate of alloys. J. Electrochem. Soc. **99**, 369–380 (1952)
47. P. Guiraldenq, P. Poyet, Influence of nickel on the volume and intercrystalline diffusion in austenitic Fe–Cr–Ni alloys—grain boundary energy evolution as a function of composition. Mem. Sci. Rev. Metall. **70**, 715–723 (1973)
48. W. Wang, X. Cao, Y. Lu, X. Ding, T. Shoji, TEM study on oxide mechanism of the dissimilar welds between 316 stainless steels/Inconel 182 in high temperature and high-pressure water. Mater. Charact. **131**, 339–347 (2017)
49. M. Sun, X. Wu, Z. Zhang, E.H. Han, Analyses of oxide films grown on Alloy 625 in oxidizing supercritical water. J. Supercrit. Fluids **47**, 309–317 (2008)
50. S.E. Ziemniak, P.A. Guilmette, R.A. Turcotte, H.M. Tunison, Oxidative dissolution of nickel metal in hydrogenated hydrothermal solutions. Corros. Sci. **50**, 449–462 (2008)
51. S.H. Jeon, E.H. Lee, D.H. Hur, Effects of dissolved hydrogen on general corrosion behavior and oxide films of alloy 690TT in PWR primary water. J. Nucl. Mater. **485**, 113–121 (2017)
52. M. Sennour, L. Marchetti, F. Martin, S. Perrin, R. Molins, M. Pijolat, A detailed TEM and SEM study of Ni-base alloys oxide scales formed in primary conditions of pressurized water reactor. J. Nucl. Mater. **402**, 147–156 (2010)
53. L. Ren, Y. Cheng, S. Wang, X. Meng, Q. Qin, J. Yang, Oxidation behavior of the supercritical water on the ternary Ni-W-P coating. Chem. Eng. J. **370**, 1388–1406 (2019)
54. P. Ampornrat, G.S. Was, Oxidation of ferritic–martensitic alloys T91, HCM12A and HT-9 in supercritical water. J. Nucl. Mater. **371**, 1–17 (2007)
55. D. Rodriguez, A. Merwin, D. Chidambaram, On the oxidation of stainless steel alloy 304 in subcritical and supercritical water. J. Nucl. Mater. **452**, 440–445 (2014)
56. C.J.O. Brien, Z. Rak, D.W. Brenner, Free energies of (Co, Fe, Ni, Zn)Fe_2O_4 spinels and oxides in water at high temperatures and pressure from density functional theory: results for stoichiometric NiO and $NiFe_2O_4$ surfaces. J. Phys.: Condens. Matter **25**, 445008 (2013)
57. J. Ehlers, D.J. Young, E.J. Smaardijk, A.K. Tyagi, H.J. Penkalla, L. Singheiser, W.J. Quadakkers, Enhanced oxidation of the 9%Cr steel P91 in water vapour containing environments. Corros. Sci. **48**, 3428–3454 (2006)
58. L. Ai, Y. Zhou, M. Chen, Role of dissolved oxygen in iron oxidation in supercritical water: insights from reactive dynamics simulations. J. Phys. Chem. **123**, 15009–15016 (2019)
59. Y. Li, S. Wang, P. Sun, J. Yang, X. Tang, D. Xu, Y. Guo, J. Yang, D.D. Macdonald, Investigation on early formation and evolution of oxide scales on ferritic–martensitic steels in supercritical water. Corros. Sci. **135**, 136–146 (2018)
60. X. Tang, S. Wang, L. Qian, Y. Li, Z. Lin, D. Xu, Y. Zhang, Corrosion behavior of nickel base alloys, stainless steel and titanium alloy in supercritical water containing chloride, phosphate and oxygen. Chem. Eng. Res. Des. **100**, 530–541 (2015)
61. M. Atashfaraz, M. Shariaty Niassar, S. Ohara, K. Minami, M. Umetsu, T. Naka, T. Adschiri, Effect of titanium dioxide solubility on the formation of BaTiO3 nanoparticles in supercritical water. Fluid Phase Equilib. **257**, 233–237 (2007)

62. E.J. Opila, Volatility of common protective oxides in high-temperature water vapor: current understanding and unanswered questions. Mater. Sci. Forum **461**, 765–773 (2004)
63. Z. Zhu, H. Xu, H.I. Khan, D. Jiang, N. Zhang, Effect of exposure temperature on oxidation of austenitic steel HR3C in supercritical water. Oxid. Met. **91**, 77–93 (2019)
64. P. Kritzer, N. Boukis, E. Dinjus, Factors controlling corrosion in high-temperature aqueous solutions: a contribution to the dissociation and solubility data influencing corrosion processes. J. Supercrit. Fluids **15**, 205–227 (1999)
65. Z. Ma, D. Xu, S. Guo, Y. Wang, S. Wang, Z. Jing, Y. Guo, Corrosion properties and mechanisms of austenitic stainless steels and Ni-base alloys in supercritical water containing phosphate, sulfate, chloride and oxygen. Oxid. Met. **90**, 599–616 (2018)
66. D. Rodriguez, D. Chidambaram, Oxidation of stainless steel 316 and Nitronic 50 in supercritical and ultrasupercritical water. Appl. Surf. Sci. **347**, 10–16 (2015)
67. D.D. Macdonald, X. Guan, Volume of activation for the corrosion of type 304 stainless steel in high subcritical and supercritical aqueous systems. NACE Int. **65**, 427–437 (2009)
68. Y. Li, D.D. Macdonald, J. Yang, J. Qiu, S. Wang, Point defect model for the corrosion of steels in supercritical water: part I, film growth kinetics. Corros. Sci. 108280 (2019)
69. X. Tang, S. Wang, D. Xu, Y. Gong, J. Zhang, Y. Wang, Corrosion behavior of Ni-based alloys in supercritical water containing high concentrations of salt and oxygen. Ind. Eng. Chem. Res. **52**, 18241–18250 (2013)
70. J. Robertson, The mechanism of high temperature aqueous corrosion of stainless steels. Corros. Sci. **32**, 443–456 (1991)
71. Z. Zhang, J. Wang, E.H. Han, W. Ke, Analysis of surface oxide films formed in hydrogenated primary water on alloy 690TT samples with different surface states. J. Mater. Sci. Technol. **30**, 1181–1192 (2014)
72. R.E. Lobnig, H.P. Schmidt, K. Hennesen, H.J. Grabke, Diffusion of cations in chromia layers grown on iron-base alloys. Oxid. Met. **37**, 81–93 (1991)
73. B. Stellwag, The mechanism of oxide film formation on austenitic stainless steels in high temperature water. Corros. Sci. **40**, 337–370 (1998)
74. Y. Li, T. Xu, S. Wang, J. Yang, J. Li, T. Xu, D. Xu, Characterization of oxide scales formed on heating equipment in supercritical water gasification process for producing hydrogen. Int. J. Hydrogen Energ. **44**, 29508–29515 (2019)
75. Y. Yi, B. Lee, S. Kim, J. Jang, Corrosion and corrosion fatigue behaviors of 9Cr steel in a supercritical water condition. Mat. Sci. Eng. A **429**, 161–168 (2006)
76. X. Gao, X. Wu, Z. Zhang, H. Guan, E.h. Han, Characterization of oxide films grown on 316L stainless steel exposed to H2O2-containing supercritical water. J. Supercrit. Fluids **42**, 157–163 (2007)
77. R. Viswanathan, J. Sarver, J.M. Tanzosh, Boiler materials for ultra-supercritical coal power plants-stearnside oxidation. J. Mater. Eng. Perform. **15**, 255–274 (2006)
78. J. Bischoff, A.T. Motta, Oxidation behavior of ferritic-martensitic and ODS steels in supercritical water. J. Nucl. Mater. **424**, 261–276 (2012)
79. T. Fujii, K. Sue, S. Kawasakia, Effect of pressure on corrosion of Inconel 625 in supercritical water up to 100 MPa with acids or oxygen. J. Supercrit. Fluids **95**, 285–291 (2014)
80. S. Xu, F. Long, S. Y. Persaud, N. Guo, Z. Yao, M.R. Daymond, W. Gao, L. Zhang, Z. Zhou, Oxidation behavior of 9Cr-4.5Al ODS steel in 600 °C supercritical water and the effect of pre-oxidation. Corros. Sci. **165**, 108380 (2020)
81. Y. Li, S. Wang, J. Yang, D. Xu, Y. Guo, L. Qian, W. Song, Corrosion characteristics of a nickel-base alloy C-276 in harsh environments. Int. J. Hydrogen Energ. **42**, 19829–19835 (2017)
82. Y. Li, S. Wang, P. Sun, Z. Tong, D. Xu, Y. Guo, J. Yang, Early oxidation of Super304H stainless steel and its scales stability in supercritical water environments. Int. J. Hydrogen Energy **41**, 15764–15771 (2016)
83. D.D. Macdonald, The point defect model for the passive state. J. Electrochem. Soc. **139**, 3434–3449 (1992)

84. E. Ghanbari, A. Saatchi, X. Lei, D.D. Macdonald, Studies on pitting corrosion of Al-Cu-Li alloys part III: passivation kinetics of AA2098-T851 based on the point defect model. Materials **2019**, 12 (1912)
85. L. Zhang, D.D. Macdonald, Segregation of alloying elements in passive systems—II. Numerical simulation. Electrochim. Acta. **43**, 2673–2685 (1998)
86. L. Zhang, D.D. Macdonald, Segregation of alloying elements in passive systems—I. XPS studies on the Ni–W system. Electrochim. Acta. **43**, 2661–2671 (1998)
87. M. Urquidi, D.D. Macdonald, Solute-vacancy interaction model and the effect of minor alloying elements on the initiation of pitting corrosion. J. Electrochem. Soc. **132**, 555–558 (1985)
88. Y. Li, S. Wang, L. Wang, P. Yu, J. Yang, Corrosion behavior of austenitic steel 304 in nearcritical aqueous solutions. Adv. Eng. Res. **45**, 189–192 (2015)
89. Y. Watanabe, T. Adschiri, K. Sue, K. Arai, Estimation of metal oxide solubility and understanding corrosion in supercritical water. NACE Int. **107**, 702–705 (2001)
90. S. Chevalier, F. Desserrey, J.P. Larpin, Oxygen transport during the high temperature oxidation of pure nickel. Oxid. Met. **64**, 219–234 (2005)
91. J. Yang, S. Wang, D. Xu, Y. Guo, C. Yang, Y. Li, Effect of ammonium chloride on corrosion behavior of Ni-based alloys and stainless steel in supercritical water gasification process. Int. J. Hydrogen Energy **42**, 19788–19797 (2017)
92. H. Kim, D.B. Mitton, R.M. Latanision, Corrosion behavior of Ni-base alloys in aqueous HCl solution of pH 2 at high temperature and pressure. Corros. Sci. **52**, 801–809 (2010)
93. J. Li, W. Zheng, S. Penttilä, P. Liu, O.T. Woo, D. Guzonas, Microstructure stability of candidate stainless steels for Gen-IV SCWR fuel cladding application. J. Nucl. Mater. **454**, 7–11 (2014)
94. A. Kimura, R. Kasada, N. Iwata, H. Kishimoto, C.H. Zhang, J. Isselin, P. Dou, J.H. Lee, N. Muthukumar, T. Okuda, M. Inoue, S. Ukai, S. Ohnuki, T. Fujisawa, T.F. Abe, Development of Al added high-Cr ODS steels for fuel cladding of next generation nuclear systems. J. Nucl. Mater. **417**, 176–179 (2011)
95. M. Payet, L. Marchetti, M. Tabarant, J.P. Chevalier, Corrosion mechanism of a Ni-based alloy in supercritical water: Impact of surface plastic deformation. Corros. Sci. **100**, 47–56 (2015)
96. W.G. Cook, R.P. Olive, Pourbaix diagrams for chromium, aluminum and titanium extended to high-subcritical and low-supercritical conditions. Corros. Sci. **58**, 291–298 (2012)
97. K. Chen, J. Wang, Z. Shen, D. Du, X. Guo, B. Gong, J. Liu, L. Zhang, Comparison of the stress cracking growth behavior of cold worked Alloy 690 in subcritical and supercritical water. J. Nucl. Mater. **520**, 235–244 (2019)
98. P. Kritzer, N. Boukis, E. Dinjus, The corrosion of nickel-base alloy 625 in sub- and supercritical aqueous solutions of HNO3 in the presence of oxygen. J. Mater. Sci. Lett. **18**, 771–773 (1999)
99. M. Sun, X. Wu, Z. Zhang, E.H. Han, Oxidation of 316 stainless steel in supercritical water. Corros. Sci. **51**, 1069–1072 (2009)
100. S. Ahn, H. Kwon, D.D. Macdonald, Role of chloride Ion in passivity breakdown on iron and nickel. J. Electrochem. Soc. **152**, B482–B490 (2005)
101. P. Kritzer, N. Boukis, E. Dinjus, The corrosion of alloy 625 (NiCr22Mo9Nb; 2.4856) in high-temperature, high-pressure aqueous solutions of phosphoric acid and oxygen. Corrosion at sub- and supercritical temperatures. Mater. Corros. **49**, 831–839 (1998)
102. M. Wagner, V. Kolarık, B. Michelfelder, M.d.M. Juez-Lorenzo, T. Hirth, N.E.u.P. Eyerer, Materials for SCWO processes for the oxidation of chlorine containing residues in supercritical water. Mater. Corros. **50**, 523–526 (1999)

Chapter 5
Corrosion Behavior of Alloys in Supercritical Water Oxidation Processes

5.1 High-Temperature Corrosion of Fe–Ni-Based and Ni-Based Alloys in SCW

5.1.1 Introduction

Ni-based superalloys and Fe–Ni-based alloys have attracted attention in recent years to meeting service requirements of advanced ultrasupercritical (A-USC) boiler superheater and reheater, and supercritical CO_2 (sCO_2) turbines [1, 2]. Among the candidate materials, the low-cost Fe–Ni-based HR6W alloy with excellent workability has been developed [3], which possesses high resistance to high-temperature corrosion in the atmosphere of combustion gases and to oxidation in steam [4]. Haynes 282, which has excellent creep strength, fabricability and high temperature strength, is designed for applications in engines for aircraft and land-based gas turbines [5]. Inconel 740 is promising to be used to make superheater and reheater pipes, and the tube material for A-USC steam boiler unit due to its unique performance [6, 7].

Some studies have been done regarding the corrosion resistance and oxidation behavior of HR6W, Haynes 282 and Inconel 740 in vapor and high-temperature air. For example, Liang et al. [8] studied the influences of temperature, SO_2, water vapor and simulated ash on the high-temperature corrosion of HR6W at 750 and 810 °C for 250–1000 h. They found that a Cr_2O_3-rich scale formed on the surface of HR6W exhibited excellent corrosion resistance, while SO_2 promoted the cracking of the Cr-rich oxide scales at 750 and 810 °C. Nnaji et al. [9] conducted an oxidation experiment on Haynes 282 under laboratory air from 0 to 10 h at 1050 and 1100 °C to investigate its overall oxidation kinetics, and confirmed that the protective properties of Haynes 282 oxide film were more stable with the increase of exposure time at an appropriate temperature. Moreover, Liang and Zhao [10] explored the high temperature oxidation of Haynes 282 in steam at 800–1000 °C for 12.5–336 h. Pérez-González et al. [11]

D. Xu and S. Guo, *Corrosion Characteristics, Mechanisms and Control Methods of Candidate Alloys in Sub- and Supercritical Water*,
https://doi.org/10.1007/978-981-16-5525-8_5

examined the isothermal oxidation resistance of Haynes 282 in the temperature range of 800–1000 °C for exposure times up to 24–150 h.

Abbasi et al. [12] investigated the effects of surface deformation and grain size on the humid air oxidation of Inconel 740 in humid air consisted of dried airand approximately 20% water vapor at 800 °C for 1021 h. They alsoconsidered the effect of alloying elements on the oxidation of Inconel 740 to improve its oxidation resistance, and conducted the oxidation experiments at 800 °C for 301 h in the presence of dried air and 20% water vapor at 1.2 atm. [13]. Besides, the oxidation of Inconel 740 in gas environment has been thoroughly discussed by the research group of Zhao et al. They investigate the oxidation and hot corrosion behavior of Inconel 740 with and without Na_2SO_4 deposit in air within 850–1000 °C [14], the corrosion of Inconel 740 exposed to synthetic coalash/flue gas environmentsat 550 and 700 °C for times up to 5008 h [15], and the coal ash corrosion performance of Inconel 740 under the simulated coal ash/flue gas condition at 704–760 °C for times up to 5000 h [16].

High-temperature corrosion of Ni-based and Fe–Ni-based alloys is very complex because many factors including temperature, pressure, exposure time, oxygen content and salt deposition affect the corrosion processes. Previous research regarding the corrosion of HR6W, Haynes 282 and Inconel 740 mainly focused on the oxidation and in high-temperature vapor in the temperature above 700 °C or the open air at room temperature. However, supercritical water (SCW) with significantly different corrosion properties as compared to liquid water is very aggressive for candidate materials, which brings great challenges to the selection and production of candidate materials. It is very important to study the corrosion performance of candidate materials used in such an aggressive environment based on equipment safety considerations. Nevertheless, little information is documented concerning the corrosion process in supercritical water environment. The influences of various factors on corrosion behavior of alloys mentioned above in supercritical water are not understood deeply. Therefore, this workaims to explore corrosion behavior of typical Fe–Ni-based alloy HR6W, Ni-based alloy Haynes 282 and Inconel 740 in SCW at 450 °C and 25 MPa for 30–200 h. The gravimetry, plan-view surface morphologies, 3D surface morphologies, structures and compositions of oxide layers of the samples after exposure were analyzed. It proposes direct comparisons of corrosion behavior of different Fe–Ni-based alloy and Ni-based alloys, and expands the corrosion research of HR6W, Haynes 282 and Inconel 740 to supercritical water environment. This information is valuable for understanding corrosion mechanisms and control means of these alloys in actual application.

5.1.2 Experiment

Tested alloys were supplied by Shanghai Boiler Works Co., Ltd., and their nominal chemical compositions are listed in Table 5.1. HR6W and Inconel 740 have higher Cr content, while Haynes 282 has higher Mo content, which may affect the corrosion

Table 5.1 Chemical compositions of tested alloys (wt%)

Alloys	Ni	Cr	Fe	Mo	Al	Co	Mn	Nb	Ti	Si	W	C
HR6W	45	23	Balance	–	–	–	1	0.2	0.1	0.5	7	0.08
Haynes 282	Balance	19.7	0.2	8.4	1.5	10.1	0.03	–	0.01	<0.05	–	0.06
Inconel 740	Balance	25	0.7	0.5	0.9	20	0.3	2	1.8	0.5	–	0.03

characteristics and will be discussed in the next section. The as-received materials were cut into cubic pieces with a size of 11 mm × 6 mm × 3 mm and a small hole in its upper part. The samples were mechanically polished by the SiC abrasive paper to 1200 grits, rinsed in acetone for 0.5 h and then ultrasonically cleaned by deionised water. Then, each samplewas polished by SiC sandpapers with particle sizes of $400^{\#}$, $600^{\#}$, $1200^{\#}$, $2000^{\#}$ and diamond polishing agent in turn. Finally, the samples were cleaned with acetone solution by an ultrasonic cleaner to remove possible contaminants on surface.

The corrosion behaviors of HR6W, Haynes 282 and Inconel 740 samples were investigated in SCW at 450 °C and 25 MPa for exposure times of 30, 60, 100, 150, and 200 h. The experimental equipment used in this workis manufactured by Beijing Century SenLong experimental apparatus Co., Ltd, which is mainly consisted of a PID temperature controller, electric heaters with 1.2 kW power, an autoclave, and asound and light alarm protection module. The reactor is made of Hastelloy C-276, which has a volume of 250 mL with 58.5 mm of inner diameter. Before experiments, alloy samples were suspended in the reactor center with silver wire, deionized water and hydrogen peroxide solution were added into the reactor. When the exposure duration was over, the tested samples were taken out from the reactor for following analyses. More details on experiment procedures had been described in our previous report [17].

The corrosion behaviors of HR6W, Haynes 282, and Inconel 740 samples were analyzed by a variety of analytical methods. Gravimetry was measured by an analytical balance (Mettler-toledo XA105) with an accuracy of 0.1 mg. A JEOL 7800F field emission scanning electron microscopy (SEM) incorporated with an energy dispersive X-ray spectroscopy (EDS) was employed for plan-view surface morphology and elemental analyses. 3D surface analyses of tested samples were carried out by aVeeco Innova atomic force microscopy (AFM). Phase compositions were identified via A Bruker D8 ADVANCE X-ray diffraction (XRD) with an X-ray line focus of 0.04 mm × 12 mm, an X-ray spot focal of 0.4 mm × 1.2 mm, a scanning range from 10° to 100°. Scale phases were analyzed with a laser Raman spectroscopy (LRS, Horiba JOBIN YVON). The elemental bonding state as a function of depth in the oxide scale was determined by an X-ray photoelectron spectrometer (XPS) microprobe (ESCALAB™ XI+) equipped with an EX06 ion source and 200–900 μm of X-ray spot sizes.

5.1.3 Results and Discussion

5.1.3.1 Gravimetry

Figure 5.1 shows the variations of corrosion weight changes of HR6W, Haynes 282 and Inconel 740 samples exposed to SCW. It is obvious that the corrosion weight change (i.e., corrosion weight gain) increased as exposure duration prolonging. However, at the end of the exposure duration, the corrosion weight gain of Haynes 282 decreased after 150 h because of oxide exfoliation. Another explanationmay be due to the samplesensitivity to the variation of operation conditions. Generally, Haynes 282 samples showed the higher weight gains while HR6W samples exhibited the lowest weight gains throughout exposure duration in Fig. 5.1a. Equation 5.1 can be used to fit experimental data and describe the oxidation process,

$$\Delta w = kt^n \tag{5.1}$$

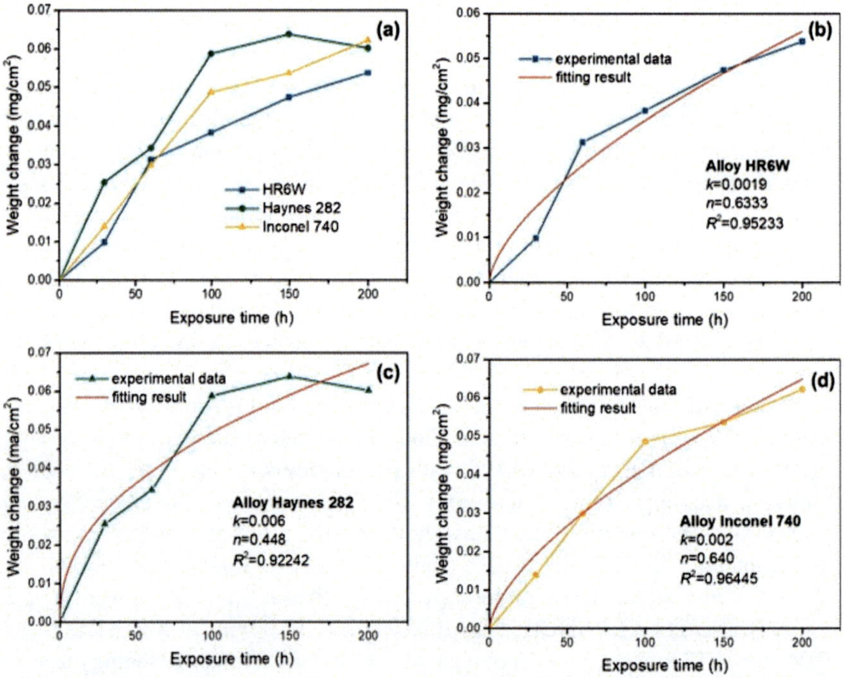

Fig. 5.1 Corrosion weight changeas a function of exposure time for **a** three tested samples, **b** HR6W, **c** Haynes 282, and **d** Inconel 740 samples exposed to SCW at 450 °C and 25 MPa for 20, 40, 60, 100 and 150 h, respectively. The solid curves in **(b–d)** represent the fitting results of data points in

where Δw(mg/cm^2) is corrosion weight change, t (h) is exposure time, k (mg/(cm^2 h)) is rate constant, and n is reaction order. The fitting parameter R^2 with a value approaching 1.00 denotes the increasing consistence of the data with the fitting results. All data in Figs. 5.1b–d produced good fitting results with the n value approaching 0.5 ($n = 0.633$, 0.448, and 0.640 for HR6W, Haynes 282 and Inconel 740, respectively). This shows that oxidation kinetics follows the parabolic law ($n = 0.5$), and three kind of alloys will have smaller weight gain trend for longer exposure time.

The oxidation kinetics mentioned above also indicates that the corrosion process is driven by transport processes and tested HR6W, Haynes 282 and Inconel 740 samples exhibit self-protection during the subsequent test [18]. The degradation of tested samples at the steady-state stage may be dominated by the outward diffusion of metal cations across the inner and outer oxide layers [19]. The rate constant is similar to that reported previously by Nie et al. [20]. After exposed to SCW at 500 °C and 25 MPa for 1300–3000 h with a dissolved oxygen content of 25 wppb, alumina-forming austenitic steel Fe–20Ni–14Cr–3Al–0.6Nb–0.1Ti exhibits a rate constant of 0.00197 while that of Incoloy 800H is 0.00285, and the oxidation kinetics of these two alloys also obey the parabolic law [20]. Therefore, three alloys tested in this workmay have good corrosion resistance comparable to alumina-forming austenitic steel and Ni-based alloy 800H. Overall, the corrosion weight gain of HR6W sample was relatively lower than that of other two samples, suggesting a higher corrosion resistance under experimental conditions.

5.1.3.2 Plan-View Surface Morphology and EDS Analyses

Figures 5.2a–c separately indicate the surface morphologies of HR6W, Haynes 282 and Inconel 740 samples after SCW exposure. The SCW-exposed HR6W sample showed a fine compact oxide scale with some oxide particles distributed on its surface

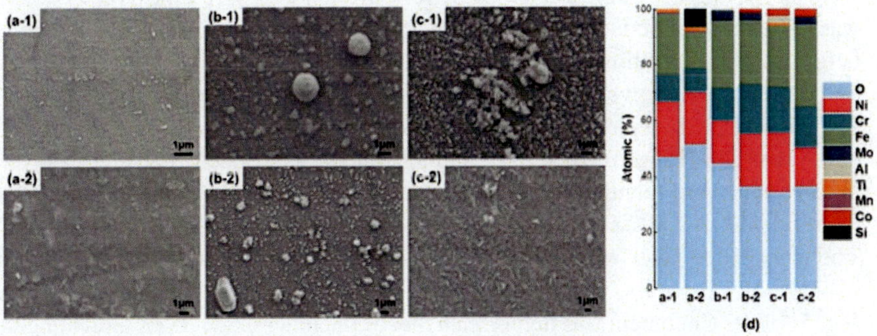

Fig. 5.2 Plan-view SEM morphologies of oxide scales grown on **a** HR6W, **b** Haynes 282 and **c** Inconel 740 after exposure to SCW at 450 °C and 25 MPa for 100 h ((**a-1**), (**b-1**) and (**c-1**)) and for 200 h ((**a-2**), (**b-2**) and (**c-2**)), **d** EDS analyses of the regions in (**a–c**)

(see Fig. 5.2a-1, a-2). In contrast, the Haynes 282 samples after 100 h of SCW exposure was covered with different sizes of oxide particles (see Fig. 5.2b-1). As exposure time increased, small particles gradually grew up, and the samplesurface was irregularly covered with lager particles (see Fig. 5.2b-2). This phenomenon indicates that big oxide nodules are subjected to more severe oxidation after corrosion in SCW, and the homogeneous oxide layer is not generated [21]. The presence of irregularly distributed particles on Haynes 282 increases the possibility of oxide exfoliation during the longer exposure time, which may be a direct reason for its highest weight gain (see Fig. 5.1a).

In the case of Inconel 740, larger oxide particles accumulated on the sample surface to form bubble-shape oxides within 100 h of exposure time. When the sample was exposed to SCW for more than 200 h, accumulated oxides were dispersed and the surface was compact without noticeable particles. This evolution of oxide crystallites with exposition time increasing can be attributed to the consequence of the dissolution [22]. Compared with HR6W sample (see Fig. 5.2a-2), Inconel 740 sample (see Fig. 5.2c-2) seems to form a layer composed of network-shaped oxides in its sample. This highly uniform oxide layer is more conducive to avoiding the contact between the alloy matrix and corrosive environment. It is noteworthy that oxide particles formed on HR6W and Inconel 740 sampleswere much denser than that on Haynes 282 sample after exposure to SCW for 200 h, suggesting the better corrosion resistance of HR6W and Inconel 740.

The surface element composition of sample after SCW exposure was analyzed by EDS analysis and the atomic percentages of elements are shown in Fig. 5.2d.Obviously, Dominant elements on tested samples were O, Fe, Ni and Cr, meaning that the oxide scale on the sample surface is mainly composed of (Fe, Ni, Cr) oxides. Small amounts of Ti, Si, Mn were observed on the HR6W sample in addition to the dominant elements. Moreover, Cu and Cowere present on the surface of the Haynes 282 sample in trace. Al, Cu, Co, and Ti were also detected on the Inconel 740 sample with atomic percent less than 5%. The finding shows that these elements have permeated through the oxide layer mainly composed of Fe, Ni and Cr to the sample surface. Among the three tested alloysafter different exposure times, the O content was lowest on the Inconel 740 surface, in accordance with the discussion mentioned before. As exposure time increased, large oxide particles on the surface of Inconel 740 samplewas dissolved and a flat network-shaped oxide layer was formed.

Figures 5.3a–c depict the plan-view morphologies of SCW-exposed samples at 450 °C and 25 MPa for 100 h. Each scanning region was scanned three times and their average value was calculated as the final atomic percentage to ensure the accuracy of the results. The oxide scale seems to be composed of a double-layer oxide structure. The inner oxide layer was relatively flat and dense, and the outer oxide layer was composed of particles. For the oxide particles on the HR6W sample surface (see Fig. 5.3a), its Cr content was higher than that of the inner oxide layer, indicating that Cr penetrates the inner oxide layer and forms Cr oxide on the HR6W surface. The average size of oxide particles on Haynes 282 was significantly larger than that on HR6W and Inconel 740. Larger oxide particles may fall off during the subsequent exposure process and ultimately result in reduced weight gain (see Fig. 5.1a). A

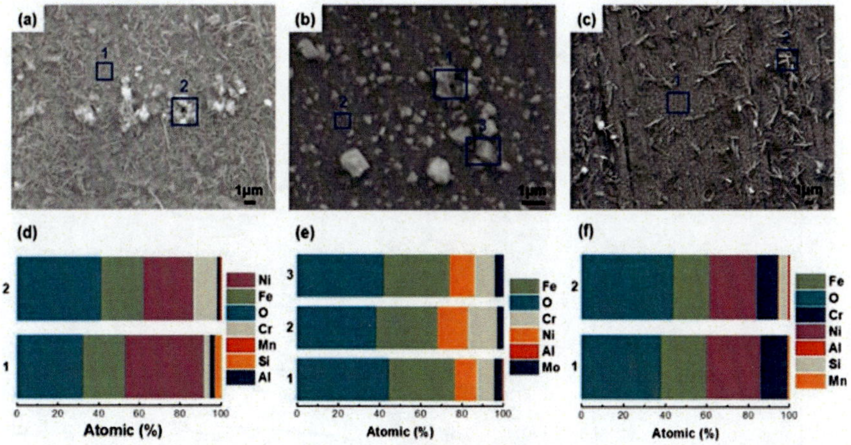

Fig. 5.3 Plan-view SEM morphologies of oxide scales grown on **a** HR6W, **b** Haynes 282 and **c** Inconel 740 after exposure to SCW at 450 °C and 25 MPa for 100 h, **d–f** EDS analyses of the regions marked with blue rectangles in (**a–c**), respectively

certain amount of Mo was detected on the three detection areas on the surface of Haynes 282 samplewhile there was no Mo detected on the HR6W and Inconel 740 surface, due to the higher Mo content (8.4 wt%) in the as-received Haynes 282 sample. Although both Cr and Mo can improve corrosion resistance of Ni-based alloys in sub- and supercritical water environments, Mo cannot exhibit good corrosion resistance in absence of Cr [23, 24]. Therefore, the presence of Cr and Mo on the Haynes 282 surface contributes to the improvement of its corrosion resistance. It can be seen from Fig. 5.3c that needle-like oxides particles were formed on the Inconel 740 sample, and the scratches on the flat region are caused by the polishing process at the pretreatment stage. The needle-like oxides were mainly composed of O, Ni, Fe, Cr and a small amount of Si, while the inner oxide layer majorly containedO, Ni, and Fe according to the EDS results. The content of O was the lowest on the HR6Wsurface among the three tested alloys, suggesting the less oxide content on the sample surfaceand thus a lower weight gain. This is in accordance with the result of corrosion weight gain shown in Fig. 5.1.

5.1.3.3 3D Surface Analyses

Figure 5.4 illustrates 2D surface topographies and particle height analyses of HR6W, Inconel 740 and Haynes 282 samples exposed to SCW at 450 °C and 25 MPa for 60 and 120 h. The scanning area is a square with a side length of 10 μm on sample surface. Figure 5.4a expresses particle height statistics for the entire scanning areain the form of frequency distribution. Line scanning was performed on each 2D image (see Fig. 5.4b) and the changes in height with distance are reflected in Fig. 5.4c. As shown in Fig. 5.4a, as exposure time increased from 60 to 200 h, the particle

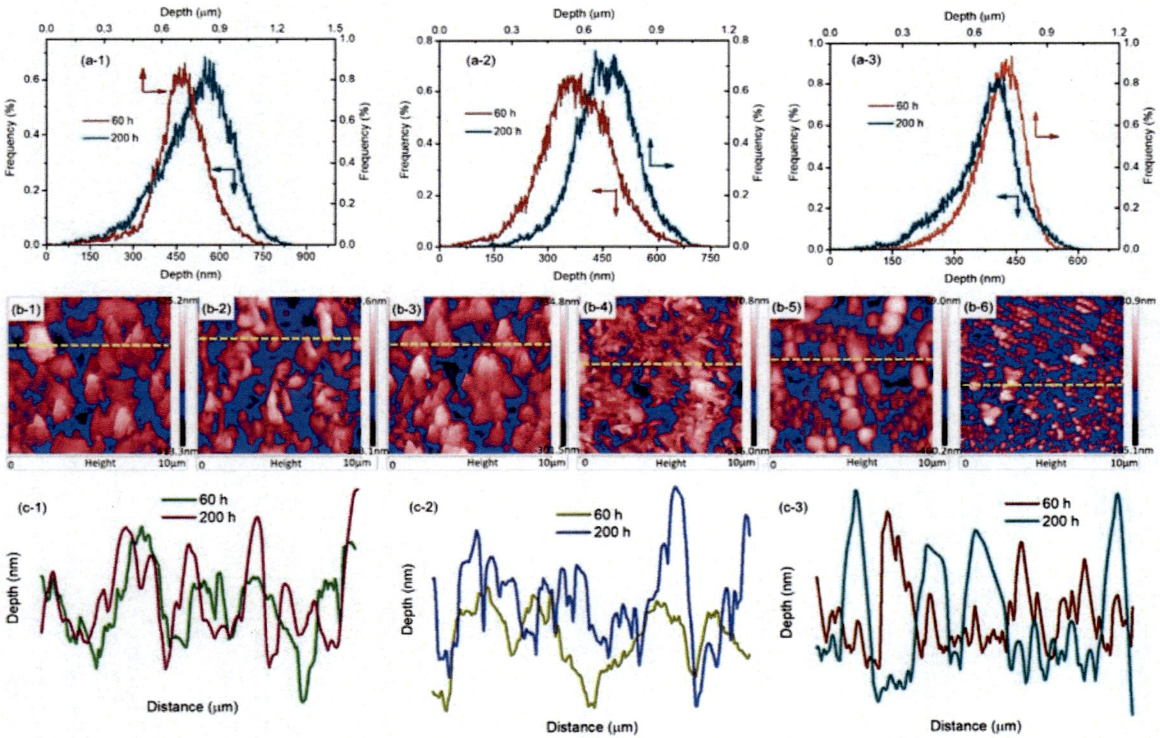

Fig. 5.4 2-D AFM analyses of HR6W, Inconel 740 and Haynes 282 samples exposed to SCW at 450 °C and 25 MPa. Frequency distribution curves of the heights ofoxide particles on **a-1** HR6W, **a-2** Inconel 740, and **a-3** Haynes 282 samples exposed to SCW for 60 and 120 h. 2-D morphologies of **b-1**, **b-2** HR6W, **b-3**, **b-4** Inconel 740, and **b-5**, **b-6** Haynes 282 samples exposed to SCW for 60 h ((**b-1**), (**b-3**) and (**b-5**)) and for 200 h ((**b-2**), (**b-4**) and (**b-6**)). Line scanning of the heights of oxide particles corresponding to the yellow dotted lines in (**b-1**)–(**b-6**)

height on the surface of HR6W and Haynes 282 samples gradually decreased, while that of Inconel 740 sample increased. Consistently, an evident particle precipitation phenomenon occurs on Inconel 740 after 120 h exposure [25]. The curve of particle height versus distance also shows that the particle height on the Inconel 740 surface was greatly affected by exposure time, and the particle height increased remarkably within 140 h. More tiny oxide particles with heights varying within 0–600 and 0–1 nm were detected on Haynes 282 surface exposed for 60 h and 200 h, respectively. Especially, the maximum particle height on the Haynes 282 surface reduced from 0.74 μm to 403 nm, suggesting that larger oxides may fall off as exposure time increases. This is in good agreement with the discussion of gravimetry and plan-view surface morphology. Notably, the maximum particle height here refers to the height value with the highest frequency. For example, the maximum particle height of the HR6W sample decreased from 0.69 μm to 566 nm as exposure time increased, so resulting in a decrease in the average particle height. It can be clearly seen from Fig. 5.4b that as the exposure time increases from 60 to 120 h, the shape and distribution of oxide particles on the HR6W sample does not change much, indicating that the oxide growth on the surface of HR6W is relatively stable. Thus, for HR6W and Haynes 282 samples, smaller surface oxide particles help to provide better protection during longer exposure duration. In addition, the larger oxide particles may be Fe–Ni–Cr oxides according to the EDS results in Figs. 5.2 and 5.3, however, this requires further examination by XRD and Raman spectra.

To visualize the surface of samples exposed to SCW, geometric characteristics of 3D topographies of oxide scales obtained from Fig. 5.4 are further provided in Fig. 5.5. The surface morphology of the HR6W sample had undergone interesting changes after exposure for 200 h, and a cluster of mountain-like oxides with similar height and size were formed on the surface. As shown in Fig. 5.5a, the heights of these mountain-like oxides are almost half of that on sample surface exposed for 60 h. The advantage of these highly similar oxides is that they help to form a dense oxide film without obvious defects. As exposure time increased from 60 to 200 h, more oxides were formed on the Inconel 740 sample surfaceand filled the gaps among initial large oxide particles (see Fig. 5.5b). This is consistent with the plan-view SEM morphologies in Fig. 5.2c. Additionally, the particles formed homogenously on the Inconel 740 sample surface may be the network-shaped oxide layer mentioned in Sect. 5.1.3.2. The gaps on the Inconel 740 sample surface (see Fig. 5.5b-1) is likely polishing scratches, which are also found in our previous research [26]. However, even if exposure time reached up to 200 h, the heights of the oxide particles on the Haynes 282 sample surface still did not show a good consistency compared with that of HR6W and Inconel 740. There were large vacancies among oxide clusters, which increases the risk of the substrate being exposed to SCW. This has good agreement with the highest corrosion weigh gain of the Haynes 282 sample in Sect. 5.1.3.1.

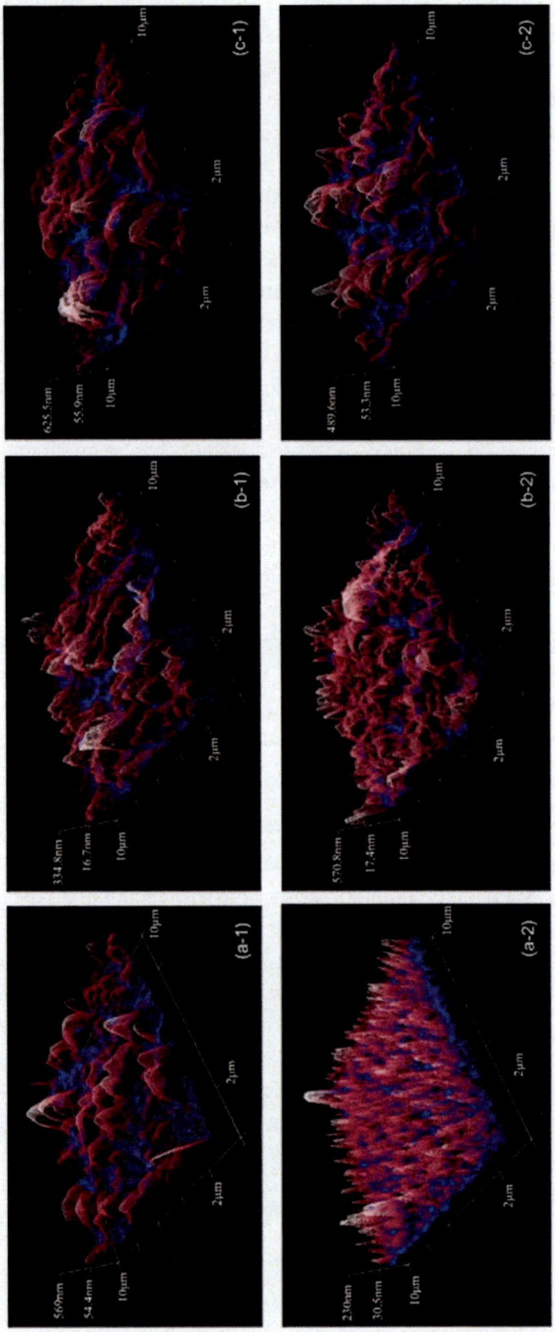

Fig. 5.5 3-D morphologies of **a** HR6W, **b** Inconel 740, and **c** Haynes 282 samples exposed to SCW at 450 °C and 25 MPa for 60 h (**(a-1)**, **(b-1)**, and **(c-1)**) and for 200 h (**(a-2)**, **(b-2)**, and **(c-2)**)

5.1.3.4 Oxide Layer Phase

Raman Analyses

Figure 5.6 elucidates the Raman spectra (in the range of 200–6000 cm^{-1}) of HR6W, Haynes 282 and Inconel 740 after exposure to SCW for 20, 60, and 150 h. A He Ne laser of 532 nm was used as the excitation source. To ensure the accuracy of the test results, five lines in each figure represent five test points randomly selected on sample surface. The existing probability of the substance is higher in the case of more overlapping curves at characteristic peaks. Overall, $FeCr_2O_4$, $NiFe_2O_4$ and Cr_2O_3 were detected on HR6W, Fe_3O_4, Cr_2O_3, $NiFe_2O_4$ and $FeCr_2O_4$ were found on Haynes 282. The oxides detected on the Inconel 740 surface contained $NiFe_2O_4$, Fe_2O_3 and Fe_3O_4.

For HR6W within the first 20 h of exposure to SCW, all the characteristic peaks in the range of 200–6000 cm^{-1}were unobvious. As exposure time increased, $FeCr_2O_4$, $NiFe_2O_4$ and Cr_2O_3were formed. In Fig. 5.6a, the peak at approximately 697 cm^{-1} may be Cr_2O_3, $FeCr_2O_4$, or $NiFe_2O_4$, and the existence of Cr_2O_3 on HR6W requires to be further confirmed by XRD detection. In the initial stage (within 50 h) of exposure in SCW, the oxides on the Haynes 282 surface were mainly Fe_2O_3, Fe_3O_4 and Cr_2O_3. As exposure time increased to 150 h, $FeCr_2O_4$and $NiFe_2O_4$were formed. For Inconel 740, $NiFe_2O_4$ existed on its sample surface within the exposure time of 20–150 h, but the peak of $Fe_3O_4$was not detected. It can be seen from Fig. 5.6c that Fe_2O_3 was found at the test point1 on the samples withdifferent exposure times, suggesting that a small amount of Fe_2O_3 exists on Inconel 740.

To further clarify oxide compositions at different locationson sample surface, Fig. 5.7 displays the Raman spectra (within 200–2000 cm^{-1}) of Haynes 282 samples after exposure to SCW for 20, 60 and 150 h. There were granular particles on the surface of three samples with different exposure times, in accordance with oxide particles on the Haynes 282 surface shown in Fig. 5.2b. Among them, irregular particles were found on the 20 and 60 h samples, while blister-shaped islands composed of $NiFe_2O_4$ and Fe_2O_3 were observed on the 150 h sample. The similar blister-shaped oxides are also found on a fine-grained compact oxide layer of alloy SS 347H after exposed to SCW at 800 °C for 12 h, which are composed of iron oxide phases such as magnetite (Fe_3O_4) [27].

The particles on three samples with different exposure times contained Fe_2O_3 and $NiFe_2O_4$, while $FeCr_2O_4$ and Fe_3O_4 only appeared on the exposure samples of 20 h and 50 h, respectively. The brief appearance of $FeCr_2O_4$ and Fe_3O_4 means that they easily react and decompose with the increase of exposure time, compared with Fe_2O_3 and $NiFe_2O_4$. It can be noted that Cr_2O_3 is not detected on three samples with different exposure times, while Fig. 5.6 shows that Cr_2O_3 existed on the surface of Haynes 282. On the one hand, this indicates that Cr_2O_3 may exist in the flat region on the alloy surface, which can reduce the effect of fluid erosion and avoid falling off from alloy surface during subsequent exposure. On other hand, the detection depth of laser Raman spectroscopy can be around dozens of nanometers, and thus Cr_2O_3 may exist in the inner oxide layer instead of on the oxide surface.

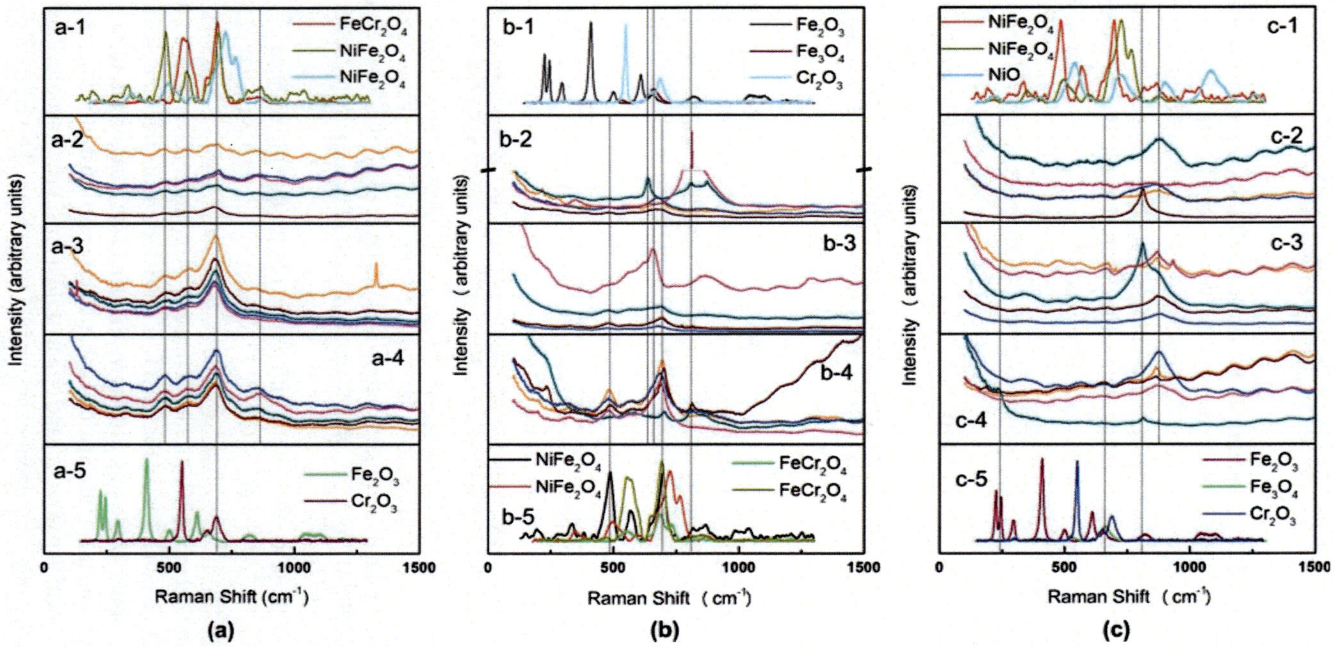

Fig. 5.6 Raman spectra of different samples obtained with 532 nm of excitation wavelength. **a-1**, **a-5**, **b-1**, **b-5**, **c-1** and **c-5** are separately Raman spectra of Fe_2O_3 (ID #R040024), Fe_3O_4 (ID #R060191), Cr_2O_3 (ID #060,892), $FeCr_2O_4$ (ID #R110059 and 110,060), $NiFe_2O_4$ (ID #R140860 and 150,038), and NiO (ID #R080121) derived from the RRUFF database. **a-2**, **a-3** and **a-4** are Raman spectra of HR6W samples exposed to SCW for 20, 60 and 150 h. respectively. **b-2**, **b-3** and **b-4** are separately Raman spectra of Haynes 282 samples exposed to SCW for 20, 60 and 150 h. **c-2**, **c-3** and **c-4** are Raman spectra of Inconel 740 samples exposed to SCW for 20, 60 and 150 h, respectively.

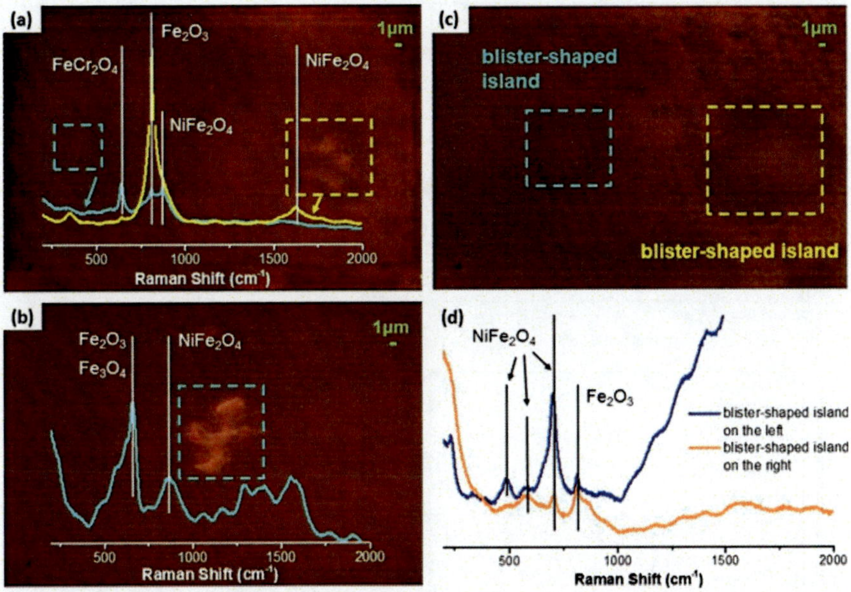

Fig. 5.7 Local Raman spectra of Haynes 282 samples after exposure to SCW for 20, 60, and 150 h. **a** Surface morphology and local Raman spectra for the 20 h Haynes 282 sample. **b** Surface morphology and local Raman spectra for the 60 h Haynes 282 sample. **c** Surface morphology for the 150 h Haynes 282 sample. **d** Local Raman spectra for the selected areas in (**c**). The dotted rectangles in (**a–c**) represent the selected Raman test points, and the curves with the same color in (**a**) and (**b**) represent corresponding Raman curves. Raman spectra were obtained with 532 nm of excitation wavelength, and the data for $FeCr_2O_4$, $NiFe_2O_4$, Fe_2O_3 and Fe_3O_4 were derivedfrom the RRUFF Project website

XPS Analyses

Figure 5.8 displays the Ni 2p, Cr 2p and Fe 2p core level spectra for HR6W, Haynes 282 and Inconel 740 after exposure in SCW at 450 °C and 25 MPa for 150 h. This helps to determinethe oxide scales and chemical states of Ni, Cr and Fe on alloy surface, Fig. 5.8a illustrates the XPS survey scan of the surface of three samples after exposure to SCW. The result indicates the presence of Cr, Fe, Ni, O, C and Ag on the sample surface, where C probably comes from surface contaminants and Ag is derived from the silver wire used to suspend samples.

As for HR6W and Inconel 740 samples in Fig. 5.8b, the signals at the binding energies of about 879.8 and 873.1 eV were assigned to NiOOH, the two peaks at around 861.8 and 855.5 eV were consistent with $Ni(OH)_2$. However, only the peak of $Ni(OH)_2$ occurred on the Ni 2p spectrum of Haynes 282 sample. Figure 5.8c shows the Cr 2p core level spectra for three samples after exposure in SCW. The binding energy signals of near 576.5 and 588.1 eV were definitely attributed to Cr_2O_3 on all sample surface. The signal at a binding energy of approximately 586.5 eV showed the presence of $NiFeCrO_4$, which appeared on the surface of HR6W and Haynes 282

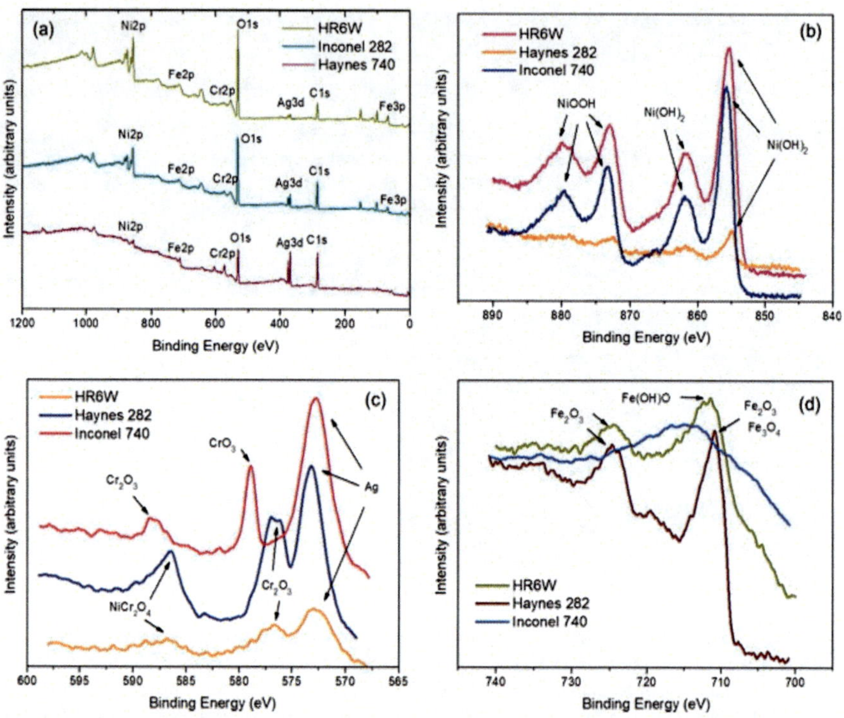

Fig. 5.8 XPS analyses of oxide films formed on the SCW-exposure HR6W, Haynes 282, and Inconel 740 samples at 450 °C and 25 MPa for 150 h. **a** Surface chemistry on different samples, **b** Ni 2p core level spectra, **c** Cr 2p core level spectra, **d** Fe 2p core level spectra

samples, while CrO_3 at the binding energy of 578.9 eV was detected on the Inconel 740 surface. In the Fe 2p core level spectra, main peaks were assigned to Fe_2O_3 and Fe_3O_4 for HR6W and Haynes 282 samples, whereas there are not obvious peaks for Inconel 740 sample.

Cr_2O_3 is generally regarded as a barrier layer on alloysurface, and can be used to prevent the outward diffusion of metal cations and the inward diffusion of oxygen as so to achieve the purpose of protecting the alloy matrix. This kinetic stability phenomenon of thermodynamically reactive metal is called "passivity" by many researchers [28–30]. However, passivity may break down and the dissolution rate of the metal increases dramatically at higher potentials [31]. This can be caused by the oxidative dissolution of the barrier layer (Cr_2O_3 in this work).

$$Cr_2O_3 + 5H_2O \rightarrow 2CrO_4^{2-} + 10H^+ + 6e^- \tag{5.2}$$

Then, the cations released by the destruction of the barrier layer react with environment and continue to generate Cr vacancies via the reaction (5.3) [31]

$$Cr(III) + 4H_2O \rightarrow CrO_4^{2-} + V_{Cr}^{3'} + 8H^+ + 3e^- \tag{5.3}$$

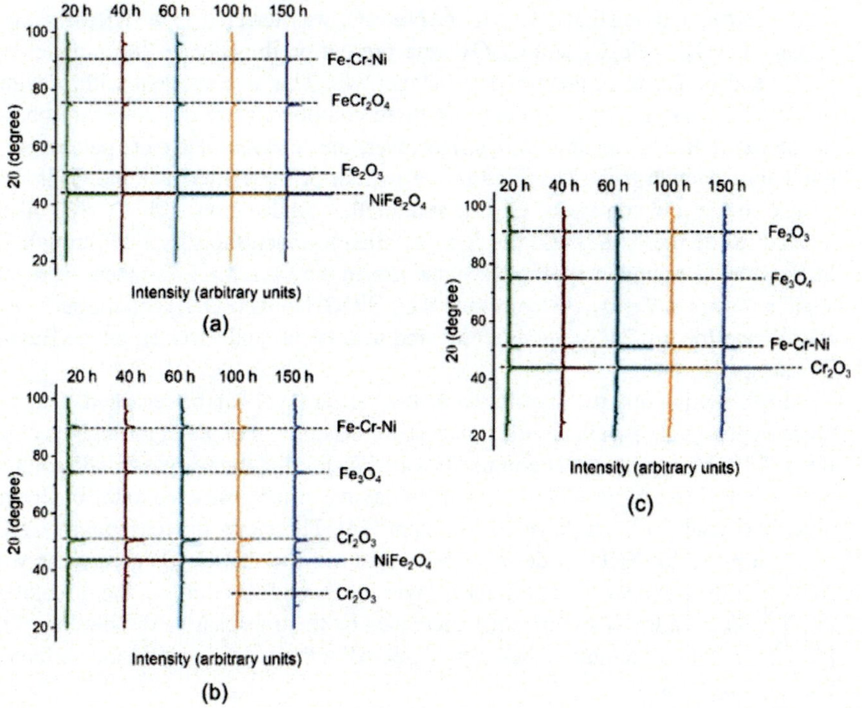

Fig. 5.9 XRD patterns of SCW-exposed **a** HR6W, **b** Haynes 282, and **c** Inconel 740 samples at 450 °C and 25 MPa for 20, 40, 60, 100 and 150 h

where $V_{Cr}^{3'}$ represents a cation vacancy. After the generation of metal vacancies, the interstitials (i.e. metal vacancies caused by the abstraction of metal positions) are dissolved and oxygen vacancies react with adsorbed water. Finally, oxygen vacancies transport outwards and oxygen ions move inwards, so resulting in the growth of the passive film [32].

XRD Analyses

Figure 5.9 indicates the surface X-ray diffraction patterns of HR6W, Haynes 282, Inconel 740 samples after exposure to SCW for different exposure times. The characteristic peaks of Fe–Cr–Ni represent the substrate and Fe–Cr–Ni oxides, and other peaks are corresponding to Fe oxides and Cr oxides. Thus, the oxide layer was mainly composed of hematite, magnetite and spinel. The peak intensitiesrelated to the content of phasesin the oxide film increased with exposuretimeincreasing especially after 100 h. When exposure time increased from 60 to 100 and 150 h, the number of Cr_2O_3 peaks increased in Haynes 282 pattern (see Fig. 5.9b), showing that Cr_2O_3 is gradually generated and increased after 100 h exposure.

It is worthynoticing that $FeCr_2O_4$ and $NiFe_2O_4$ was detected on the HR6W sample surface, Fe_3O_4, $NiFe_2O_4$ and Cr_2O_3 were present on the Haynes 282 sample, and Fe_2O_3 and Fe_3O_4 were identifiedon Inconel 740. This is consistent with previous results of Raman spectra, proving that these compounds are the dominant phases in the oxide layers of three alloys. However, the features of Cr_2O_3 pattern were unobvious enough to be observed in the pattern of the HR6W sample. This may be due to the low content of Cr_2O_3, and another explanation is that Cr_2O_3 likely located inside the oxides and the X-rays failed to reach the substrate through the thick scale. Combined with the XPS analyses in the Sect. 5.1.3.4.2, there were still small amounts of Cr_2O_3, Ni hydroxide (i.e., NiOOH and $Ni(OH)_2$) on the surface of HR6W and Inconel 740 samples, compared with Fe_2O_3 and $Ni(OH)_2$ on the Haynes 282 sample surface.

The formation and the dissolution of the barrier layer can be regarded as a slow reaction process in parallel to the transpassive dissolution of metal through its oxide layer [33]. It is reported that compositional changes due to transpassive dissolution (e.g., the decrease of the Cr content) of the passive film is restricted to the outermost layer in contact with corrosive environment [34]. Therefore, the reason why Cr_2O_3 is not detected by XRD on the HR6W surface may be that Cr_2O_3 contacting with SCW dissolves and Cr_2O_3 in the inner layer is retained. In addition, the dissolution of Ni from the alloy is significantly increased by the transpassive dissolution of Cr [35]. Thus, Ni-containing oxides were detected in the oxide layer on the surface of the three samples.

5.1.4 Conclusions

The oxidation kinetics of HR6W, Haynes 282 and Inconel 740 followed the parabolic law. The corrosion weight gain of Haynes 282 decreased at the end of exposure duration due to oxide exfoliation. Haynes 282 samples showed higher weight gains while HR6W samples exhibited the lowest weight gains throughout exposure duration. A layer composed of network-shaped oxides was formed on the Inconel 740 surface. This highly uniform oxide layer can avoid the contact between alloy substrate and corrosive environment. The heights of oxide particles on the Haynes 282 sample surface were not consistent enough compared with that of HR6W and Inconel 740 samples in SCW for 200 h. There were large vacancies among oxide clusters, which increases the risk of substrate destruction. [$FeCr_2O_4$, $NiFe_2O_4$], [Fe_3O_4, $NiFe_2O_4$ and Cr_2O_3], and [Fe_2O_3, Fe_3O_4] were the dominant phases in the oxide layers of HR6W, Haynes 282 and Inconel 740, respectively. The blister-shaped islands on the Haynes 282 surface after exposure for 150 h were composed of $NiFe_2O_4$ and Fe_2O_3. $FeCr_2O_4$ and Fe_3O_4 were separately detected on the oxide particles of the Haynes 282 sample after exposure for 20 and 50 h, and disappeared after longer corrosion duration due to reaction and decomposition with the increase of exposure time.

5.2 Corrosion Properties of Alloys in SCW Containing Phosphate, Sulfate, Chloride and Oxygen

5.2.1 Introduction

Sewage sludge and some organic wastewaters commonly contain high concentration of refractory and toxic organic matters and certain amounts of inorganic salts (e.g., phosphates, chlorides and sulphates). Compared with traditional treatment approaches, supercritical water oxidation (SCWO) has unique economic and technical advantages in harmlessly disposing these complex pollutants. A SCWO plant with 7 $m^3 h^{-1}$ of treatment capacity has been built by Chematur Engineering AB (Sweden) to deal with sewage sludge. However, a key obstacle for the extensive commercialization of SCWO is the reactor corrosion caused by harsh reaction conditions, such as high temperature and pressure, oxygen, aggressive inorganic salts [36].

Till now, researchers have tested corrosion resistances of some reactor candidate materials such as Ni-base alloy, ceramic, stainless steel (SS), titanium alloy and noble metal in supercritical water (SCW) [37–42]. The effects of aggressive species on the corrosion behaviors of Ni-base alloys are explored in SCW containing alkaline, HCl, HBr, H_2SO_4, HNO_3, or H_3PO_4, etc. [40–43]. For example, the corrosion properties of the Ni-base Alloy 625, Alloy 686 and pure chromium were studied at 500 °C and 46.5 MPa in the model fluid consisting of $HCl + H_2O + NaCl +$ methanol with H_2O_2 as oxidant [44]. It is confirmed that acid radical ion, active free radical, and salt deposition in SCW accelerate equipment material corrosion [45, 46]. Chloride can penetrate and destroy protective oxide layers, so resulting in severe localized corrosion, pitting corrosion, and stress corrosion cracking [47]. Each material shows an individual corrosion resistance property under the same reaction conditions, and can not withstand the corrosions of all conditions in SCWO [39]. Nonetheless, presently existing corrosion tests are typically conducted in a relatively simple SCW environment, such as with or without oxygen and/or one kind of inorganic salt. The corrosion characteristics of some alloys in SCW containing two types of inorganic salts had been examined in our previous research [40, 41], and the results showed that the synergism of chloride and oxygen caused severe corrosion than the single factor [40], and a low corrosion rate was exhibited under the condition of oxidized and chlorinated SCW in the presence of phosphate [41].

As an extension, this work explored the corrosion behaviors of austenitic stainless steels (316 SS and 316L SS) and Ni-base alloys (Inconel 625, Incoloy 825 and Hastelloy C276) in SCW simultaneously containing high concentrations of phosphate, chloride, sulphate and oxygen. Corresponding corrosion rates, corrosion morphologies, elements and products components of oxide films, corrosion layer structures were evaluated by weight measurement, X-ray diffraction, Raman spectroscopy, scanning electron microscopy, and energy dispersive spectroscopy, etc. The formation mechanisms of oxide films on 316 SS, Incoloy 825 and Hastelloy C276 were proposed as well. This information is valuable for guiding the reactor material

Table 5.2 Chemical compositions of test alloys

Alloy	Chemical composition (wt%)						
	C	Ni	Cr	Mo	Ti	Fe	Others
316 SS	0.080	9.69	16.87	1.13	0.12	69.21	0.35Si, 1.05Cu
316L SS	0.030	10.02	17.83	2.06	–	65.90	0.80Si, 2.20Mn
Inconel 625	0.025	56.19	20.20	10.30	0.36	3.38	0.35Si, 0.08Mn, 0.32Ai, 3.63Nb
Incoloy 825	0.030	42.00	21.20	3.00	0.68	30.40	0.20Si, 0.50Mn, 2.20Cu, 0.10Ai
Hastelloy C276	0.003	58.97	16.94	15.15	–	5.75	0.06Si, 0.01Mn, 2.50Co

selection and corrosion control in the SCWO of sewage sludge or organic wastewater containing phosphate, sulphate and chloride.

5.2.2 Experiment Section

5.2.2.1 Materials

30 wt% of H_2O_2 solution was used as oxidant and obtained from Tianjin Tianli Chemical Agent Co., Ltd. NaCl (>99.8 wt% purity) and Na_2SO_4 (>99.5 wt% purity) were chosen to represent corrosive chloride and sulphate, and purchased from Tianjin Kemiou Chemical Agent Co., Ltd and Tianjin Hongyan Chemical Agent Co., Ltd, respectively. Na_3PO_4 (>99.5 wt% purity) was adopted to represent phosphates and gained from Tianjin Hongyan Chemical Agent Co., Ltd. Deionized water was prepared by a water purification plant (ULUP-II) in our lab and used throughout. Five kinds of tested alloys (i.e., 316 SS, 316L SS, Inconel 625, Inconel 825 and Hastelloy C276) were supplied by Shenzhen Metal Alloy Material Co., Ltd. Their chemical compositions are listed in Table 5.2.

Each alloy was cut into several small samples with a size of 10 mm × 10 mm × 7 mm. They were polished with #80, #400, #600, #800, #1000, #1200, #1500 and #2000 abrasive papers and then with aluminum polishing powder in turn to form slick mirror surfaces by a metallographic specimen polishing machine (P-1). Before corrosion tests, the samples were cleaned by an ultrasonic cleaner in acetone for 10 min and then dried in an oven (Electro-thermostatic blast) at 60 °C for 30 min.

5.2.2.2 Experimental Procedures

The corrosion experiments were performed in a 415 ml batch reaction autoclave. Before the experiments, the autoclave was opened and the specimen were placed in its center, and then certain amounts of deionized water, Na_3PO_4, NaCl, Na_2SO_4

Table 5.3 Experimental conditions set for corrosion tests

Con.[a]	Temperature(°C)	Pressure(MPa)	Added corrosive species ($mg \cdot L^{-1}$)[b]				Exposure time (h)
			O_2	Cl^-	SO_4^{2-}	PO_4^{3-}	
1	450	25	31,600	0	0	5800	70
2	450	25	31,600	0	2700	5800	70
3	450	25	31,600	4000	0	5800	70
4	450	25	31,600	4000	2700	5800	70
5	450	25	0	4000	2700	5800	70

[a]Experimental condition
[b]At ambient temperature

and/or 30 wt% H_2O_2 solution were added. Their individual amounts were confirmed according to the desired test conditions, and the detailed method was described in our previous study [40]. Table 5.3 lists five types of experimental conditions (Cons. 1–5) used in this research. Herein, each corrosion test was conducted in SCW (at 450 °C, 25 MPa) with an amount of Na_3PO_4. The differences among the five Cons. depend on whether NaCl, Na_2SO_4 and/or oxidant were added. Notably, these test conditions were chosen roughly according to the SCWO parameters of sewage sludge or organic wastewater containing inorganic salts. After all materials were loaded, the autoclave was sealed and heated with 3.5 °C/min of heating rate by a temperature controller, and thus the autoclave pressure increased with temperature accordingly. Finally, the autoclave reached up to 450 °C and 25 MPa and was maintained there for 70 h. Further, we shut off the temperature controller and naturally cooled the autoclave down to ambient temperature, and then opened the autoclave and took out corrosion samples, and further cleaned them by deionized water for 3 min and then by acetone for 3 min in an ultrasonic cleaner to remove possible salt deposition on the sample surfaces. Subsequently, we naturally dried the corrosion samples at least for 2 h and weighed them, and then kept the samples in soft filter papers for following analyses.

5.2.2.3 Analysis Methods

The corrosion rate of the alloy was evaluated by its sample weight variation before and after corrosion test as well as by the corrosion layer thickness of the sample. An electronic scale with an accuracy of 0.0001 g, was used to measure the sample weight. The surface morphology and element composition of oxide scale were analyzed by a scanning electron microscope (SEM, JSM–6390A) equipped with an energy dispersive spectrometer (EDS). A field emission scanning electron microscope (FE-SEM, Gemini500) equipped with an EDS was adopted to analyze the sample cross-section morphology and corresponding element distribution. The corrosion products of sample surface were determined by an X-Ray Diffractomer (XRD, D8 Advanced). A laser Raman spectrometer (LabRAM HR Evolution) equipped with a powerful laser (1.5 W of maximum power) was applied to identify the oxide composition in the oxide film. The chemical state and relative content of the P element were analyzed by an X-ray photoelectron spectroscopy (XPS, ESCALAB250Xi).

5.2.3 Results and Discussion

5.2.3.1 Corrosion Rate

Figure 5.10 illustrates the corrosion rates of tested alloys exposed to SCWs at different conditions. Apparently, the highest corrosion rates of 316 SS, 316L SS, Inconel 625, Incoloy 825 and Hastelloy C276 appeared at Con. 4, corresponding to around 0.0397, 0.0357, 0.0252, 0.0330 and 0.0231 mg cm^{-2} h^{-1}, respectively, in good agreement with our previous findings [41]. That is, the most severe corrosion in this study occurred in SCW containing phosphate, chloride, sulphate and oxygen simultaneously. Overall, the corrosion rate of each alloy presented an increase tendency from Con. 1 to Con. 4, implying that the corrosion rates of these alloys in SCW containing oxygen and phosphate are increased by adding sulphate, chloride and especially both them. The coexistence of sulphate and chloride resulted in much more severe corrosion in comparison to their individual presence, especially for 316 SS, 316L SS and Incoloy 825. This is likely due to the synergistic effect between sulphate and chloride on accelerating the alloy corrosion in SCW containing phosphate and oxygen. Apparently, oxygen played a crucial role in promoting the corrosions of all tested alloys according to the comparison between Cons. 4 and 5. This reminds us that it is significant to reduce the oxygen content in SCWO in the presence of

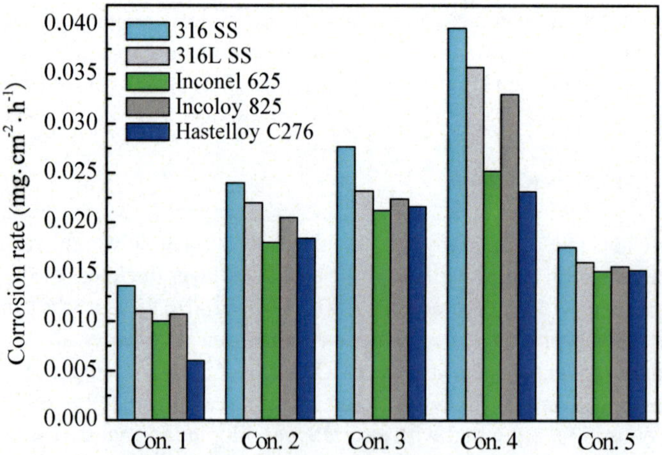

Fig. 5.10 Corrosion rates of tested alloys at different conditions

complex inorganic salts (e.g., chlorides, sulphates and phosphates) so as to lower the corrosion rate of the reactor.

5.2.3.2 Corrosion Morphology

Figure 5.11 elucidates the surface corrosion morphologies of alloy specimens exposed to SCW (at 450 °C and 25 MPa) for 70 h at various conditions. Apparently, each alloy showed the slightest corrosion morphology at Con. 1 (in the presence of phosphate and oxygen) among all Cons. The addition of sulphate or chloride accelerated alloy corrosion based on the comparisons of Fig. 5.11a–c, and their coexistence resulted in more severe corrosion than the single presence (see Fig. 5.11d). Similarly, it appears to have certain synergistic effect between chloride and sulphate on accelerating the alloy corrosion in SCW containing phosphate and oxygen. The more severe corrosion morphologies at Con. 4 than at Con. 5 also exhibits that oxygen plays a significant effect on enhancing the alloy corrosion in SCW containing chloride, sulphate and phosphate at the same time. Overall, the corrosion morphologies of the tested alloy specimens agree well with the corresponding corrosion rates reflected in Fig. 5.10.

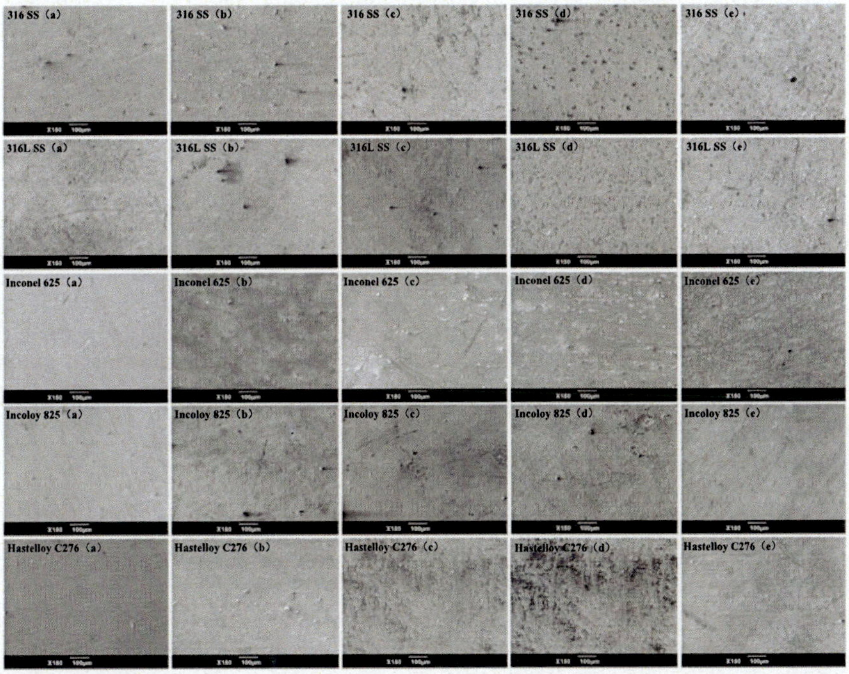

Fig. 5.11 SEM images (×150) of tested alloys under various corrosion conditions: **a** Con. 1, **b** Con. 2, **c** Con. 3, **d** Con. 4, and **e** Con. 5

It seems that Inconel 625 exhibited the best corrosion resistance at all test conditions (Cons. 1–5) since all of its oxide films were dense, uniform and smooth. 316 SS showed a severe spallation of oxide film after exposed to SCW containing sulfide, chloride and oxygen in our prior reports [9, 10], and the inconsistence should be due to corrosion condition differences. Hastelloy C276 is commonly regarded as a suitable alloy adoptable to SCWO, but it seemingly displayed relatively severe corrosion at Cons. 4 (see Fig. 5.11d). The presence of oxygen, phosphate, chloride and sulphate likely destroyed local oxide film by exfoliation or dissolution. This destroying easily makes local alloy substrate be exposed to harsh reaction medium and suffer from further corrosion, so lowering its corrosion resistance.

The most severe corrosion phenomena occurred at Con. 4 (see Fig. 5.11d), so high magnification (×1500) SEM images of tested alloy specimens at this condition are further displayed in Fig. 5.12. Apparently, numerous loose big particle corrosion products adhered to the substrates of 316 SS and 316L SS (see Fig. 5.12a, b). This is in accordance with the finding that a loose corrosion layer is formed when 316 SS is exposed to SCW at 450 °C, 25 MPa for 100 h with NaCl. Regular corrosion products uniformly distributed on the surface of Incoloy 825 (see Fig. 5.12d). Small amounts of corrosion products with much smaller particles evenly adhered to the surface of Inconel 625 (see Fig. 5.12c), suggesting its slightest corrosion and the best corrosion resistance among these alloys. Compared with the four alloys above, Hastelloy C276 presented an irregular and uneven corrosion morphology (see Fig. 5.12e), probably due to the exfoliation or dissolution of oxide film at Con. 4. Consistently, Hastelloy C276 surface displays a severe exfoliation of corrosion products after exposed to SCW (at 500–600 °C, 25 MPa) [48]. It can be concluded that uniform and dense corrosion product distribution helps to inhibit further corrosion by avoiding the contact of

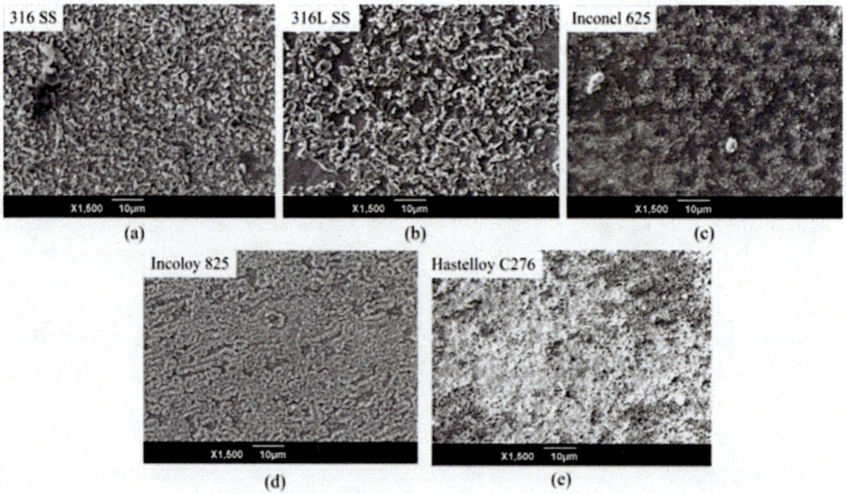

Fig. 5.12 SEM images (×1500) of corroded alloys in SCW at Con. 4: **a** 316 SS, **b** 316L SS, **c** Inconel 625, **d** Incoloy 825, and **e** Hastelloy C276

between alloy substrate and corrosive species. Certainly, oxide film will gradually thicken with corrosion products increasing and likely further take place corrosion layer exfoliation with exposure time prolonging.

5.2.3.3 Element Composition of Oxide Film

Figure 5.13 illustrates the EDS analysis results of oxide films of all tested alloys exposed to various conditions. Major elements (i.e., Fe, Ni, Cr, Mo, O and P) accounted for more than 99.5 wt% of all elements, and other elements in trace were ignored herein. On the whole, compared with the substrates of 316 SS and 316L SS, their oxide scales had indiscernible change in the Fe content, but remarkable decrease in Cr and Ni contents and obvious increase in the O content after SCW exposure at Cons. 1–4. The disappearance or below measurement limit of Mo (probably due to diffusion and dissolution) in the oxide films of 316 SS at Cons. 3 and 4 is likely related to the relatively severe corrosion of 316 SS in SCW containing phosphate, chloride and oxygen. After corrosion at Cons. 1–4, Ni-base alloys (Inconel 625 and Hastelloy C276) had a significant increase in the O content, a remarkable decrease in the Cr or Mo content, and an unobvious variation in the Ni element. Incoloy 825 exhibited a certain decrease in the Ni, Cr or Mo content, a obvious increase in the O content after corrosion exposures at Cons. 1–4. High content of O in each oxide scale should be caused by the oxygen diffusion inward alloy substrate at Cons. 1–4. The element composition of alloy substrate was similar to that of its test specimen after

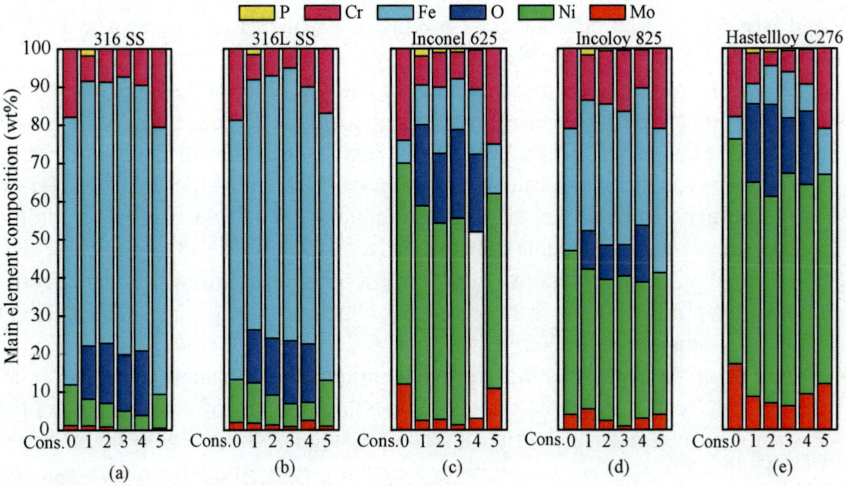

Fig. 5.13 Major element compositions of oxide films of alloys exposed in SCWs at different conditions (Con. 0 represents alloy substrate): **a** 316 SS, **b** 316L SS, **c** Inconel 625, **d** Incoloy 825, and **e** Hastelloy C276

corrosion at Con. 5, suggesting much slight corrosion occurrence in SCW containing phosphate, sulphate and chloride in absence of oxygen.

Notably, the P content in the oxide film of each alloy at Con. 1 was relatively high, showing the presence of water-insoluble (at ambient temperature) P-containing compounds (e.g., some phosphates) as corrosion products in the oxide film. The P content roughly reduced in Ni-base alloy (i.e., Inconel 625, Incoloy 825 or Hastelloy C276) from Con. 1 to Con. 4, suggesting that sulphate, chloride, and especially both them destroyed the formation of the water-insoluble P-containing compounds in O-containing SCW. One possible explanation is that some alloy elements diffuse outward to combine with O and Cl^- and SO_4^{2-} rather than PO_4^{3-} to form oxides and water-soluble chlorides and sulphates. The loss of these formed compounds is likely beneficial to oxygen diffusion inward at Con. 4, especially for 316 SS and Incoloy 825 (see Fig. 5.13a, d). Moreover, it can be observed that all oxide films did not contain Cl and S elements, implying that possibly relevant corrosion products dissolve in SCW or are water-soluble at ambient temperature. Overall, the content changes of Fe, Cr, Ni, Mo, O and P significantly determined the corrosion results of each alloy in this work.

5.2.3.4 Corrosion Product

To further clarify corrosion product compositions, Fig. 5.14 provides the XRD analysis results of corrosion products formed on 316 SS, Incoloy 825 and Hastelloy C276 samples after exposure to SCWs of various Cons. At Con. 1, the oxide films of these alloys included some identical products (i.e., Fe_2O_3, Cr_2O_3, NiO, $CrPO_4$ and $Ni_3(PO_4)_2$). The differences of product components rested on extra $FePO_4$ in the oxide film of 316 SS, $FePO_4$ and $NiCr_2O_4$ in the oxide film of Incoloy 825, and $NiCr_2O_4$ and MoO_2 in the oxide film of Hastelloy C276. Similarly, Cr_2O_3, NiO and $NiCr_2O_4$ are reported to be three major components in the oxide film of Incoloy 825 exposed to SCW at 520 °C and 25 MPa [42], and the formation of Cr_2O_3 and $NiCr_2O_4$ was observed on the oxide scale of Hastelloy C276 in oxidizing SCW [49]. Herein, it can be further confirmed that the P content in Fig. 5.14 existed as corrosion products in the form of water-insoluble phosphates (e.g., $CrPO_4$, $Ni_3(PO_4)_2$ and/or $FePO_4$). The second passivation of phosphates forms a phosphate film, which can effectively prevent the alloy substrate from contacting corrosive substances, so decreasing the alloy corrosion rate in SCWO [43].

As shown in Fig. 5.14, the corrosion products components and their relative contents of each alloy took place certain variations at different test conditions. Particularly, except for the alloy substrate peaks, there were almost no other peaks found in the XRD pattern at Con. 5 for each alloy, suggesting the much weak corrosion in SCW without oxygen in spite of containing phosphate, sulphate and chloride. This is in good agreement with the variation of the elemental composition in Fig. 5.13 at Con. 5. Overall, in combination with Figs. 5.13, 5.14, the formation and action of water-insoluble phosphates (e.g., $FePO_4$, $CrPO_4$ and $Ni_3(PO_4)_2$) on alloy

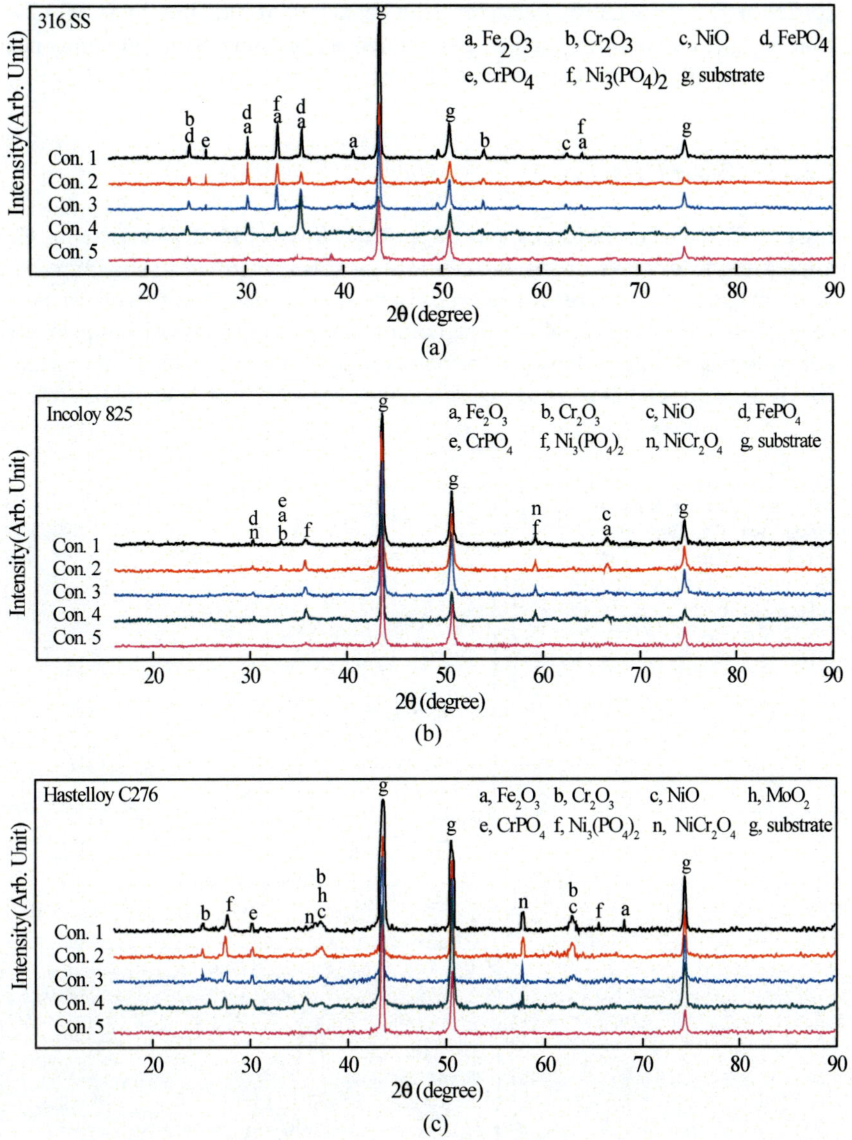

Fig. 5.14 XRD patterns of oxide scales of alloy specimens under different conditions: **a** 316 SS, **b** Incoloy 825, and **c** Hastelloy C276

surface may significantly affect corrosion behaviors of the alloy in SCW. Corrosion product compositions should mainly depend on alloy chemical components, corrosion resistance and corrosion conditions.

5.2.3.5 Oxide Phase

Figure 5.15 indicates the Raman spectra of oxide films on 316 SS and Hastelloy C276 after exposure to SCWs at various conditions to further determine oxide composition. As shown in Fig. 5.15a, the Raman spectrum of the oxides identified on ecah oxide scale of 316 SS at Cons.1–4 consisted of a major peak and several sub-peaks. The peaks corresponding to Fe_2O_3 were observed at 225 and 297 cm^{-1}, the peaks of Cr_2O_3 appeared at 304 and 633 cm^{-1}, and the peak at 480 cm^{-1} could be attributed

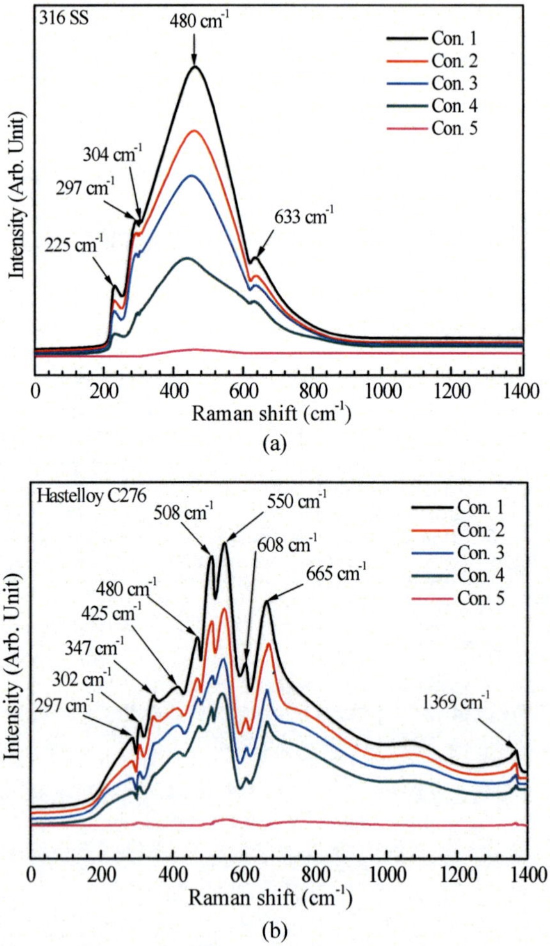

Fig. 5.15 Raman spectra of oxide scales of 316 SS and Hastelloy C276 after exposure to various conditions: **a** 316 SS and **b** Hastelloy C276

to NiO. Similarly, for Hastelloy C276, the Raman peaks corresponding to Fe_2O_3, Cr_2O_3, NiO, $NiCr_2O_4$ and MoO_2 were detected at Cons.1–4. The peaks at 297 and 480 cm^{-1} separately represented Fe_2O_3 and NiO, and the peaks at 302, 347 and 550 cm^{-1} were consistent with the characteristics of Cr_2O_3 [50]. The characteristic peaks of $NiCr_2O_4$ were at 425, 508 and 665 cm^{-1}, and the MoO_2 peakwas confirmed at 1369 cm^{-1} as well. No signal on the Raman spectra of oxide films of 316 SS and Inconel 625 at Con. 5 shows that the corrosion in SCW with high concentrations of phosphate, sulphate and chloride and without oxygen was negligible within 70 h of exposure time, in good agreement with the findings in Figs. 5.13 and 5.14.

On the whole, identified oxides are completely consistent with the findings by the XRD analysis in Fig. 5.14. In spite of common elemental distribution rule [51], chromium oxides can provide the alloy sample with a good corrosion resistance. Herein, the absence of $NiCr_2O_4$ in the oxide film of 316 SS in comparison to Hastelloy C276 (as well as Incoloy 825 in Fig. 5.14) might result from its high thermodynamics instability in SCW at Con. 4 [52]. The $NiCr_2O_4$ presence was one significant difference in the components of oxide scales, and may be one reason for the corrosion resistance differences between 316 SS and Ni-base alloy (e.g., Incoloy 825 and Hastelloy C276).

5.2.3.6 Phosphate Analysis

Phosphates were not only the significant corrosion products of affecting alloy corrosion behaviors but also the important fractions of potentially improving alloy corrosion resistance. Thus, Fig. 5.16 further displays the XPS and SEM analyses of oxide films of alloy specimens (i.e., 316 SS, Incoloy 825 and Hastelloy C276) at Con. 1 to determine the chemical state of P 2p. The P 2p characteristic peaks on the surfaces of 316 SS, Incoloy 825 and Hastelloy C276 were identified at the binding energies

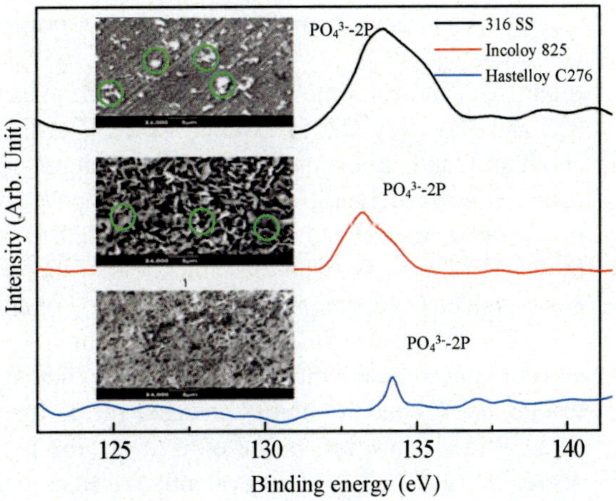

Fig. 5.16 XPS patterns of phosphates and SEM images (×4000) of 316 SS, Incoloy 825 and Hastelloy C276 samples after exposure to Con. 1

of 133.9, 133.3 and 134.0 eV, respectively. Depending on the peak area of the P 2p [40], the formed phosphate content was in the order: 316 SS>Incoloy 825>Hastelloy C276 here. This is consistent with the results of the EDS analysis in Fig. 5.13, and detailed phosphate components can be observed by the XRD analysis results (see Fig. 5.14). Hence, it can be confirmed that the Fe, Ni and Cr contents in the alloys above can react with PO_4^{3-} derived from feedstock to form some water-insoluble phosphate products (e.g., $FePO_4$, $CrPO_4$ and/or $Ni_3(PO_4)_2$). This is a significant factor contributing to alloy corrosion since there were only two types of corrosion products (i.e., metal oxides and phosphates) in the oxide films of alloys exposed in SCW containing oxygen and phosphate (see Fig. 5.14). However, the formation and uniform distribution of phosphates can isolate corrosive species and alloy substrate, so reducing the further corrosion of halogen ions (e.g., Cl^-) to the tested alloy [47].

As shown by the high magnification (×4000) of SEM images in Fig. 5.16, big particle phosphates (i.e., corrosion products) unevenly dispersed on the 316 SS surface, but much small and dense corrosion products adhered on the Hastelloy C276 surface evenly. This compact oxide layer "passivation" on the Hastelloy C276 surface should help to prevent the alloy substrate from further contacting corrosive substances such as O, Cl^- and SO_4^{2-} in this research. Overall, the corrosion product distribution property of Incoloy 825 was between those of 316 SS and Hastelloy C276. Note that the alloy corrosion resistance was in the order: Hastelloy C276>Incoloy 825>316 SS on the basis of the corrosion rate analysis (see Fig. 5.10) at Con. 1, and this is consistent with the judgment by the formed phosphate contents and their uniform distribution degrees (see Fig. 5.16). Certainly, apart from the formed phosphates, the chemical compositions and corrosion conditions of the tested alloy should crucially affect the corrosion properties of the alloy. It is worth noticing that phosphates (as corrosion products) were stably formed at Con. 1 (in SCW containing oxygen and phosphate), perhaps the condition can be adopted in the pretreatment of reactor material to form a protective phosphate film before SCWO in actual run.

5.2.3.7 Corrosion Layer Structure and Composition

Figure 5.17 indicates the cross-section SEM images of alloys (i.e., 316 SS, Incoloy 825 and Hastelloy C276) exposed to SCW at Con. 4 (the most severe corrosion condition), and corresponding EDS line scanning profiles. As shown in Fig. 5.17a, the oxide scale of 316 SS was distinctly composed of two layers, and the thicknesses of the outer and inner layers were identified to be separately 62.5 and 53.1 μm by the SEM analysis. The total thickness of the oxide scale was 115.6 μm, which corresponded to approximately 11.89 $mm \cdot a^{-1}$ of corrosion rate based on the average value in the exposure time (i.e., 115.6 μm for 70 h) for 300 days in one year. The Cr and O contents remarkably increased, the Ni content slightly reduced, and meanwhile the Fe and P contents hardly changed in the inner layer compared with the 316 SS substrate. However, in the outer layer, the Fe, Cr and O contents remarkably decreased, and the P content evidently raised, and the Ni content slightly increased. Note that the corrosion products of 316 SS were composed of Fe_2O_3, Cr_2O_3, NiO,

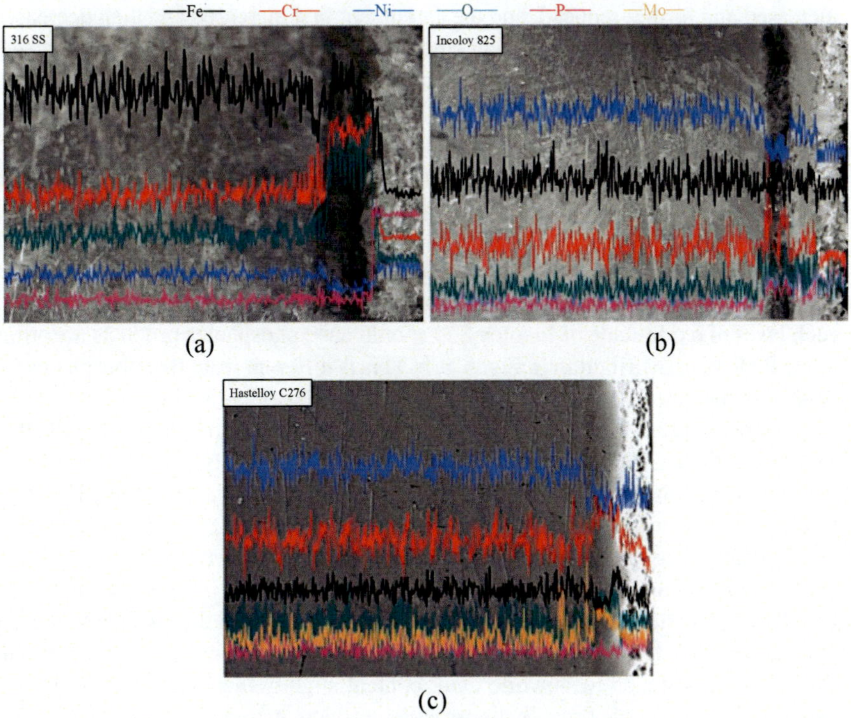

Fig. 5.17 SEM images (×150) and EDS line scanning profiles of cross-sections of alloy samples after exposure to Con. 4: **a** 316 SS, **b** Incoloy 825, and **c** Hastelloy C276

$FePO_4$ and $Ni_3(PO_4)_2$ (see Figs. 5.14a, 5.15a), so the oxide film of 316 SS should consist of an inner layer with the Cr-rich oxide (i.e., Cr_2O_3), and an outer layer with the Fe-, Cr- and Ni-rich phosphates (e.g., $FePO_4$) as well as oxides, based on the element content changes above. The Cr-rich oxide layer has a better corrosion resistance than the Fe/Ni oxide layer due to its lower dissolution and diffusion [51], implying that the corrosion rate of the alloy has a reduction tendency. In fact, the alloy element Fe exhibits high dissolving and diffusing rates in SCWO [52], so resulting in a great deal of Fe loss in the outer layer of oxide scale of 316 SS here. Similarly, a two-layer structure of oxide scale, mainly consisting of Fe_2O_3/Fe_3O_4 + spinel/Cr_2O_3/Ni-enrichment from outer to inner layer, is developed on 316 SS after exposure to SCW containing oxygen [53]. The composition differences should be attributed to corrosion condition variations, especially for the presence of Na_3PO_4 in the feedstock.

The oxide film growing on Incoloy 825 in SCW at Con. 4 displayed a three-layer structure (see Fig. 5.17b). The thicknesses of each layer from the outer to inner layer were identified to be 21.9, 43.8 and 31.3 μm, respectively. Similarly, the total 97.0 μm thickness corresponded to around 9.97 mm·a^{-1} of corrosion rate. Overall, from the alloy substrate to its outmost surface, the Ni content first decreased and then increased and further reduced, and the Cr content first increased and then decreased, and the P content raised, and the O content first increased and then was relatively stable, and the Fe content was always stable. The presence of the three-layer structure is likely attributed to the differences of the diffusion, oxidation and/or dissolution rates of Cr, Ni, Fe, O and PO_4^{3-}. For instance, the inner oxide layer might be formed by the diffusions and oxidations of Ni and Cr, so forming some oxides and phosphates by reacting with O and PO_4^{3-}. The previous XRD analysis (see Fig. 5.14) confirms that the products included Fe_2O_3, NiO, $FePO_4$, $Ni_3(PO_4)_2$ and $NiCr_2O_4$ in the oxide scale of Incoloy 825. Apparently, the content variations of these corrosion products determined the element content distributions in the whole corrosion layer. Notably, each layer of oxide scale of Incoloy 825 should have phosphates products according to the P element distribution in Fig. 5.17b. This is different from that phosphates just existed in the outer layer of the 316 SS oxide scale (see Fig. 5.17a).

As shown in Fig. 5.17c, it seems to only have one corrosion layer tightly adhering to the Hastelloy C276 substrate at Con. 4, but the EDS analysis clearly confirmed that the oxide film of Hastelloy C276 also exhibited a two-layer structure. The thicknesses of its inner and outer layers were around 28.5 and 30.9 μm, respectively. It is calculated to be approximately 6.11 mm·a^{-1} of corrosion rate based on the total 59.4 μm of thickness. Apparently, from the alloy substrate to its outmost surface, the Cr, Mo and O contents first increased and then decreased, and the Ni content initially reduced and then tended to be stable, and the Fe content declined and then raised and further decreased, and meanwhile the P content slightly increased. The XRD analysis (see Fig. 5.14) confirms that corrosion products included Fe_2O_3, Cr_2O_3, NiO, MoO_2, $FePO_4$, $CrPO_4$, $Ni_3(PO_4)_2$ and $NiCr_2O_4$. Obviously, based on the corrosion rates, corrosion layer structures and compositions, the order of corrosion resistances of the three tested alloys was Hastelloy C276 > Incoloy 825 > 316 SS. Moreover, the much high corrosion rates above may be explained by that the high average corrosion rate in the initial 70 h was used to evaluate the corrosion rate of one year (unit: mm a^{-1}). Hence, the variation law of the corrosion rate of each alloy with exposure time increasing should be further explored in subsequent work.

5.2.3.8 Formation Mechanism of Oxide Film

Generally, element and ion diffusion and oxide film formation mainly consists of potential corrosion mechanisms of the tested alloys exposed to SCW containing high concentrations of phosphate, sulphate, chloride and oxygen in this work. According to the EDS and XRD analyses, there were majorly Fe, Ni, Cr, O and P in the corrosion layers of the tested alloys (i.e., 316 SS, Incoloy 825 and Hastelloy C276). Based on the Cr distributions in the oxide films of 316 SS, Incoloy 825 and Hastelloy C276 (see Fig. 5.17), Cr should diffuse outward by formingCr_2O_3, $NiCr_2O_4$ and $CrPO_4$ (on Incoloy 825 and Hastelloy C276). The Cr diffusion properties in Incoloy 825 and Hastelloy C276 were similar, but different from that in 316 SS. The opposite variation trends of the Cr and Ni contents in the inner layer of oxide film of each alloy (see Fig. 5.17) suggest the outward diffusing rate of Cr was higher than that of

Ni. Cr_2O_3 should be the single Cr-containing product in the inner layer of 316 SS oxide film, and the major fractions in the inner layer of the Hastelloy C276 oxide film should be Cr oxides based on the P, O and Cr contents changes (see Figs. 5.14, 5.17). The significant decrease of the Cr content in the outer oxide layer of each alloy (see Fig. 5.8) can be attributed to the high diffusing rate of Cr^{3+}, because its oxides are easy to be converted into Cr^{6+} to further dissolve in SCWO [54].

The alloy element Ni diffused outward (see Fig. 5.17) to react with oxygen and PO_4^{3-} to produce either NiO, $Ni_3(PO_4)_2$ and $NiCr_2O_4$ on Incoloy 825 and Hastelloy C276 or NiO and $Ni_3(PO_4)_2$ on 316 SS, as shown in Fig. 5.14. The Ni content reduction in the oxide film of Incoloy 825 or Hasterlloy C276 (see Fig. 5.17b, c) may be due to its high diffusion and dissolution rates, and thus some formed Ni-containing oxides are unstable and easy to dissolve in SCW. The diffusion rate of Ni is higher than that of Cr, so the corrosion layer mainly consists of Ni-rich oxides in SCW [28]. The formed NiO likely further reacts with Cr_2O_3 through oxidation reaction to form Ni–Cr spinel ($NiCr_2O_4$) on the Ni-base alloys (Incoloy 825 and Hastelloy C276). This can be confirmed by the much weak NiO signal intensity for Incoloy 825 or Hastelloy C276 at Con. 4 (see Fig. 5.14b, c). The results about the Cr oxides above have good agreement with the findings that the oxide scale of Hastelloy C276 has a two-layer structure, mainly consisting of NiO in the outer layer and Cr_2O_3 and $NiCr_2O_4$ in the inner layer.

As indicated in Fig. 5.17c, Cr and Mo were rich in the inner corrosion layer, and their contents in the outer layer had no obvious differences from those of the Hastelloy C276 substrate at Con. 4. This is in agreement with corresponding corrosion products including MoO_2, $CrPO_4$, Cr_2O_3 and $NiCr_2O_4$ (see Figs. 5.14c, 5.15b). Notably, the synergistic actions of Cr and Mo are confirmed to improve the corrosion resistance of the Ni-base alloy in SCWO [40]. Apart from that, the high Mo content in the Hastelloy C276 substrate and relatively stable existence in the oxide scale may also be significant reasons for the better corrosion resistance of the alloy. It is reported that Fe has a high diffusing rate and leads to the formations of Fe_2O_3 and Fe_3O_4 on the specimen surface of austenitic stainless steel exposed to SCW (especially higher than 400 °C) [55]. Herein, Fe_3O_4 was not found in the corrosion products of 316 SS, Incoloy 825 and Hastelloy C276 at Con. 4, likely due to excessive oxygen presence.

In this research, the P element in the oxide film of the tested alloy existed in the form of phosphates, in which PO_4^{3-} was derived from the initial test solution. Apparently, these formed phosphates were stable water-insoluble corrosion products including $FePO_4$ and $Ni_3(PO_4)_2$ on 316 SS and Incoloy 825, $CrPO_4$ and $Ni_3(PO_4)_2$ on Hastelloy C276 (see Fig. 5.14). Notably, as shown in Fig. 5.17, PO_4^{3-} displayed a good diffusion performance toward the substrate of Incoloy 825 but a weak diffusion ability on Hastelloy C276 in SCW at Con. 4. Overall, the diffusing rate ofPO_4^{3-} toward alloy substrate was far lower than that of O in the three alloys above in SCW at Con. 4.

On the whole, in SCW containing high concentrations of phosphate, sulphate, chloride and oxygen, Fe, Ni and Cr diffused outward and O and P (in the form of PO_4^{3-}) diffused inward continuously, and reacted with each other to produce metal oxides and phosphates, such as Fe_2O_3, Cr_2O_3, NiO, $CrPO_4$ and $Ni_3(PO_4)_2$

(see Figs. 5.14, 5.15). A two-layer structure of oxide scale on 316 SS was formed, and consisted of an inner layer with the Cr-rich oxide (i.e., Cr_2O_3) and an outer layer with rich phosphates (e.g., $FePO_4$ and $Ni_3(PO_4)_2$). A three-layer oxide scale structure appeared on Incoloy 825 and each layer had phosphate products such as $FePO_4$ and $Ni_3(PO_4)_2$. A two-layer structure of oxide scale was formed on Hastelloy C276, and its inner layer was composed of metal oxides while the outer layer contained metal oxides as well as phosphates. The formed oxides and phosphates in the outer layer of oxide scale of each alloy probably react with corrosion species (e.g., Cl^- and SO_4^{2-}), and thus further dissolve and diffuse in water.

5.2.4 Conclusion

The corrosion rates of 316 SS, 316L SS, Inconel 625, Incoloy 825 and Hastelloy C276 in SCW containing oxygen and phosphate were accelerated by adding sulphate, chloride, and especially both them. The Fe, Ni and Cr contents in 316 SS, Incoloy 825 or Hastelloy C27 diffused outward and reacted with PO_4^{3-} derived from the feedstock to form some water-insoluble phosphate products (e.g., $FePO_4$, $CrPO_4$ and $Ni_3(PO_4)_2$) existing in the oxide film. These phosphates seemingly helped to inhibit further alloy corrosion by avoiding the contact between corrosive species (e.g., Cl^-, SO_4^{2-} and O) and the alloy substrate. At the coexistence condition of high concentrations of phosphate, sulphate, chloride and oxygen in SCW, the corrosion rates of 316 SS, Incoloy 825 and Hastelloy C276 were approximately 11.89, 9.97 or 6.11 $mm \cdot a^{-1}$, respectively. A two-layer structure of oxide scale on 316 SS consisted of an inner layer with the Cr-rich oxide (i.e., Cr_2O_3) and an outer layer rich in phosphates (e.g., $FePO_4$ and $Ni_3(PO_4)_2$). A three-layer oxide scale structure was formed on Incoloy 825, and corresponding corrosion products included Fe_2O_3, NiO, $FePO_4$, $Ni_3(PO_4)_2$ and $NiCr_2O_4$. A two-layer structure of oxide scale was formed on Hastelloy C276, with metal oxides in the inner layer and phosphates in the outer layer. Corrosion products were Fe_2O_3, Cr_2O_3, NiO, MoO_2, $FePO_4$, $CrPO_4$, $Ni_3(PO_4)_2$ and $NiCr_2O_4$.

References

1. K.A. Rozman, G.R. Holcomb, C.S. Carney, N. Doan, J.A. Hawk, Effect of 730 °C supercritical fluid exposure on the fatigue threshold of Ni-based superalloy haynes 282. J. Mater. Eng. Perform. **28**, 4335–4347 (2019)
2. M. Grudzien-Rakocz, L. Rakoczy, R. Cygan, F. Kromka, O. Milkovi, Fabrication and characterization of the newly developed superalloys based on inconel 740. Materials (Basel) **13** (2020)
3. A. Iwamaru, H. Hisazawa, Y. Terada, Microstructure evolution of Fe–Ni-based alloy HR6W during isothermal aging. Mater. Trans. **60**, 824–829 (2019)

4. A. Zieliński, J. Dobrzański, H. Purzyńska, R. Sikora, M. Dziba-Kaluza, Z. Kania, Evaluation of creep strength of heterogeneous welded joint in hr6w alloy and sanicro 25 steel. Arch. Metall. Mater. **62**, 2057–2064 (2017)
5. J. He, R. Sandström, S. Notargiacomo, Low-cycle fatigue properties of a nickel-based superalloy Haynes 282 for heavy components. J. Mater. Eng. Perform. **26**, 2257–2263 (2017)
6. M. Wang, C. Sun, M. W. Fu, Z. Liu, C. Wang, Experimental investigations and constitutive modeling of the dynamic recrystallization behavior of Inconel 740 superalloy. Mater. Sci. Eng. A **793** (2020)
7. M.J. Wang, C.Y. Sun, M.W. Fu, Z.L. Liu, L.Y. Qian, Study on the dynamic recrystallization mechanisms of Inconel 740 superalloy during hot deformation. J. Alloys Compd. **820** (2020)
8. Z. Liang, T. Guo, S. Deng, Q. Zhao, High-temperature corrosion of an Fe–Ni-based alloy HR6W under various conditions at 750 °C and 810 °C: Effect of the temperature, water vapor, simulated ash and SO_2. Mater. Chem. Phys. **256** (2020)
9. R.N. Nnaji, M.A. Bodude, L.O. Osoba, O.S.I. Fayomi, F.E. Ochulor, Study on high-temperature oxidation kinetics of Haynes 282 and Inconel 718 nickel-based superalloys. Int. J. Adv. Manuf. Tech. **106**, 1149–1160 (2019)
10. Z. Liang, Q. Zhao, High temperature oxidation of Fe–Ni-base alloy HR120 and Ni-base alloy HAYNES 282 in steam. Mater. High Temp. **36**, 87–96 (2018)
11. F.A. Pérez-González, N.F. Garza-Montes-de Oca, R. Colás, High temperature oxidation of the haynes 282© nickel-based superalloy. Oxid. Met. **82**, 145–161 (2014)
12. M. Abbasi, B.K. Kim, D.I. Kim, W.S. Jung, Effects of surface deformation on the oxidation behavior of inconel 740 superalloy in humid air. J. Alloys Compd. **683**, 212–220 (2016)
13. M. Abbasi, D.I. Kim, J.H. Shim, W.S. Jung, Effects of alloyed aluminum and titanium on the oxidation behavior of INCONEL 740 superalloy. J. Alloys Compd. **658**, 210–221 (2016)
14. S. Zhao, X. Xie, G.D. Smith, The oxidation behavior of the new nickel-based superalloy Inconel 740 with and without Na2SO4 deposit. Surf. Coat. Technol. **185**, 178–183 (2004)
15. S. Zhao, X. Xie, G.D. Smith, S.J. Patel, The corrosion of INCONEL alloy 740 in simulated environments for pulverized coal-fired boiler. Mater. Chem. Phys. **90**, 275–281 (2005)
16. S. Zhao, X. Xie, G.D. Smith, S.J. Patel, Research and improvement on structure stability and corrosion resistance of nickel-base superalloy INCONEL alloy 740. Mater. Des. **27**, 1120–1127 (2006)
17. S. Guo, D. Xu, Y. Liang, Y. Gong, J. Yang, Corrosion characterization of ZrO_2 and TiO_2 ceramic coatings via air plasma spraying on 316 stainless steel in oxygenated sub- and supercritical water. J. Supercrit. Fluids **157** (2020)
18. M. Fulger, M. Mihalache, D. Ohai, S. Fulger, S.C. Valeca, Analyses of oxide films grown on AISI 304L stainless steel and Incoloy 800HT exposed to supercritical water environment. J. Nucl. Mater. **415**, 147–157 (2011)
19. Y. Li, S. Wang, J. Yang, D. Xu, Y. Guo, L. Qian, W. Song, Corrosion characteristics of a nickel-base alloy C-276 in harsh environments. Int. J. Hydrogen Energy **42**, 19829–19835 (2017)
20. S.H. Nie, Y. Chen, X. Ren, K. Sridharan, T.R. Allen, Corrosion of alumina-forming austenitic steel Fe–20Ni–14Cr–3Al–0.6Nb–0.1Ti in supercritical water. J. Nucl. Mater. **399**, 231–235 (2010)
21. H. Zhao, T. Liu, Z. Bai, L. Wang, W. Gao, L. Zhang, Corrosion behavior of 14Cr ODS steel in supercritical water: The influence of substituting Y_2O_3 with $Y_2Ti_2O_7$ nanoparticles. Corros. Sci. **163** (2020)
22. M. Payet, L. Marchetti, M. Tabarant, J. Chevalier, Corrosion mechanism of a Ni-based alloy in supercritical water: Impact of surface plastic deformation. Corros. Sci. **100**, 47–56 (2015)
23. H. Kim, D.B. Mitton, R.M. Latanision, Corrosion behavior of Ni-base alloys in aqueous HCl solution of pH 2 at high temperature and pressure. Corros. Sci. **52**, 801–809 (2010)
24. H. Kim, D.B. Mitton, R.M. Latanision, Effect of pH and temperature on corrosion of nickel-base alloys in high temperature and pressure aqueous solutions. J. Electrochem. Soc. **157**, C194–C199 (2010)

25. J. Yang, S. Wang, D. Xu, High temperature oxidation of alloy 617, 740, HR6W and Sanicro 25 in supercritical water at 650 °C. Corros. Eng. Sci. Technol. **55**, 196–204 (2020)
26. J. Yang, S. Wang, Y. Li, D. Xu, Under-deposit corrosion of Ni-based alloy 825 and Fe–Ni based alloy 800 in supercritical water oxidation environment. Corros. Sci. **167** (2020)
27. Y. Behnamian, A. Mostafaei, A. Kohandehghan, B.S. Amirkhiz, L.L. Jing, A comparative study of oxide scales grown on stainless steel and nickel-based superalloys in ultra-high temperature supercritical water at 800 °C. Corros. Sci. **106**, 188–207 (2016)
28. D.D.M. Elzbieta Sikora, Defining the passive state. Solid State Ionics **94**, 141–150 (1997)
29. D.D. Macdonald, The passive state in our reactive metals-based civilization. Arab. J. Sci. Eng. **37**, 1143–1185 (2012)
30. M. Bojinov, S. Cattarin, M. Musiani, B. Tribollet, Evidence of coupling between film growth and metal dissolution in passivation processes. Electrochim. Acta **48**, 4107–4117 (2003)
31. D.D. Macdonald, A. Sun, An electrochemical impedance spectroscopic study of the passive state on Alloy-22. Electrochim. Acta **51**, 1767–1779 (2006)
32. M. Bojinov GF, T. Laitinen, K. Mäkelä, T. Saario, G. Sundholm, Coupling between ionic defect structure and electronic conduction in passive films on iron, chromium and iron–chromium alloys. Electrochim. Acta **45**, 2029–2048 (2000)
33. M.B. Iva Betova, L. Timo, M. Karia, P. Pekka, S. Timo, The transpassive dissolution mechanism of highly alloyed stainless steels I. Experimental results and modelling procedure. Corros. Sci. **44**, 2675–2697 (2002)
34. I. Betova MB, P. Kinnunen, T. Laitinen 1, P. Pohjanne, T. Saario, Mechanism of transpassive dissolution of nickel-based alloys studied by impedance spectroscopy and rotating ring-disc voltammetry. Electrochim. Acta **47**, 2093–2107 (2002)
35. I. Betova, M. Bojinov, T. Tzvetkoff, Oxidative dissolution and anion-assisted solubilisation in the transpassive state of nickel–chromium alloys. Electrochim. Acta **49**, 2295–2306 (2004)
36. P. Kritzer, E. Dinjus, An assessment of supercritical water oxidation (SCWO) existing problems, possible solutions and new reactor concepts. Chem. Eng. J. **83**, 207–214 (2001)
37. E. Asselin, A. Alfantazi, S. Rogak, Corrosion of nickel–chromium alloys, stainless steel and niobium at supercritical water oxidation conditions. Corros. Sci. **52**(1), 118–124 (2010)
38. X. Tang, S. Wang, L. Qian, M. Ren, P. Sun, Y. Li, J. Yang, Corrosion properties of candidate materials in supercritical water oxidation process. J. Adv. Oxid. Technol. **19**, 141–157 (2016)
39. P.A. Marrone, G.T. Hong, Corrosion control methods in supercritical water oxidation and gasification processes. J. Supercrit. Fluid **51**, 83–103 (2009)
40. X. Tang, S. Wang, D. Xu, Y. Gong, J. Zhang, Y. Wang, Corrosion behavior of Ni-based alloys in supercritical water containing high concentrations of salt and oxygen. Ind. Eng. Chem. Res. **52**, 18241–18250 (2013)
41. X. Tang, S. Wang, L. Qian, Y. Li, Z. Lin, D. Xu, Y. Zhang, Corrosion behavior of nickel base alloys, stainless steel and titanium alloy in supercritical water containing chloride, phosphate and oxygen. Chem. Eng. Res. Des. **100**, 530–541 (2015)
42. Y. Li, S. Wang, X. Tang, D. Xu, Y. Guo, J. Zhang, L. Qian, Effects of sulfides on the corrosion behavior of Inconel 600 and Incoloy 825 in supercritical water. Oxid. Met. **84**, 509–526 (2015)
43. P. Kritzer, N. Boukis, E. Dinjus, The corrosion of alloy 625 (NiCr22Mo9Nb; 2.4856) in high-temperature, high-pressure aqueous solutions of phosphoric acid and oxygen. Corrosion at sub- and supercritical temperatures. Mater. Corros. **49**, 831–839 (1998)
44. M. Wagner, V. Kolarik, B. Michelfelder, M.d.M. Juez-Lorenzo, T. Hirth, N.E.u.P. Eyerer, Materials for SCWO processes for the oxidation of chlorine containing residues in supercritical water. Mater. Corros. **50**, 523–526 (1999)
45. L. Tan, X. Ren, K. Sridharan, T.R. Allen, Effect of shot-peening on the oxidation of alloy 800H exposed to supercritical water and cyclic oxidation. Corros. Sci. **50**, 2040–2046 (2008)
46. M. Sun, X. Wu, Z. Zhang, E. H. Han, Analyses of oxide films grown on Alloy 625 in oxidizing supercritical water.J. Supercrit. Fluids **47**, 309–317 (2008)
47. P. Kritzer, Corrosion in high-temperature and supercritical water and aqueous solutions: A review. J. Supercrit. Fluids **29**, 1–29 (2004)

48. Q. Zhang, R. Tang, K. Yin, X. Luo, L. Zhang, Corrosion behavior of Hastelloy C-276 in supercritical water. Corros. Sci. **51**, 2092–2097 (2009)
49. D.B. Mitton, N. Eliaz, J.A. Cline, R.M. Latanision, An overview of the current understanding of corrosion in SCWO systems for the destruction of hazardous waste products. Mater. Technol. **16**, 30–53 (2001)
50. J.E. Maslar, W.S. Hurst, J.H. Hendricks, M.I. Aquino, In situ Raman spectroscopic investigation of nickel hydrothermal corrosion. Corrosion **58**, 225–231 (2002)
51. Y. Daigo, Y. Watanabe, K. Sugahara, T. Isobe, Compatibility of Nickel-based alloys with supercritical water applications: Aging effects on corrosion resistance and mechanical properties. NACE Int. **62**, 174–181 (2006)
52. X. Gao, X. Wu, Z. Zhang, H. Guan, E.-H. Han, Characterization of oxide films grown on 316L stainless steel exposed to H2O2-containing supercritical water. J. Supercrit. Fluids **42**, 157–163 (2007)
53. M. Sun, X. Wu, Z. Zhang, E.-H. Han, Oxidation of 316 stainless steel in supercritical water. Corros. Sci. **51**, 1069–1072 (2009)
54. S.H. Son, J.H. Lee, Corrosion phenomena of alloys by subcritical and supercritical water oxidation of 2-chlorophenol. J. Supercrit. Fluids **44**(3), 370–378 (2008)
55. B. Stellwag, The mechanism of oxide film formation on austenitic stainless steels in high temperature water. Corros. Sci. **40**, 337–370 (1998)

Chapter 6
Corrosion Control Methods in Supercritical Water Oxidation Processes

6.1 Corrosion of Ceramic Coatings on 316 SS in Oxygenated Sub- and Supercritical Water

6.1.1 Introduction

Supercritical water (SCW) possesses some unique properties (e.g., benign transportation and solubility properties) and thus can provide an opportunity to carry out chemical reactions in a single phase. Supercritical water oxidation (SCWO) is a promising technology of harmlessly treating organic pollutants (e.g., organic wastewater and sewage sludge). It has many advantages such as a high destruction efficiency (>99.9%), a high reaction rate, no NO_x formation, and good reaction confining [1]. However, high temperature, high pressure and high concentrations of oxygen and ions cause severe reactor corrosion in a SCWO system. Therefore, corrosion control is of great significance for the scale-up and commercialization of SCWO.

In recent years, surface modification techniques (e.g., coating) have developed rapidly and drawn much attention to corrosion control [2]. Coating can avoid corrosive attack by a thin and protective layer tightly bonding onto metal substrate, and is versatile and applicable to SCWO facilities [3]. Till now, corrosion behavior of many types of coatings on different metal substrates have been evaluated in SCW extensively, such as ceramic-coated ZIRLO™ via cathodic arc physical vapor deposition [4], NiCrAlY-coated Alloy 214 [5], NiCrAlY-coated Inconel 625 by plasma spraying [6], FeCrAlY- and NiCrAl-coated model alloy via plasma spraying [7], TiN-based-coated Zircaloy-4 by pulsed laser deposition [8], Al slurry-coated ferritic 9% Cr T92 steel and Ni-based alloy A263 [9], Cr-coated ferritic-martensitic steel CNS-I (9Cr) and modified CNS-II (12Cr) by magnetron sputtering [10], FeAl-slurry-coated ferritic-martensitic steel (P91) [11], Al_2O_3-and TiO_2-coated 316L SS via atmospheric plasma spraying [12]. The plasma spraying process shows its versatility and cost-efficiency in the coating field, and plasma spraying ceramic coating

D. Xu and S. Guo, *Corrosion Characteristics, Mechanisms and Control Methods of Candidate Alloys in Sub- and Supercritical Water*,
https://doi.org/10.1007/978-981-16-5525-8_6

Table 6.1 Elemental compositions (wt%) of 316 SS substrate

Elements	Fe	Cr	Ni	Mo	Cu	Si	Ti	C	Others
316 SS	69.21	16.87	9.69	1.13	1.05	0.35	0.12	0.08	1.50

has great potential in corrosion prevention and control [13]. Ceramic materials (e.g., zirconia, titania, and diamond) have been proposed as candidate coatings for SCWO reactors [3]. Hui et al. [14] confirmed that Al_2O_3 ceramic coating on the P91 substrate could effectively inhibit the oxidation of P91 in SCW at 500 °C and 25 MPa. As an important candidate material of SCWO preheaters and reactors, 316 SS has good corrosion resistance, mechanical properties and a relatively low cost. Nonetheless, to the best of our knowledge, there is now no research on the protective effect of ceramic coating on the 316 SS substrate in oxygenated sub- and supercritical water.

Therefore, in this work, ZrO_2 and TiO_2 were coated on 316 SS substrates via an air plasma spraying process. Then, corrosion experiments were carried out in oxygenated high temperature water (300–480 °C) at 24 MPa with 3 wt% of oxygen content, based on commonly operating conditions in SCWO. The effects of ceramic coatings (i.e., ZrO_2 and TiO_2) and temperature on corrosion behavior (e.g., corrosion weight changes, morphologies, and chemical compositions and structures of oxide films) of coated samples were explored and compared systematically. This information is documented here for the first time, and helps to get insight into 316 SS corrosion control in SCWO by ZrO_2 and TiO_2 coatings via air plasma spraying.

6.1.2 Experimental Section

6.1.2.1 Materials

Alloy 316 SS, supplied by Shenzhen Metal Alloy Material Co., Ltd., was used as substrate material. Each sample has a size of 20 × 7 × 3 mm and a small hole in its upper part. As shown in Table 6.1, the chemical compositions of 316 SS included Fe, Cr, Ni, Mo, Cu, Si, Ti, C and others (i.e., P, Mn, S and Co). Oxidant was 30 wt% of hydrogen peroxide solution, which was purchased from Tianjin Tianli Chemical Agent Co., Ltd. Deionized water was prepared by a water purification plant (ULUP-II) in lab and used throughout. Two types of commercial ZrO_2 and TiO_2 powders were selected as coating materials, and their physical properties can be seen from Table 6.2.

6.1.2.2 Air Plasma Spraying Processes

Prior to air plasma spraying, each 316 SS sample surface was polished by SiC sand-papers with particle sizes of $400^{\#}$, $600^{\#}$, $1200^{\#}$, $2000^{\#}$ and diamond polishing agent

Table 6.2 Physical properties of two types of coating powders

Physical properties	Melting point (°C)	Rockwell hardness	Density (g/cm^3)	Particle size (μm)
ZrO_2	2535	25	5.3	10–53
TiO_2	1920	53	4.1	45–106

Table 6.3 Major air plasma spraying conditions for ZrO_2- and TiO_2-coated 316 SS samples

Parameters	Powder feeding rate (g/min)	Spray distance (mm)	Voltage (V)	Current (A)	Plasma gas (m^3/h)	
					Ar	H_2
ZrO_2-coated sample	45	100	70	500	2.72	0.57
TiO_2-coated sample	53					

in turn. Then, the sample was cleaned in acetone and in deionized water to remove possible contaminants from the sample surface. Ceramic powers (i.e., ZrO_2 and TiO_2) were sprayed onto each dried 316 SS sample via an air plasma spraying plant. The nozzle and electrode of the spray gun were connected to positive and negative anodes of a power source, respectively. A mixed gas of argon and hydrogen (with a flow ratio of 24:5) was used as the plasma gas, which was introduced to the gap between the nozzle and the electrode. A high frequency spark was adopted to ignite a plasma arc, which heated and ionized the plasma gas and caused its thermal expansion. A high velocity plasma jet was then ejected from the nozzle. Further, the ZrO_2 or TiO_2 powder was fed from the nozzle into the plasma jet, heated to the molten state, accelerated by the plasma jet, and then sprayed onto the sample surface to form the desired coating finally. Detailed air plasma spraying conditions are listed in Table 6.3. After the spraying processes were completed, the samples were placed in a vacuum environment at 100 °C for 2 h.

6.1.2.3 Experimental Procedures

A batch autoclave (409 mL), made of Hastelloy C-276, was used as the corrosion experimental device under sub- and supercritical water conditions. Before experiments, as-received 316 SS, ZrO_2- and TiO_2-coated samples were suspended in the reactor center with silver wire. Deionized water and hydrogen peroxide solution were added into the reactor. Their individual amounts were determined according to desired test temperature and pressure, and the detailed method had been described in our previous report [15]. The reactor was heated with 3.5 °C/min of heating rate by a temperature controller, and its temperatures were set at separately (300, 350, 420, 450 and 480) °C. The reactor was constantly maintained at 24 MPa and 3 wt% of oxygen content, according to common SCWO conditions. After 100 h of exposure time, the heater was turned off to naturally cool the reactor to ambient temperature.

Subsequently, tested samples were taken out and rinsed in deionized water. They were further dried, weighed and kept in sealed bags for following analyses. Notably, the experimental uncertainties of weight, temperature and pressure were 0.0038%, 0.42% and 0.69%, respectively.

6.1.2.4 Analytical Methods

Sample weight was measured by an analytical balance (Mettler-toledo XA105) with an accuracy of 0.1 mg. Corrosion weight change is defined as the weight change before and after corrosion per unit surface area. Corrosion morphologies of sample surfaces were analyzed by a scanning electron microscope (SEM, JEOL JSM-6390) equipped with an energy dispersive spectrometer (EDS). Phase compositions of formed surface oxides after SCW exposure were identified via X-ray diffraction (XRD, Bruker D8 ADVANCE) with an X-ray line focus of 0.04 × 12 mm, an X-ray spot focal of 0.4 × 1.2 mm, a scanning range of 10°–100°, and a minimum step size of 0.0001°. X-ray photoelectron spectroscopy (XPS) was performed by an X-ray photoelectron spectrometer microprobe (ESCALAB™ XI+) equipped with an EX06 ion source and (200–900) μm of X-ray spot sizes. Electron back-scatter diffraction (EBSD) was employed to determine phase distributions via a tungsten filament scanning electron microscope (HITACHI, SU3500) with (0.3–30) kV of acceleration voltage, incorporated with the 3rd generation image acquisition and analysis processor (MICs F+). In order to obtain high quality of EBSD images, each sample was further electropolished to minimize surface distortion before testing. More specially, an ion beam milling device (LEICA EM RES102) that has two saddlefield ion sources with variable ion energy was used to electropolish herein.

6.1.3 Results and Discussions

6.1.3.1 Morphology and Element Analyses

Figures 6.1 and 6.2 show the SEM images of the cross-sections of ZrO_2- and TiO_2-coated samples before oxygenated SCW exposure, respectively. The average coating thickness was around 185 μm (see Figs. 6.1a and 6.2a). The coating combination with the 316 SS substrate exhibited a serration morphology, which is beneficial to the integrity of coating and metal substrate and to the reduction of coating exfoliation tendency. However, some small holes were found inside the ZrO_2 coating (see Fig. 6.1a). A possible explanation is that spray material heterogeneously melts and thus causes molten powder particles to aggregate at the nozzle outlet, and thus deposits in a large spherical shape in the coating. There were also some small impurity particles inside the TiO_2 coating (see Fig. 6.2b), which may be attributed to fine particles doped during the spraying process. Notably, air plasma spraying is a nonequilibrium process, during which raw powders undergo fast melting and cooling

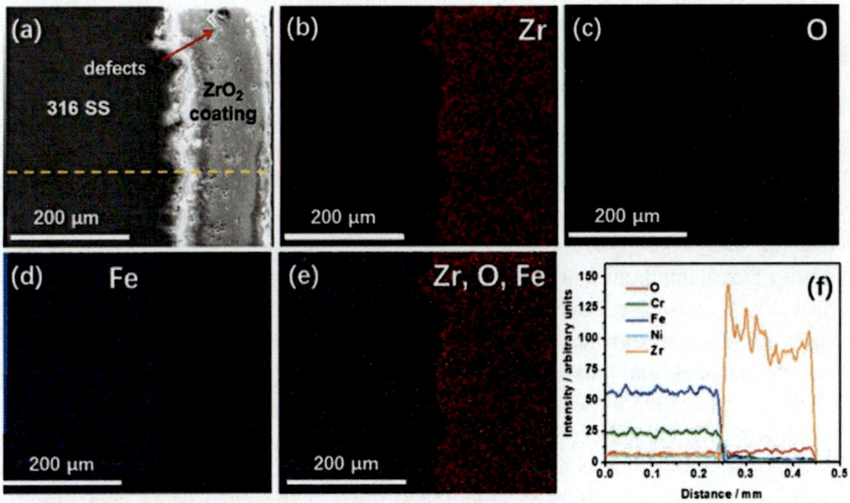

Fig. 6.1 Polished cross-section morphologies of **a** the ZrO_2-coated sample, **b–e** the distributions of Zr, O and Fe from (**a**), **f** EDS analysis along the dotted line shown in (**a**)

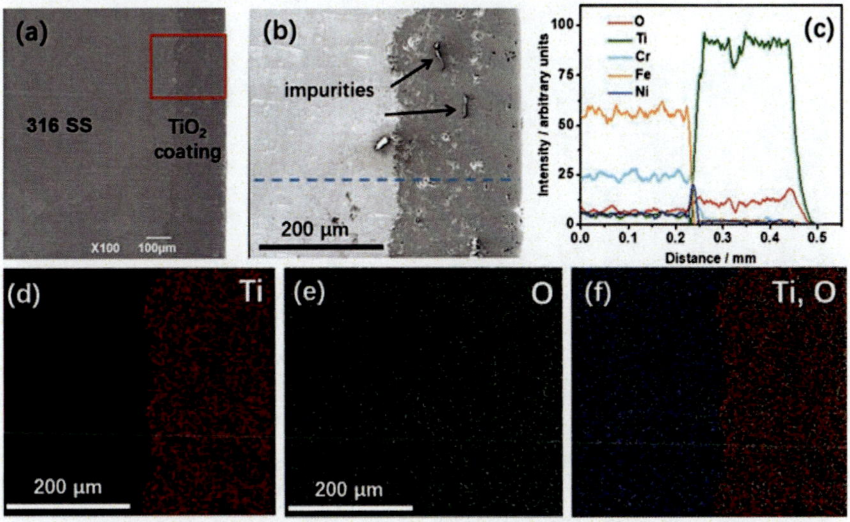

Fig. 6.2 Polished cross-section morphologies of **a** the TiO_2-coated sample and **b** the high magnification region marked with a red rectangle in (**a**), **c** EDS analysis along the dotted line shown in (**b**), **d–f** the distributions of Ti and O from (**b**)

processes [16]. Therefore, more attention should be paid to the process optimization of air plasma spraying to avoid the aforementioned defects.

Apparently, element distribution images of cross-sections display that the main elements in the ZrO_2 coating were O and Zr, compared with O and Ti in the TiO_2 coating (see Figs. 6.1b–e, and 6.2d–f). The intensity of Zr indicated a decrease trend while that of Ti was relatively stable in the individual coating from the 316 SS substrate to the sample outmost surface (see Figs. 6.1f and 6.2c), suggesting that the TiO_2 coating has a better quality than does the ZrO_2 coating. It is worth noting that a small amount of O was detected inside each coated sample substrate, probably because the 316 SS substrate is lightly oxidized during high temperature spraying. Plasma surrounding atmosphere can react with the metal substrate and molten particles, and thus leads to metal oxidation and carbide decomposition [17]. Hence, a controlled atmosphere chamber for spraying under industrial conditions should be fully considered during air plasma spraying.

Figure 6.3 illustrates coating microstructures and compositions via SEM and EDS analyses. As shown in Figs. 6.3a–c, the surface of the ZrO_2 or TiO_2 coating was smooth and flat, but there were some small pits distributing on each surface. This may be attributed to inconsistent thicknesses of sprayed material when the spray gun moved above the 316 SS substrate. Another possible explanation is that the

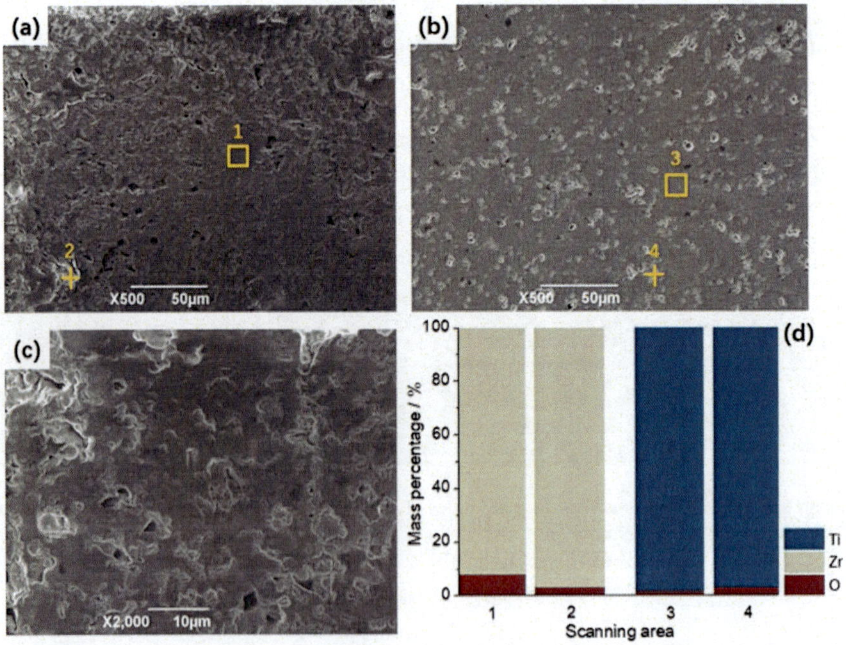

Fig. 6.3 Surface morphologies of **a** the ZrO_2-coated sample, **b** the TiO_2-coated sample, **c** the high magnification image of (**a**), **d** EDS analyses of the regions marked with yellow crosses and rectangles in (**a**, **b**)

unstable flow rate of protective gas results in these small structural defects. The high magnification of image (see Fig. 6.3c) displays that pit depths were different, but they were much lower than the coating thickness (see Figs. 6.1 and 6.2). Notably, element composition changes of the flat and pit regions of the ZrO_2-coated sample were more remarkable according to the EDS spectrum (see Fig. 6.3d). This implies that the interior microstructure of the ZrO_2 coating is rather complex, exhibiting anisotropic behavior induced by defects such as pores, cracks and embedded unmelted particles [18]. It is believable that these defects negatively affect the corrosion resistance of coating in SCW and should be avoided as possible.

6.1.3.2 Corrosion Weight Changes

Figure 6.4 indicates the weight changes of as-received 316 SS, ZrO_2- and TiO_2-coated samples exposed to oxygenated SCW. They were mostly reflected as corrosion weight gains under different temperature conditions. However, the ZrO_2-coated sample at 300 and 350 °C and the TiO_2-coated sample at 450 °C showed weight losses, which may be caused by the dissolutions of some metal elements and oxides on the sample surface. Apparently, the total dissolution rate was higher than the formation rate of oxide film during the whole test time (i.e., 100 h) at these temperatures. Besides, possible oxide film exfoliation can result in a decrease in weight. Consistently, corrosion weight loss is confirmed by the studies on Alloy 33 exposed to oxygenated SCW at 550 °C, 25 MPa for 600 h [19] and on Alloy 718, 617 and 625 exposed to SCW at 550 °C, 25 MPa within 150–600 h [12]. Overall, as-received 316

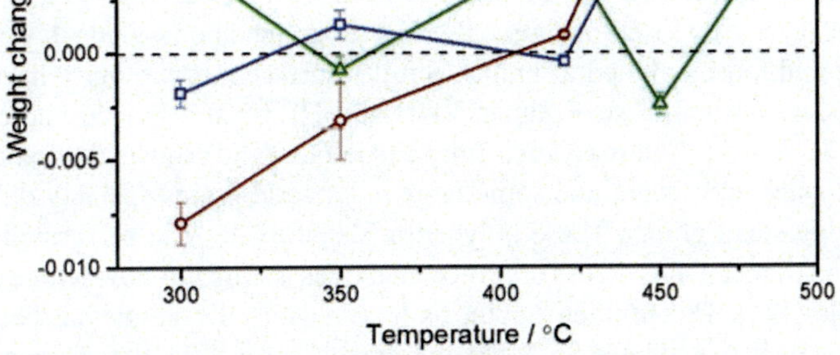

Fig. 6.4 Weight changes of as-received 316 SS sample (square), ZrO_2-coated sample (circle) and TiO_2-coated sample (triangle) exposed to high temperature water at 24 MPa with 3 wt% of oxygen content for 100 h

SS, TiO_2- and ZrO_2-coated samples showed much smaller absolute values of weight changes at 350 °C, corresponding to (0.676×10^{-3}, 0.143×10^{-3}, 1.332×10^{-3}) mg/cm^2, respectively. This can be attributed to the approximate balance between the total dissolution rate of elements and the formation rate of oxides on the sample surface.

The weight change values of as-received 316 SS and ZrO_2-coated samples increased with temperature raising, indicating that increasing temperature accelerates corrosion reactions and promotes oxide formation. This is detrimental to the protection effect of coating on the metal substrate, in good agreement with the report that high temperature results in high corrosion rates for alloys in SCW [20]. Notably, temperature variations along circuits also provoke the local precipitation and accumulation of oxidation products (i.e., Ni- and Fe-rich oxides and Cr_2O_3) [21], so increasing corrosion weight gain. The phenomena above show that the corrosion processes of as-received and ZrO_2-coated samples were gradually dominated by the formation of oxides with an increase in temperature. Especially, when temperature went across the supercritical point of water (i.e., 374 °C) from 350 to 420 °C, corrosion weight changes took place dramatic variations. This implies that water property changes near its critical point have significant effects on sample corrosion. Although there was not a discernible trend for the corrosion weight change of the TiO_2-coated sample with temperature increasing, this sample showed lower absolute values of weight changes than did the ZrO_2-coated sample under most temperature conditions. This suggests that the TiO_2 coating is relatively stable and plays a better protection effect compared with the ZrO_2 coating, likely by restraining the oxidation or dissolution of the metal substrate in oxygenated sub- and supercritical water.

6.1.3.3 SEM and EDS Analyses

Figures 6.5 and 6.6 separately illustrate low-magnification and high-magnification SEM images of three samples in oxygenated SCW at 450 and 480 °C. It can be seen that the surface oxides on as-received 316 SS were not dense enough, and some grooves appeared in the case of 450 °C (see Fig. 6.5a, d). A double-layer oxide film structure on as-received 316 SS was found at 480 °C (see Fig. 6.6a, d), including an inner layer composed of fine and compact grains and an outer layer composed of large and loose polyhedral grains. Similar surface morphologies have also been observed on austenitic steel Super 304H and HR3C and ferritic-martensitic steel T91 in SCW [22]. The magnified image (see Fig. 6.6d) shows that oxide particles varied in size and shape, and some large polyhedral grains sparsely distributed on underlying small grains. These polyhedral shape oxides can be formed by the fast migration of metal ions across unprotected oxides and by the slow lattice diffusion of Cr in alloy [23]. This implies that the as-received 316 SS sample has been damaged by corrosion in oxygenated SCW to some extent, and is difficult to form a continuous uniform protective oxide film.

In contrast, the delamination phenomenon could not observed on the coated samples. Two types of coatings were compact and maintained high integrity without

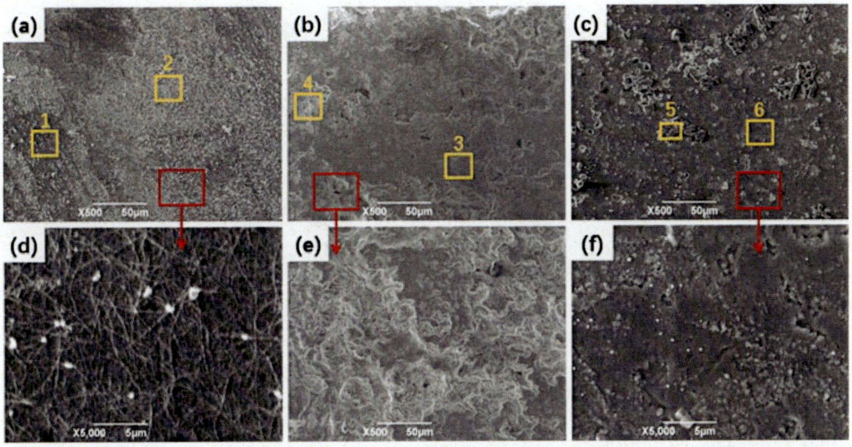

Fig. 6.5 SEM images of **a** the as-received 316 SS sample, **b** the ZrO_2-coated sample, and **c** the TiO_2-coated sample exposed to SCW at 450 °C, 24 MPa with 3 wt% of oxygen content for 100 h, **d–f** are high magnification of morphologies of red rectangles in (**a–c**), respectively

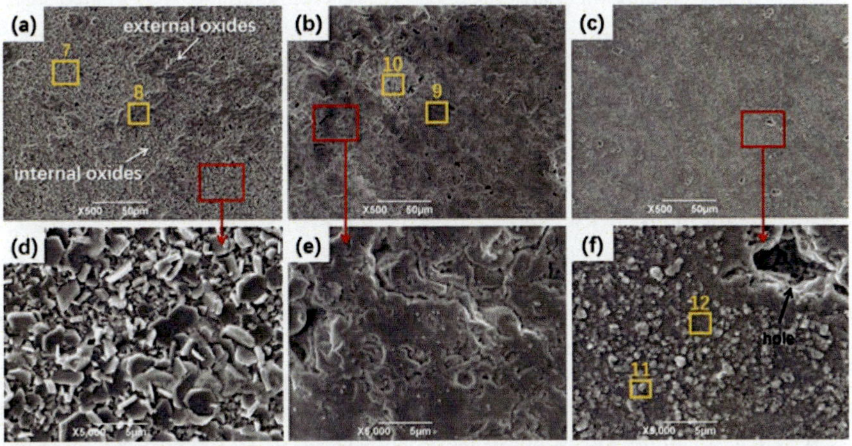

Fig. 6.6 SEM images of **a** the as-received 316 SS sample, **b** the ZrO_2-coated sample, and **c** TiO_2-coated sample exposed to SCW at 480 °C, 24 MPa with 3 wt% of oxygen content for 100 h, **d–f** are high magnification of morphologies of red rectangles in (**a–c**), respectively

exfoliation in spite of having different corrosion morphologies. Notably, increasing temperature from 450 to 480 °C accelerated each coating corrosion, in agreement with the findings in Fig. 6.4. Obviously, finer grain sizes and smoother oxide surface help the coating to protect the 316 SS substrate. However, some tiny holes were found on the ZrO_2- and TiO_2-coated samples at 450 and 480 °C (see Figs. 6.5 and 6.6), in accordance with the findings in Fig. 6.3. This is mainly due to original defects (on the coating surface) formed during air plasma spraying. These holes probably

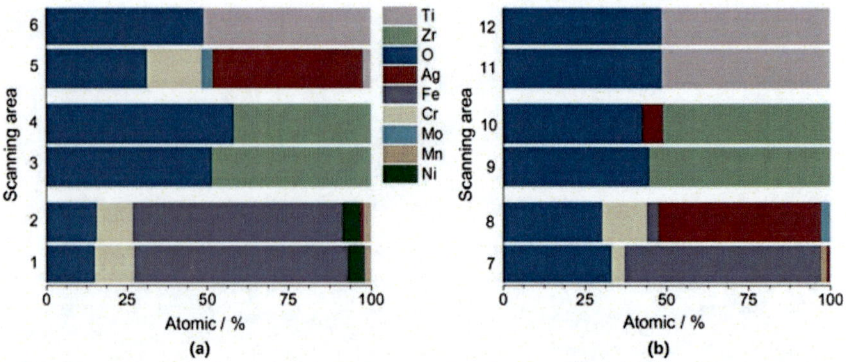

Fig. 6.7 Elemental analyses of the regions marked with yellow rectangles in Figs. 6.5 and 6.6. The numbers of scanning areas in (**a**) and (**b**) here correspond to the numbers of yellow rectangles in Figs. 6.5 and 6.6, respectively

become some channels, from which O diffuses into the coating interior, so causing a risk of accelerating corrosion.

It is worth noticing that particle deposition was found on the TiO_2-coated sample at 450 °C (see Fig. 6.5c). The EDS analysis (see the scanning area 5 in Fig. 6.7a) indicates that these particles were mainly composed of Ag, O, and Cr, likely in the forms of Ag_2O and Cr_2O_3. It can be inferred that Ag in silver wire used to suspend samples took place oxidation and deposition in oxygenated SCW. Similarly, Ag was also detected on the ZrO_2-coated sample surface at 480 °C (see the scanning area 10 in Fig. 6.7b). Overall, except for particle deposition, the TiO_2-coated sample had a more uniform and denser surface than did the ZrO_2-coated sample, based on the high magnification of morphologies shown in Figs. 6.5 and 6.6, implying the better protective effect of the TiO_2 coating.

The semi-quantitative EDS detection (see Fig. 6.7) also elucidates that the surface compositions of as-received 316 SS consisted of Fe oxides, Cr oxides as well as small amounts of oxides of Ni, Mo and Mn. Nonetheless, surface elemental compositions were mainly Ti and O in the TiO_2-coated sample, compared with Zr and O in the ZrO_2-coated sample. This suggests that coating (ZrO_2 or TiO_2) can efficiently inhibit the outward diffusion of elements in the 316 SS substrate and the inward diffusion of O in SCW, so slowing down the corrosion process. For the TiO_2-coated sample, the atomic percents of Ti at three different positions (scanning areas 6, 11 and 12) were separately 50.9, 51.3 and 51.4%. Herein, the small differences show the higher stability of TiO_2 with temperature increasing from 450 to 480 °C. For the ZrO_2-coated sample, the atomic percents of Zr at the scanning areas of 3, 4 and 9 were 48.5%, 41.7% and 55.2%, respectively. The significant differences suggest that the ZrO_2 coating was neither uniform at the same temperature nor stable with temperature variation in oxygenated SCW. Therefore, it can be believable that the TiO_2 coating is more stable than is the ZrO_2 coating in oxygenated SCW at 450, 480 °C and 24 MPa. ZrO_2 may be partially converted into ionic states in oxygenated SCW and then leave the coating surface, so weakening its protective effect on the 316 SS substrate. This implies that the TiO_2-coated sample exhibits better corrosion resistance in the tested SCW than does the ZrO_2-coated sample, in agreement with the observation in Fig. 6.4.

6.1.3.4 Surface Oxide Analyses

Figure 6.8 illustrates the XRD patterns of as-received 316 SS, ZrO_2-coated and TiO_2-coated samples exposed to oxygenated SCW at different temperatures for 100 h. The peaks of magnetite (Fe_3O_4) and chromium oxide (Cr_2O_3) were detected on the three samples. In the spectra of the ZrO_2-coated sample (see Fig. 6.8b), $NiFe_2O_4$ and intrinsic ZrO_2 were main phases under each temperature condition, and magnetite

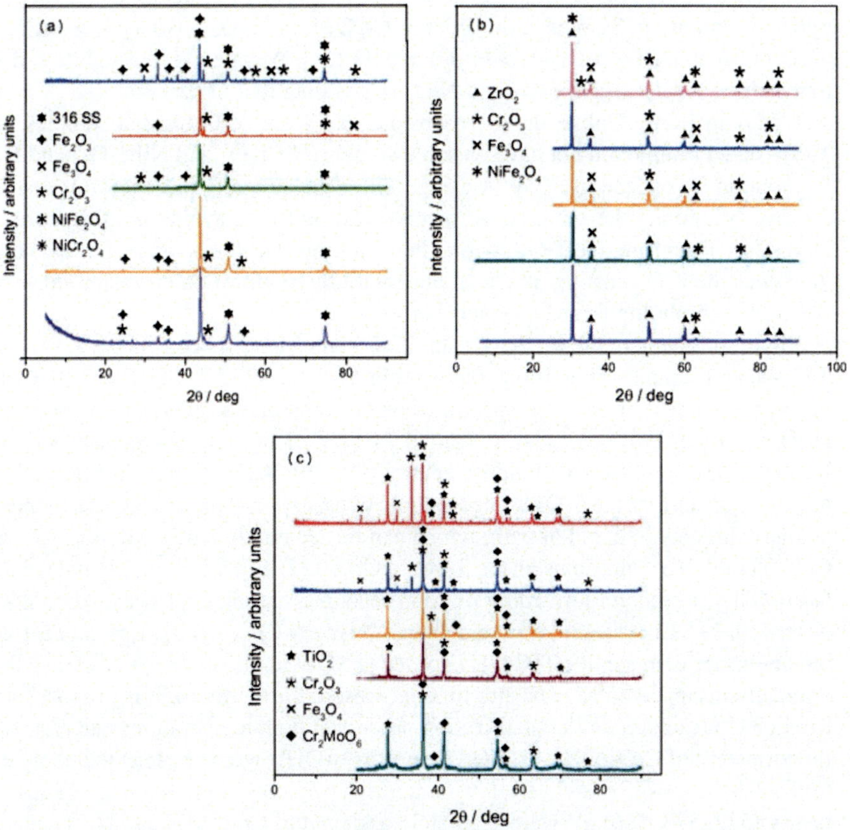

Fig. 6.8 XRD patterns of oxidized surfaces of **a** the as-received 316 SS sample, **b** the ZrO_2-coated sample, and **c** the TiO_2-coated sample exposed to high temperature water (300–480 °C) at 24 MPa with 3 wt% of oxygen content for 100 h. In **a–c**, the lines from top to bottom represent 480 °C, 450 °C, 420 °C, 350 °C and 300 °C, respectively

and Cr_2O_3 separately appeared after 350 and 420 °C. In contrast, in the spectra of the TiO_2-coated sample (see Fig. 6.8c), Cr_2O_3, Cr_2MoO_6 and intrinsic TiO_2 were primary phases. As temperature increased, the magnetite phase was detected at 450 and 480 °C. These results indicate that different temperatures induced various oxide compositions, and high temperature could lead to more oxide formation and convert some oxides to other forms. As indicated in Fig. 6.8, under different temperature conditions, the corrosion products of as-received 316 SS were composed of Cr_2O_3, Fe_2O_3, Fe_3O_4, $NiCr_2O_4$ and/or $NiFe_2O_4$. The formed oxides on the TiO_2-coated sample included Cr_2O_3, Fe_3O_4 and/or Cr_2MoO_6, while that on the ZrO_2-coated sample consisted of Cr_2O_3, Fe_3O_4 and/or $NiFe_2O_4$. This shows that ZrO_2 or TiO_2 coating could avoid the formation of some oxides, suggesting good protection effects on the 316 SS substrate.

Notably, Cr_2MoO_6 was found on the TiO_2-coated sample at each temperature condition (see Fig. 6.8c), in good agreement with the finding that the atomic percent of Mo on the TiO_2-coated sample at 450 °C was 3.5% (see Fig. 6.7a). Oxide thermodynamic stability in SCW is in the order of $Cr_2O_3 > FeCr_2O_4 > NiCr_2O_4 > NiFe_2O_4 > NiO$ [24] and $TiO_2 > Cr_2O_3 > Fe_3O_4 > NiO$ [25]. Apparently, the stability of Ni oxides (i.e., $NiCr_2O_4$, $NiFe_2O_4$ and NiO) is less than that of Cr oxide (i.e., Cr_2O_3) and TiO_2 in SCW. Unlike the as-received 316 SS and ZrO_2-coated samples, the TiO_2-coated sample did not have unstable spinels ($NiCr_2O_4$ and $NiFe_2O_4$) and NiO forming on its surface in SCW (see Fig. 6.8). This suggests that the TiO_2 coating effectively inhibits the outward diffusion of Ni and thus avoids the formation of Ni oxides. Therefore, the TiO_2 coating helps to improve corrosion resistance better than does the ZrO_2 coating, in accordance with the findings on corrosion rates and corrosion morphologies and products before.

Figure 6.9 depicts the Fe 2p, Cr 2p, Zr 3d and Ti 2p spectra of the ZrO_2- and TiO_2-coated samples after 100 h exposure in SCW at 480 °C, 24 MPa with 3 wt% of oxygen content. The Fe 2p spectra of oxide films on coated samples are shown in Fig. 6.9a, d. Fe_2O_3 located at around 710 eV of binding energy while Fe_3O_4 located at about 714 eV of binding energy [22, 26]. As displayed in Fig. 6.9b, e, the Cr 2p spectra of the TiO_2-coated sample consisted of two binding energy peaks at approximately 576 and 580 eV, which can be separately attributed to Cr_2O_3 and Cr^{6+}. Notably, Cr^{6+} may exist in the forms of CrO_3 [27, 28], $Cr_2O_7^{2-}$[29] and volatile $CrO_2(OH)_2$, which are formed by relevant reactions among Cr, Cr_2O_3, oxygen and water [30, 31]. Li et al. [26] reported that Cr^{6+} (likely $CrO_2(OH)_2$) was only formed on the surface of austenitic TP347H exposed to SCW in the very early stage and then gradually disappeared as exposure time increased. Furthermore, it is proposed that trioxide CrO_3 undergoes a volatilization process at high temperatures and results in the formation of $CrO_2(OH)_2$ [32] CrO_3 and $CrO_2(OH)_2$ were not detected likely due to their decomposition and very low contents. The peaks at the binding energies of nearly (572, 576, 586) eV were separately assigned to Cr, Cr_2O_3 and $NiCr_2O_4$ [24, 33] on the ZrO_2-coated sample surface. The Zr 3d peaks of the ZrO_2-coated sample appeared at the bond energies of approximately 181 and 184 eV, corresponding to Zr and ZrO_2, respectively (see Fig. 6.9c). Moreover, for the TiO_2-coated sample, the Ti

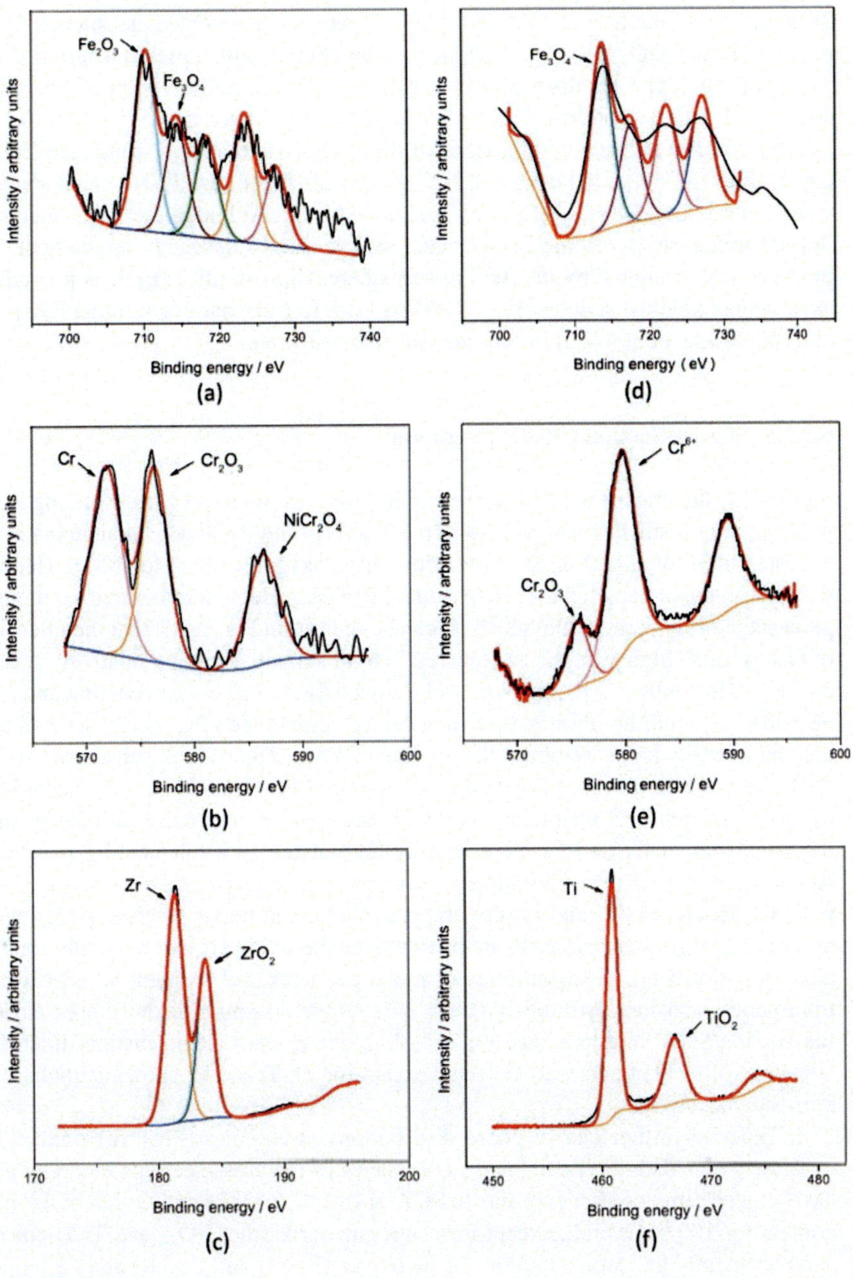

Fig. 6.9 XPS (Cr 2p, Fe 2p, Zr 3d, Ti 2p) spectra of oxidized surfaces of **a–c** the ZrO_2-coated sample and **d–f** the TiO_2-coated sample exposed to SCW at 480 °C, 24 MPa with 3 wt% of oxygen content for 100 h

2p bonding energies at about 461 and 467 eV separately represented the characteristic peaks of Ti and TiO_2 (see Fig. 6.9f), in good agreement with a previous report [34].

Overall, the XPS results are well consistent with the findings in previous XRD analyses. It can be confirmed that after exposure to oxygenated SCW at 480 °C, 24 MPa for 100 h, the surface oxides of the ZrO_2-coated sample contained ZrO_2, Cr_2O_3, Fe_3O_4, Fe_2O_3, $NiFe_2O_4$ and $NiCr_2O_4$, while that of the TiO_2-coated sample were composed of TiO_2, Cr_2O_3, Fe_3O_4 and Cr_2MoO_6. Although the XRD analyses did not found Fe_2O_3 on the ZrO_2-coated sample due to its much low content, its presence was validated by the XPS analysis here. Consistently, Fe_2O_3 is proved to exist in the oxides of stainless steel TP347H [26], ferritic-martensitic steel P92 [35] and stainless steel 304 [36] in sub- and supercritical water.

6.1.3.5 Cross-Section Characterization

Figure 6.10 illustrates the SEM images of the cross-sections and corresponding EDS line scanning results of as-received 316 SS, ZrO_2- and TiO_2-coated samples after exposure to SCW at 480 °C, 24 MPa with 3 wt% oxygen content for 100 h. Herein, the EDS analyses depicted the relative content of each element rather than its atomic percentage or mass ratio. Obviously, it can be seen from Fig. 6.10a that the intensity of O was much higher on the as-received 316 SS surface, showing relatively severe corrosion. In contrast, a distinct and well-bonded interface between coating and 316 SS substrate could be observed on each coated sample (see Fig. 6.10b, c), and the coating integrity helps to inhibit the interface failure. The Ni-rich substances at the ZrO_2 coating interface (see Fig. 6.10b) should be $NiFe_2O_4$ and $NiCr_2O_4$, according to the XRD and XPS results in Figs. 6.8b and 6.9b. Figure 6.10b, c display that the intensities of Cr and Fe were slightly high at the substrate/coating interface. According to the XRD and XPS analyses, these enriched elements are likely derived from Cr_2MoO_6, Cr_2O_3 and Fe_3O_4. Importantly, a continuous Cr-rich oxide layer, as a barrier of further oxidation, helps to reduce the oxide growth rate and remarkably slow down the diffusion rates of metal elements and oxygen, so effectively improving corrosion resistance [20, 35, 37]. As the scanning distance approached the coating/SCW interface (see Fig. 6.10b, c), it seems that no distinct intensity signals of other elements were detected except for Zr, Ti and O, showing unobvious corrosion occurrence.

In order to further clarify phase distributions in coatings, Figs. 6.11 and 6.12 separately elucidate the EBSD and EDS images of the cross-sections of ZrO_2- and TiO_2-coated samples after exposure to SCW at 480 °C, 24 MPa with 3 wt% of oxygen content for 100 h. Overall, except for coating material, the ZrO_2- and TiO_2-coated samples mainly had small amounts of Fe oxides (Fe_3O_4 and Fe_2O_3) and Cr oxides (Cr_2O_3) in the coatings, as shown by XRD and XPS analyses (see Figs. 6.11b and 6.12b). It can be found that the concentration differences of Fe and Cr between the 316 SS substrate and the coating in the ZrO_2-coated sample were significantly higher than that in the TiO_2-coated sample (see Figs. 6.11e, g and 6.12e, g). This shows that more Fe and Cr distributed in the ZrO_2 coating, meaning that more Fe and Cr in the

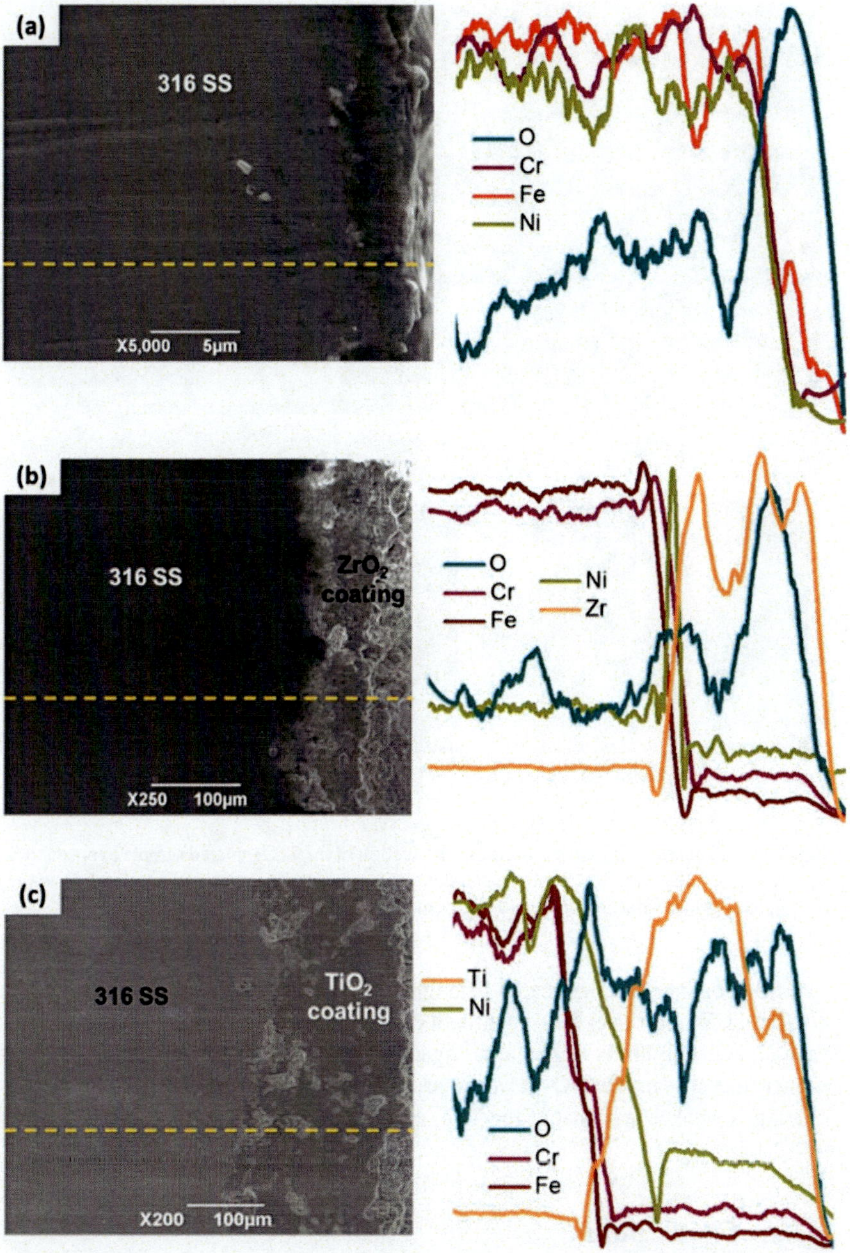

Fig. 6.10 Polished cross-section morphologies and EDS analyses along the dotted lines of **a** the as-received 316 SS sample, **b** the ZrO_2-coated sample, and **c** the TiO_2-coated sample after exposure to SCW at 480 °C, 24 MPa with 3 wt% of oxygen content for 100 h

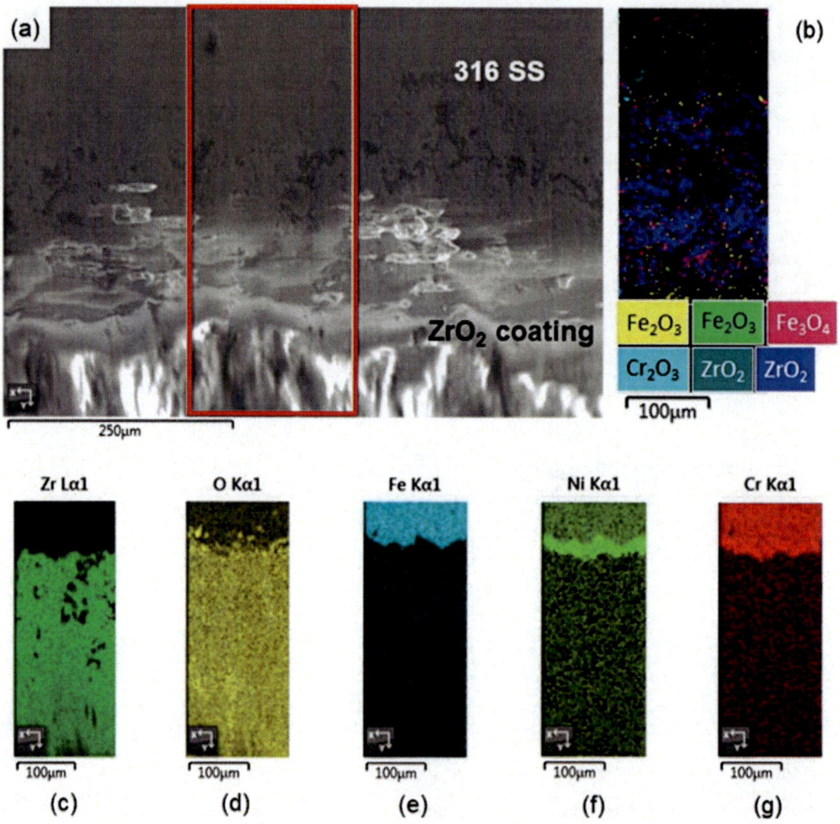

Fig. 6.11 EBSD scanning images of the cross-section of the ZrO_2-coated sample exposed to SCW at 480 °C, 24 MPa with 3 wt% of oxygen content for 100 h. The scanning area of **b–g** is the red rectangle in (**a**), and different colors in (**b**) demonstrate phase distributions

316 SS substrate had permeated through the ZrO_2 coating and diffused toward the coating/SCW interface. Consistently, it can be observed from Figs. 6.11b and 6.12b that the concentrations of hematite, magnetite and Cr_2O_3 in the ZrO_2 coating were higher than that in the TiO_2 coating, proving that the protection effect of the ZrO_2 coating is weaker than that of the TiO_2 coating.

6.1.4 Conclusions

ZrO_2 and TiO_2 were coated by an air plasma spraying method on the 316 SS substrate as protective coatings. Two coated samples showed better corrosion resistance than did as-received 316 SS after exposure to high temperature water (from 300 to 480 °C) at 24 MPa with 3 wt% of oxygen content for 100 h. High temperature promoted

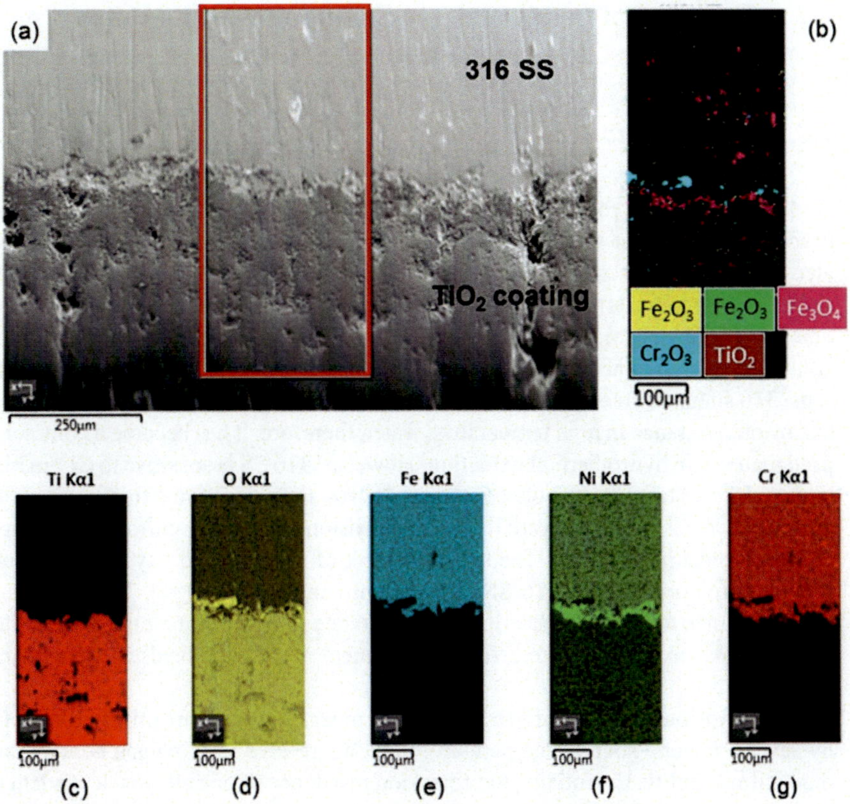

Fig. 6.12 EBSD scanning images of the cross-section of the TiO_2-coated sample exposed to SCW at 480 °C, 24 MPa with 3 wt% of oxygen content for 100 h. The scanning area of **b–g** is the red rectangle in (**a**), and different colors in (**b**) demonstrate phase distributions

corrosion and surface morphologies of coated samples were relatively dense without cracking and exfoliation phenomena. The oxide scale on the ZrO_2-coated sample was composed of Cr_2O_3, Fe_3O_4, Fe_2O_3, $NiFe_2O_4$ and $NiCr_2O_4$, while that on the TiO_2-coated sample consisted of Cr_2O_3, Fe_3O_4 and Cr_2MoO_6 at 480 °C. The concentrations of Fe_2O_3, Fe_3O_4 and Cr_2O_3 in the TiO_2 coating were lower than that in the ZrO_2 coating. Overall, the TiO_2-coated sample exhibited better corrosion resistance than did the ZrO_2-coated sample, based on the comparisons of weight changes and oxide contents after exposure in oxygenated high temperature water. The TiO_2 coating could better inhibit the outward diffusion of metal elements (especially Ni) and the inward diffusion of O. The TiO_2 coating on the 316 SS substrate via air plasma spraying is an effective method of improving its corrosion resistance in tested oxygenated SCW environment. However, it is necessary to further optimize the air plasma spraying process to eliminate some defects (e.g., pores) in the coating.

6.2 Corrosion of Passivated 316 SS in Subcritical Water Containing Oxygen and NaCl

6.2.1 Introduction

High temperature hydrothermal oxidation is an advanced oxidation technology used for the efficient treatment of high concentration refractory organic wastewater. However, relatively high temperature, pressure, oxygen content, and corrosive substances (e.g., salts) accelerate facility of material corrosion. This has become one key factor that hinders its commercial application, especially for wastewater containing a high concentration of chloride [38, 39]. Among candidate metal materials, 316 stainless steel (316 SS) (UNS S31600) has a relatively low price and good corrosion resistance in high temperature water, therefore, it has become a commonly used material in hydrothermal oxidation. However, 316 SS is sensitive to Cl^- in high temperature water. Cl^- can penetrate and destroy the protective oxide layer on the alloy surface, resulting in severe localized corrosion, and pitting corrosion and stress corrosion cracking [15, 40]. The synergy effect of chloride and oxygen can induce more severe corrosion on 316 SS in high temperature water [34]. Therefore, it is of significance to improve the corrosion resistance of 316 SS in a high temperature hydrothermal oxidation system for the treatment of organic feedstock containing chloride.

The corrosion behaviors of 316 SS in high temperature water are primarily affected by several factors, such as temperature, time, aggressive salt, oxygen content, and oxide film stability. Essentially, the corrosion resistance of one alloy is dependent on the stability of the oxide film formed on the alloy surface [41]. A dense oxide film can be formed on the alloy surface in a corrosive environment, and this passivation process can protect the alloy substrate from further corrosion. Similarly, the formed phosphate film can effectively reduce the corrosion rate of Alloy 625 (UNS N06625) in supercritical water oxidation [42]. This suggests that the passivation pretreatment of 316 SS can improve its corrosion resistance in subsequent applications. Phosphate passivation cannot inhibit 316 SS corrosion in supercritical water containing oxygen and chloride [34]. Currently, there are no relevant reports on whether the passivation pretreatment of 316 SS works in subcritical water containing oxygen and chlorides. However, this is a typical environment in high temperature hydrothermal oxidation for the treatment of some chloride-containing organic wastewaters (e.g., pesticide wastewater). Therefore, this gap in the literature motivated the present research.

In this study, two 316 SS samples were first passivated with and without sodium phosphate (Na_3PO_4) in supercritical water separately, and then their early corrosion behaviors in subcritical water containing high concentrations of sodium chloride and oxygen, at different temperature conditions for a short period of time, were explored and compared systematically. Corresponding corrosion weight gains, corrosion rates, oxide film morphologies, oxide components, element compositions, and oxide film structures were analyzed comprehensively. The effects of temperature on passivation sample corrosion are discussed in depth. To the best of our knowledge, this is documented here for the first time. The information is valuable for fully understanding the early corrosion behaviors of 316 SS passivation film and for improving 316 SS corrosion resistance in subcritical water containing chloride and oxygen.

6.2.2 Experimental Section

6.2.2.1 Materials

The oxidant is 30 wt% of hydrogen peroxide (H_2O_2) solution, that easily decomposes into oxygen (O_2) and water (H_2O) in high temperature water, and was purchased from Tianjin Tianli Chemical Reagent Co., Ltd. Sodium chloride (NaCl) (>99.8 wt% purity) was used as corrosive chloride and was obtained from Tianjin Kemiou Chemical Reagent Co., Ltd. Na_3PO_4 (>99.5 wt% purity) was used to passivate the 316 SS sample and was obtained from Tianjin Hongyan Chemical Reagent Co., Ltd. Deionized water was prepared by a water purification plant (ULUP-II) in the lab and used throughout this study. A hole of 10 × 10 × 7 mm was drilled at the edge of each sample, and its six surfaces were polished with #80, #600, #1000, #1500 and #2000 abrasive papers in turn, and then with aluminum polishing powder by a metallographic specimen polishing machine (P-1). Before the experiments, these samples were cleaned by an ultrasonic cleaner in acetone for 20 min and then dried in an oven (Electro-Thermostatic Blast Oven) at 60 °C for 40 min.

6.2.2.2 Experimental Procedures

The passivation and corrosion experiments for the 316 SS samples were conducted in a 415 mL batch reactor. Before the passivation experiments, the reactor was opened, and the sample was suspended in the reactor center by silver wire. Then, certain amounts of deionized water and Na_3PO_4were added into the reactor. Their amounts were determined according to the desired test temperature and pressure conditions. The reactor was sealed and then heated by a temperature controller up to 450 °C, and thus its pressure increased with temperature accordingly. The 316 SS sample was passivated in supercritical water containing Na_3PO_4, and then the temperature controller was shut off to naturally cool the reactor to ambient temperature. The passivated sample was taken out of the reactor, and then cleaned with deionized water for 10 min in an ultrasonic cleaner to remove possible Na_3PO_4 deposited on the sample surface. Subsequently, the passivated sample was naturally dried for at least 3 h and then weighed. Herein, the obtained sample is called the phosphate-passivated 316 SS sample. Similarly, another 316 SS sample was passivated by the method above, but no Na_3PO_4 was added. Then, the two samples were kept in soft filter papers for subsequent corrosion tests and analyses.

In the corrosion experiments, certain amounts of deionized water, NaCl, and 30 wt% of H_2O_2 solution were added into the cleaned reactor. Their amounts were also confirmed by the same method mentioned above. In four corrosion tests, the reactor was separately heated up to 100, 160, 230, and 300 °C with a 3.5 $°C{\cdot}min^{-1}$ heating rate by the temperature controller, and was maintained at 25 MPa for 50 h at each temperature condition. When each corrosion test was finished, the temperature controller of the reactor was shut off, and the sample was processed by the same procedures mentioned above.

6.2.2.3 Analytical Methods

The average corrosion rate of the sample $\nu_w(g \cdot mm^{-2} \cdot h^{-1})$ was evaluated by its weight variation in unit area and time after corrosion.

$$v_w = \frac{\Delta m}{St} \tag{6.1}$$

where S (mm^2) is the sample area, t (h) is corrosion time, and Δm (g) is the weight variation of the sample. The average corrosion thickness of the sample per unit time $\nu_d\left(mm \cdot h^{-1}\right)$ can also reflect its average corrosion rate,

$$v_d = \frac{v_w t}{\rho} \tag{6.2}$$

where ρ is the sample density (7.98×10^{-3} g mm^{-3}).

The sample weight was measured by an electronic scale with an accuracy of 0.0001 g. The surface morphology and the elemental composition of the oxide film were analyzed by a scanning electron microscope (SEM, SU3500) equipped with an energy dispersive spectrometer (EDS). The 3D morphologies of the samples were obtained by atomic force microscopy (AFM, Veeco Innova). The oxide components on the sample surface were determined by an X-ray diffraction analyzer (XRD, D8 ADVANCE). The phase compositions of corrosion products were examined by an X-ray photoelectron spectroscope (XPS, ESCALAB Xi type).

6.2.3 Results and Discussions

6.2.3.1 Gravimetry

Figure 6.13 shows the corrosion weight gains of tested samples (i.e., passivated 316 SS and phosphate-passivated 316 SS) exposed to subcritical water with oxygen and NaCl for 50 h under different temperature conditions. Herein, weight gain can be attributed to the formation of the oxide film that tightly adhered to the sample surface,

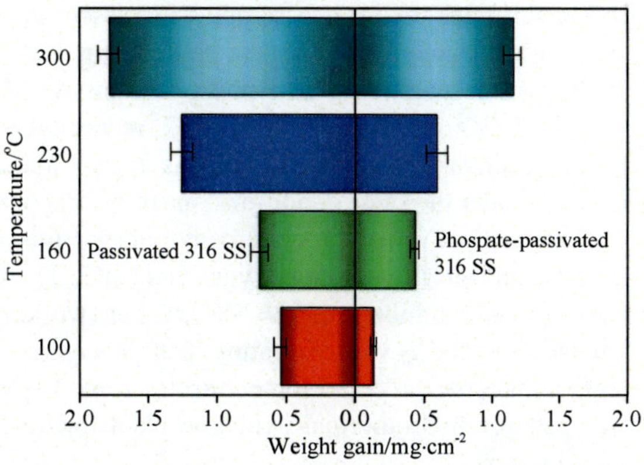

Fig. 6.13 Corrosion weight gains of the samples after exposure to subcritical water for 50 h at different temperature conditions

as exfoliation could not take place in such a short test time. Overall, the weight gains of the two samples increased with an increase in temperature, and their maximum values were observed at 300 °C. A possible explanation is that increasing temperature promotes electrochemical reactions and chemical reactions that are responsible for alloy corrosion in subcritical water. Notably, the phosphate-passivated 316 SS sample exhibited a much lower weight gain than the passivated 316 SS sample under the same conditions. This implies that Na_3PO_4 passivation pretreatment in supercritical water is beneficial to improving the corrosion resistance of 316 SS in subcritical water containing oxygen and NaCl. Similarly, a low corrosion rate of 500 μm per 100 h is observed for Inconel 625 (UNS N06625) in subcritical water at 250–300 °C and 24 MPa with the presence of PO_4^{3-} [40].

Figure 6.14 displays the corrosion rates of two passivated 316 SS samples in subcritical water under different temperature conditions. This evaluation is valuable for understanding their corrosion resistance properly. It is apparent that the

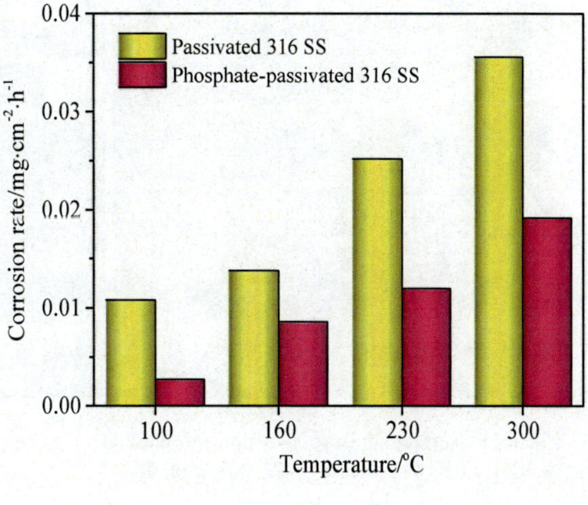

Fig. 6.14 Corrosion rates of the tested samples in subcritical water under various temperature conditions

corrosion rates of the two samples increased with increasing temperature, which is in good agreement with the findings in Fig. 6.13. The maximum corrosion rates for the passivated and phosphate-passivated 316 SS samples were 0.0356 and 0.0192 mg cm^{-2} h^{-1} at 300 °C, respectively. Overall, the corrosion rate of the phosphate-passivated 316 SS was approximately half of that of the passivated 316 SS under the same conditions, meaning that phosphate passivation pretreatment in supercritical water can effectively improve the corrosion resistance of 316 SS in subcritical water containing oxygen and NaCl. Therefore, Na_3PO_4 can be considered as a corrosion inhibitor of 316 SS facilities (working in subcritical water containing oxygen and NaCl) via passivating them in supercritical water in advance. This is of importance for the corrosion control of some key facilities (e.g., reactor) in a high temperature hydrothermal oxidation plant for treating chloride-containing organic wastewater.

6.2.3.2 Surface Morphology

Surface and cross-section morphologies of the two kinds of passivated 316 SS samples are shown in Fig. 6.15. Hence, the two samples exhibited a similar cross-sectional morphology but significantly different surface morphologies. The surface

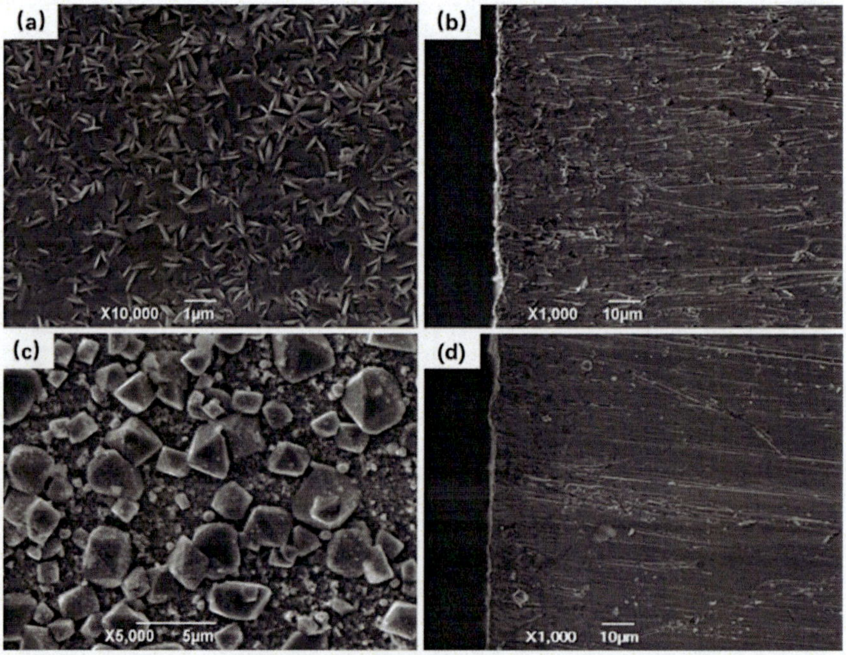

Fig. 6.15 Surface and cross-section morphologies of the phosphate-passivated 316 SS sample (**a**, **b**) and the passivated 316 SS sample (**c**, **d**)

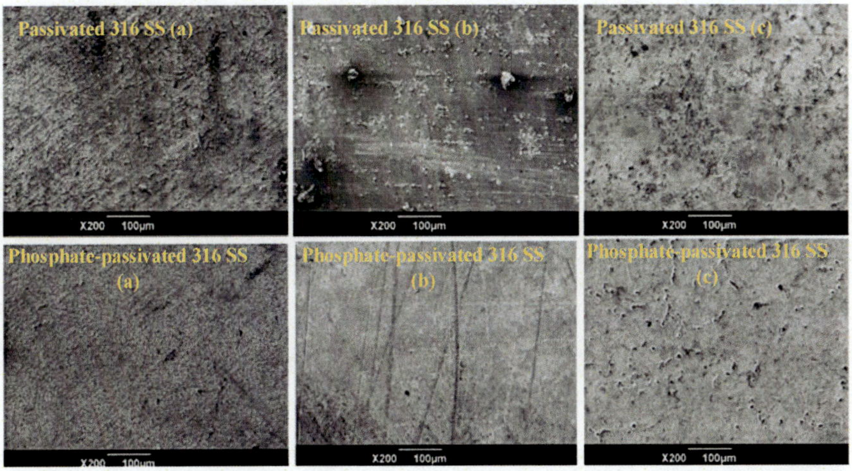

Fig. 6.16 Corrosion morphologies of the tested samples after exposure to subcritical water for 50 h under different temperature conditions: **a** 160 °C, **b** 230 °C, **c** 300 °C

of the phosphate-passivated sample was covered with acicular oxides of uniform size, while the oxide particles on the surface of the passivated sample were of different sizes and could not completely cover the 316 SS substrate. Therefore, it can be inferred that these uniformly distributed oxide particles can provide excellent protection for the substrate in subsequent corrosion tests. Specific protective effects and oxide compositions will be discussed below.

Figure 6.16 indicates the surface corrosion morphologies of the two samples exposed to subcritical water for 50 h at various temperature conditions. The two samples exhibited significantly different corrosion morphologies under the same conditions due to corrosion resistance variations. Overall, compared with the passivated 316 SS sample, the phosphate-passivated 316 SS sample had an oxide film with a much denser and more uniform morphology at each temperature condition. This suggests that the latter had better corrosion resistance in subcritical water with oxygen and NaCl, which is in good accordance with the findings in Figs. 6.13 and 6.14. This can be explained by the fact that uniform and dense corrosion products on the sample surface help inhibit further corrosion by avoiding contact of the alloy substrate with corrosive species [20, 43]. For 316 SS, severe pitting corrosion readily takes place in subcritical water containing oxygen and HCl with a concentration of 0.05 mol kg^{-1} at 160 °C and 24 MPa [40]. Nevertheless, herein pitting corrosion was indiscernible at each temperature condition, likely due to the relatively short exposure time. Consistently, the most severe corrosion appeared at 300 °C, and Cl^- may result in the local damage (e.g., the dissolution of corrosion products) of the passivation film in subcritical water with the help of oxygen at a high temperature.

Figure 6.17 further elucidates high magnification (×3000) SEM images of the passivated and phosphate-passivated 316 SS samples exposed to subcritical water with oxygen and NaCl at 300 °C for 50 h. There were coarse flaky accumulation

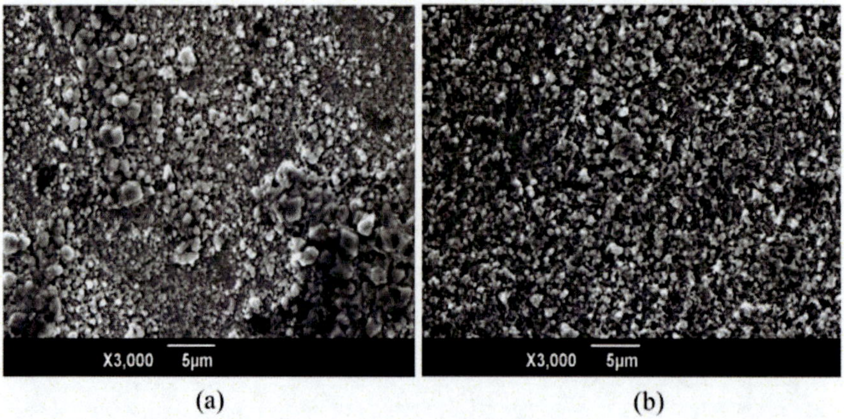

Fig. 6.17 Surface corrosion morphologies (×3000 magnification SEM images) of the tested samples after exposure to subcritical water at 300 °C for 50 h. **a** Passivated 316 SS, **b** phosphate-passivated 316 SS

regions of several multi-angle oxides on the passivated 316 SS sample surface (see Fig. 6.17a) that should be due to the formation of new oxides at defect locations such as oxide grain boundaries and pores [22]. Nevertheless, a more uniform and denser oxide film was formed on the phosphate-passivated 316 SS sample surface (see Fig. 6.17b). The stable and protective oxide film was composed of much smaller particles and thus could effectively avoid the further contact of corrosive species (e.g., Cl^- and O) with the alloy substrate. This is one reason for the improved corrosion resistance of the phosphate-passivated 316 SS. According to Figs. 6.13, 6.14, 6.15 and 6.16, it can be concluded that the phosphate passivation pretreatment of 316 SS in supercritical water can significantly improve the corrosion resistance of 316 SS in subcritical water containing NaCl and oxygen. Inevitably, the corrosion resistance of one alloy material significantly depends on the oxide stability of the alloy surface [23, 25]. The oxide film (or corrosion products) on the sample surface, in the present study, would thicken gradually and exfoliation would even take place with prolonged corrosion time, so a much longer test time is required for further examination in subsequent research.

Figure 6.18 shows the geometric characteristics of 3D topographies of oxide scales obtained from the phosphate-passivated 316 SS sample before and after exposure to subcritical water at 300 °C for 50 h. It can be noted that the 3D parameters in (c) and (d) were expanded from the roughness (2D) parameters in (a) and (b). In addition, the frequency distributions of the height of oxide particles on the phosphate-passivated 316 SS sample before and after exposure to subcritical water are expressed in Fig. 6.19 that visualizes height data acquired by AFM. For the phosphate-passivated 316 SS sample before and after exposure to subcritical water, the heights of the oxide particles varied from 0 to 600 nm, and from 0 to 250 nm, respectively. More tiny oxide particles with heights varying within 100–180 nm appeared on the alloy surface after

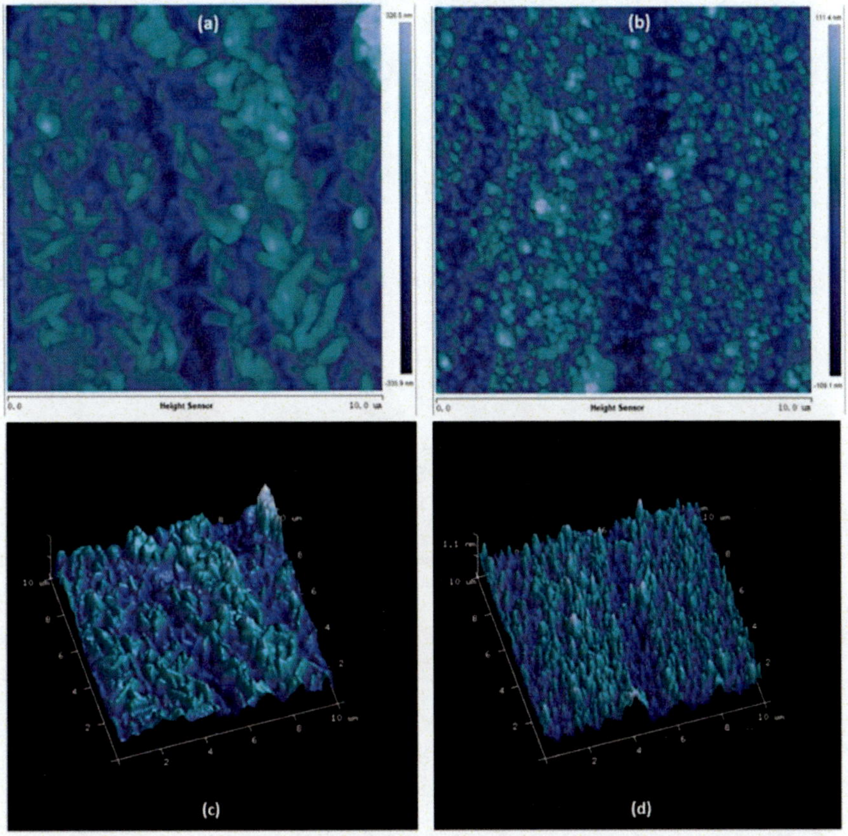

Fig. 6.18 Typical 2D and 3D morphologies of the phosphate-passivated 316 SS sample (**a**, **c**), and the phosphate-passivated 316 SS sample after exposure to subcritical water at 300 °C for 50 h (**b**, **d**)

subcritical water exposure. Compared to large oxide particles, these small oxide particles can form an oxide film with smaller gaps and thus reduce the ways that oxygen penetrates the oxide film to contact the alloy substrate, so better protecting the substrate from further corrosion damage. The composition of these oxide particles can be determined by the subsequent XRD and XPS analysis.

6.2.3.3 Oxide Phase Analysis

Figure 6.20 illustrates the XRD patterns of corrosion products on two samples exposed to subcritical water for 50 h at various temperature conditions. Thus, stable Cr_2O_3 and Fe_2O_3 were formed on the passivated 316 SS sample surface after corrosion in subcritical water with oxygen and NaCl (see Fig. 6.20a). These two oxides

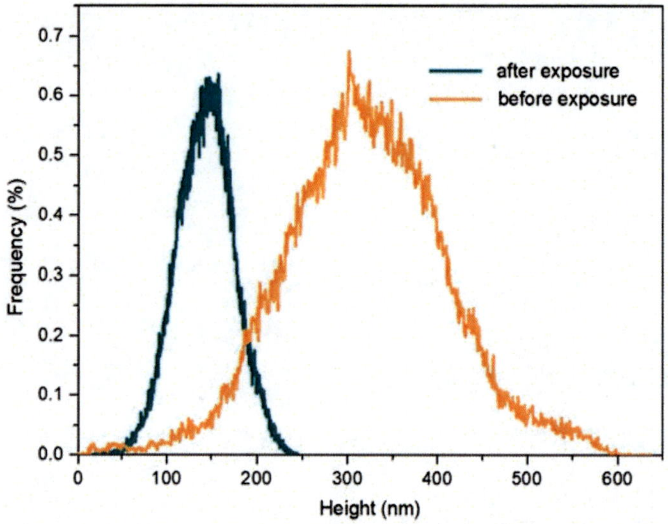

Fig. 6.19 Frequency distribution curves of the height of the oxide particles on the phosphate-passivated 316 SS sample before and after exposure to subcritical water at 300 °C for 50 h

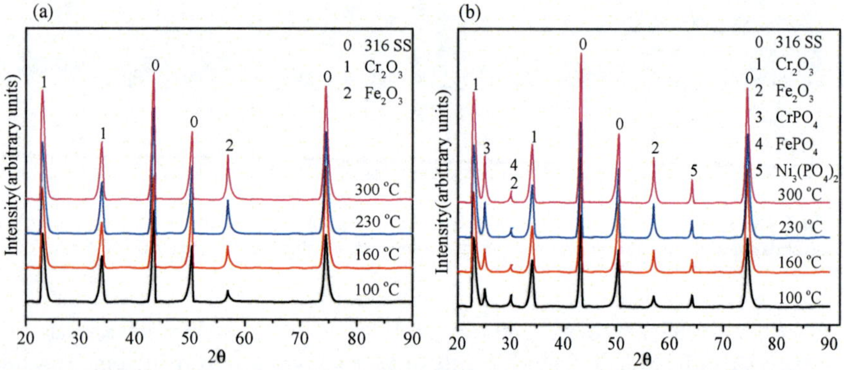

Fig. 6.20 XRD patterns of corrosion products after exposure of the sample to subcritical water for 50 h at various temperature conditions. **a** Passivated 316 SS, **b** phosphate-passivated 316 SS

were formed in the initial passivation process in supercritical water and at the corrosion stage in subcritical water. Essentially, the metal element diffuses outward and losses electrons, meanwhile oxygen diffuses inward through grain boundaries. This provides short diffusion pathways [44, 45] to form metal oxides via chemical or electrochemical reactions. Cr first reacts with dissolved oxygen to form Cr_2O_3 in the early phase of corrosion [20], because Cr is preferentially oxidized by the inward diffusion of oxygen due to its much higher oxygen affinity than Ni and Fe [43]. Cr_2O_3 commonly has a compact structure and thus can inhibit further corrosion [46]. Fe can reacts with dissolved oxygen to form Fe_2O_3 that suffers from certain dissolution by corrosive anions under high temperature and pressure conditions [34]. Notably, no NiO signal was found here, and a possible explanation is that once NiO is formed, Cl^- instantly reacts with it to form water-soluble metal chlorides.

Except Cr_2O_3 and Fe_2O_3, the phosphate-passivated 316 SS sample surface also showed the formation of $CrPO_4$, $FePO_4$, and $Ni_3(PO_4)_2$ at all temperature conditions (see Fig. 6.20b). Thus, it can be believable that in the phosphate-passivation process in supercritical water, the dissolved metal ions (i.e., Cr^{3+}, Fe^{3+}, and Ni^{2+}) in corrosion pits react with $PO_4{}^{3-}$ to form the water-insoluble phosphates above in the oxide film on the 316 SS surface. These formed stable phosphates can effectively prevent further alloy corrosion caused by corrosive species (e.g., Cl^- and O), and this should be the most significant reason for the corrosion resistance differences of two samples shown in Figs. 6.1, 6.2, 6.3 and 6.4. Overall, the intensities of XRD detection peaks of Cr_2O_3, Fe_2O_3, $CrPO_4$, $FePO_4$, and $Ni_3(PO_4)_2$ in corrosion products increased as temperature increased, implying the increase of oxide amount and oxide film thickness with temperature. Temperature is also reported to be a key factor that affects the oxide components of austenitic HR3C steel after exposure to supercritical water [23]. Hence, it is believable that temperature significantly influences the corrosion resistance of the passivated or phosphate-passivated 316 SS in subcritical water containing oxygen and chloride.

Figure 6.21 depicts the analysis results of the chemical valence states of P, Cr, and Fe after the corrosion tests of the passivated and phosphate-passivated 316 SS samples. The core level spectra of P, Cr, and Fe were P 2p, Cr 2p, and Fe 2p, respectively. The core level spectrum of P 2p consisted of two peaks located at

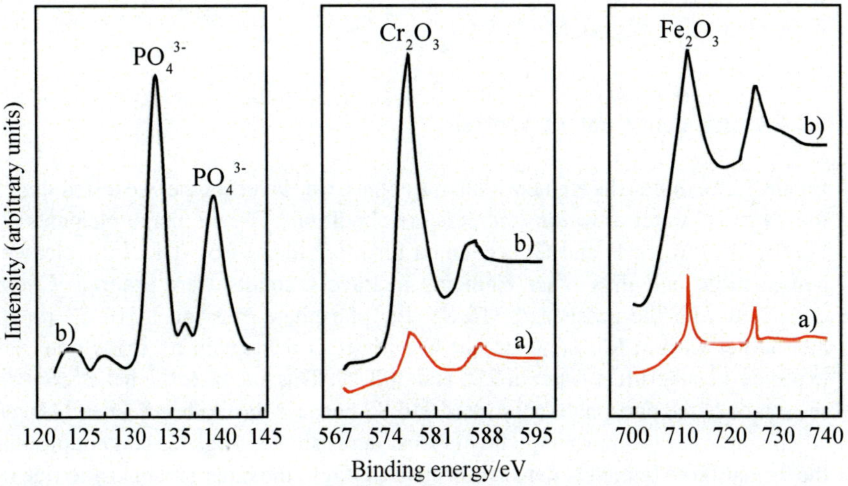

Fig. 6.21 XPS analyses of the surface oxides of the tested samples after exposure to subcritical water containing oxygen and NaCl at 300 °C for 50 h. **a** Passivated 316 SS, **b** phosphate-passivated 316 SS

134.5 and 139.7 eV, representing the characteristic peaks of PO_4^{3-}. The peaks of Cr_2O_3 and Fe_2O_3 appeared at the bond energies of 577.2 and 711.4 eV in the core level spectra of Cr 2p and Fe 2p, respectively [47]. Hence, the surface oxides of the phosphate-passivated 316 SS mainly consisted of metal phosphates, Cr_2O_3, and Fe_2O_3, whereas those of the passivated 316 SS were composed of Cr_2O_3 and Fe_2O_3 after 300 °C of corrosion testing in subcritical water with oxidant and NaCl. Fe_3O_4 was not found in corrosion products, likely due to the abundant oxygen presence that may promote the decomposition of Fe_3O_4 or inhibit the formation of Fe_3O_4. Herein, the XPS findings on corrosion products are in good agreement with the XRD results shown in Fig. 6.20.

Generally, the corrosion resistance of an alloy is dependent on corrosion product properties, such as chemical stability, strength, and compactness [15]. Chloride can penetrate and destroy the protective oxide layer and water-insoluble phosphates by exfoliation or dissolution, resulting in corrosion acceleration in high temperature water [40, 48]. The existence of sulfate, chloride, and especially both of them in combination, destroys the formation of water-insoluble phosphates on the alloy oxide film in supercritical water containing oxygen and phosphate [48]. The peak area in an XPS analysis can reflect the relative content of the corrosion product on the sample surface. Herein, the phosphate-passivated 316 SS sample had much higher peak areas of Cr_2O_3 and Fe_2O_3 on its surface than the passivated 316 SS sample (see Fig. 6.21). This suggests that the oxide film enriched in phosphates on the phosphate-passivated 316 SS surface can better protect Cr_2O_3 and Fe_2O_3 from the destruction and dissolution of corrosive anions (e.g., Cl^-) in oxidizing subcritical water. Therefore, it is believable that the formed metal phosphates (i.e., $CrPO_4$, $FePO_4$, and $Ni_3(PO_4)_2$ in Fig. 6.20b) in the 316 SS passivation process in supercritical water play crucial roles in improving the corrosion resistance of 316 SS in subcritical water containing oxygen and NaCl.

6.2.3.4 Element Content Analysis

Figure 6.22 exhibits the element contents of the oxide layers of the two tested samples in subcritical water at various temperature conditions. Herein, major elements (i.e., Fe, Cr, Ni, Mo, O, P, and C) accounted for more than 99.5 wt% of the elemental composition, and thus other elements in trace amounts were ignored. Overall, compared with the passivated 316 SS, the phosphate-passivated 316 SS roughly had higher Cr and Mo contents but had lower O content in its oxide layer after exposure to subcritical water at 230 and 300 °C. This suggests the better corrosion resistance of the phosphate-passivated 316 SS because the oxides of Cr and Mo play significant roles in improving alloy corrosion resistance. High Cr content promotes the formation of relatively dense protective Cr_2O_3 in the inner layer of the oxide film [49] (see Fig. 6.20). Cr_2O_3 can inhibit the diffusion of some species, such as O^{2-} and Ni^{2+}, and thus plays a protective role in the oxidation of 316 SS [50]. Mo has a relatively low diffusion rate in the inner layer of the oxide film and can improve the anti-pitting corrosion performance [34]. O was detected mainly due to the oxidation

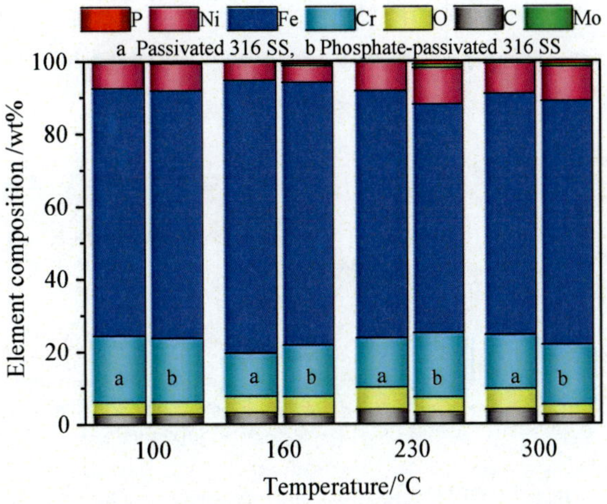

Fig. 6.22 Surface elemental mass content changes of the tested samples after exposure to different temperatures of subcritical water for 50 h

of Cr and Fe to form corresponding oxides (e.g., Cr_2O_3 and Fe_2O_3), as verified in Figs. 6.20 and 6.21.

Although the P content was much lower in the oxide layer of the phosphate-passivated 316 SS sample, it further proves that P could stably exist in the forms of $CrPO_4$, $FePO_4$, and $Ni_3(PO_4)_2$ (see Fig. 6.20b) in subcritical water containing oxygen and NaCl. As shown in Figs. 6.16b and 6.17b, these formed water-insoluble phosphates, together with Cr_2O_3 and Fe_2O_3, uniformly and compactly adhered to the sample surface. Thus, they played a significant role in isolating contact of the alloy substrate with corrosive species and in decreasing the diffusion and dissolution of metal elements (e.g., Cr). It is worth noting that the differences in element components in the two samples occurred after 100 °C, showing the important effects of temperature. It is apparent that the changes the content of Fe, Cr, Ni, Mo, O, and P significantly reflect and affect the corrosion resistance of the tested sample in subcritical water containing oxygen and NaCl.

6.2.3.5 Cross-Section Morphology and Element Analysis

Figure 6.23 provides the SEM morphology and line scanning of the cross-section of the phosphate-passivated 316 SS sample after exposure to subcritical water containing oxygen and NaCl at 300 °C for 50 h. Overall, the Fe and Cr contents discernibly decreased, the Ni content slightly reduced, the O content remarkably increased, and the Mo content was relatively stable from the metal substrate to its outmost surface. P detected in the oxide layer further verifies the formation of water-insoluble phosphates in corrosion products. The different compositions of Fe, Cr, Ni, and Mo are due to their different oxygen affinities and outward diffusion rates in the oxide layer [51]. The drop in the Cr content can be explained by the fact that oxidant induces relatively high electrochemical potential and thus converts Cr^{3+}

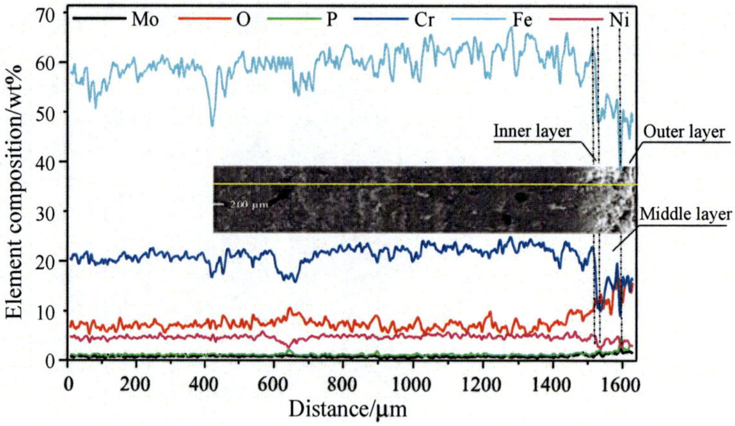

Fig. 6.23 SEM morphology and line scanning of the cross-section of the phosphate-passivated 316 SS sample after exposure to subcritical water at 300 °C for 50 h

(i.e., Cr_2O_3) to Cr^{6+} (i.e., $Cr_2O_7^{2-}$[44], CrO_3 [52, 53], and $CrO_2(OH)_2$ [45, 54]), further dissolving into the corrosion medium from the oxide layer. It is apparent that this depassivation dissolution process accelerates corrosion. However, the Cr diffusion rate in the oxide layer is lower than that of Fe and Ni, and the chromium oxide solubility is also lower than that of iron oxide in high temperature water [55]. The drop in the Ni content can be attributed to the synergy effect of chloride and oxygen on the Ni dissolution [34], because they too play a crucial role in promoting the decomposition of Ni oxides. Fe has the highest oxidation rate among Ni, Cr, Fe, and Mo, while Mo has a comparably low diffusion rate in the oxide film [34]. The information above is in good accordance with the changes in the Cr, Ni, Fe, and Mo contents (see Fig. 6.11). Moreover, the P distribution in the outer layer suggests the diffusing rate of PO_4^{3-} toward the alloy substrate is far lower than that of O. The diffusion rates of aggressive ions (e.g., O and Cl^-) in the oxide film should be the rate-determining steps for the oxidation and dissolution of metal elements.

The relative variations of the element components above reflect the depth and structure of the oxide film. Metal dissolution (e.g., Ni and Fe) and oxide film growth (e.g., Cr_2O_3 and Fe_2O_3) should be responsible for the alloy corrosion mechanism. The outer oxide layer is formed by the outward diffusion of metal cations, while the inner oxide layer is generated by the inward diffusion of oxygen anions [35], thereby determining the growth rate of the whole oxide layer [37]. Oxide film growth includes initial oxide nucleation at selected locations and the subsequent uniform growth of oxide grains till their combination to form a dense oxide layer [37]. Based on the elemental content changes in Fig. 6.11, the oxide film was composed of three-layer structures (i.e., inner layer, outer layer, and middle layer), as marked by dashed lines. Different alloy elements have different electric potentials and oxide enthalpies, resulting in complicated oxide layers formed under corrosion conditions [15]. According to the results in Figs. 6.8 and 6.11 and our latest report [48], the inner oxide layer was mainly composed of Cr_2O_3, the outer oxide layer was rich in Fe_2O_3 as well as phosphates including $CrPO_4$, $FePO_4$, and $Ni_3(PO_4)_2$, and the middle layer was the transition layer caused by diffusion action. The Cr-rich oxide layer has better corrosion resistance than the Fe-rich oxide layer, owing to the lower dissolution and diffusion rates of Cr-rich oxides. Cr-rich oxide [55].

After the metal oxides Cr_2O_3 and Fe_2O_3 were formed (see Figs. 6.20 and 6.21), Cl^- could react with them to generate dissoluble metal chlorides, but the phosphates formed in the 316 SS passivation process can inhibit these corrosion reactions. According to Figs. 6.13, 6.14, 6.20, and 6.21, phosphate formation in the outer layer of the oxide scale is an important factor for improving the corrosion resistance of 316 SS. Also, they probably react with Cl^- to produce acids [56] in subcritical water. Certainly, besides these phosphates, chemical compositions and the corrosion conditions of tested samples crucially affect final corrosion results. Notably, all obtained data in this research comes from a relatively short corrosion time (i.e., 50 h) and a low subcritical temperature (i.e., $\leq$300 °C), so it is necessary to validate the correctness of the present result with far longer times and higher subcritical temperatures.

6.2.4 Conclusions

The corrosion rate of phosphate-passivated 316 SS was approximately half of that of passivated 316 SS under the same conditions in subcritical water containing oxygen and NaCl at 25 MPa for 50 h. Compared with the passivated 316 SS, the phosphate-passivated 316 SS had a much denser and more uniform oxide film that was composed of Cr_2O_3, Fe_2O_3, $CrPO_4$, $FePO_4$, and $Ni_3(PO_4)_2$. The oxide film of the phosphate-passivated 316 SS sample after exposure to subcritical water at 300 °C consisted of three layers, that included an inner layer mainly composed of Cr_2O_3, an outer layer rich in Fe_2O_3 as well as phosphates, and a middle transition layer. The alloy elements (i.e., Cr, Fe, and Ni) in 316 SS could react with Na3PO4 in supercritical water to form stable water-insoluble protective phosphates (i.e., $CrPO_4$, $FePO_4$, and $Ni_3(PO_4)_2$). These phosphates could effectively prevent corrosive substances (e.g., Cl^- and O) from making contact with the 316 SS substrate, and thus inhibit corrosion in subcritical water containing oxygen and NaCl.

6.3 Corrosion of Coated and Uncoated 316 SS in Supercritical Water Oxidation of Municipal Sludge

6.3.1 Introduction

Municipal sludge (MS) contains a large amount of water and complex components (e.g., organic pollutants and inorganic salts), and thus is difficult to be disposed efficiently and cost-effectively via conventional methods (e.g., landfill and incineration). Supercritical water oxidation (SCWO) is a promising approach of harmlessly treating MS, with some advantages such as rapid reaction rate, no secondary pollution, and low energy consumption [57, 58]. However, harsh reaction environment (including high temperature and pressure, strong oxidizing property, and complex feedstock nature, etc.) makes the SCWO reactor suffer from corrosion problems, so severely hindering its development and application.

Till now, researchers have developed a variety of corrosion control techniques, such as the uses of high corrosion resistant material, liner, coating, transpiring wall reactor, adsorption/reaction on the fluidized solid phase, vortex/circulating flow reactor and pre-neutralization, and the optimization of operating conditions [3, 59]. Nonetheless, no one means can withstand corrosions at all feedstock conditions. Corrosion resistant coating, with some advantages in terms of operating cost, flexibility and feasibility, is an effective method of reducing the reactor corrosion rate, because it can prevent corrosive species (e.g., oxygen and inorganic salts) in the reaction fluid from contacting the reactor material in SCWO. Coating can be made of corrosion resistant materials (e.g., Ni-based alloy, Ti alloy and ceramic material) on a common metal substrate (e.g., austenitic stainless steel). For example, Agüero et al. [60] coated the P92 surface with Fe-Al compounds to enhance its corrosion resistance in ultra-supercritical steam oxidation. NiCrAlY coating on the surface of Incoloy 625 or Alloy 214 can effectively improve the corrosion resistance of the coated material in supercritical water [5, 6]. Drews et al. [61] found that TiO_2 coating could significantly increase the corrosion resistance of 316 stainless steel in oxidizing subcritical/supercritical water containing HCl. Hui et al. [14]conducted ceramic coating (e.g., Al_2O_3, CeO_2 and ZrO_2) on the P91 substrate by spray pyrolysis deposition to reduce its oxidation in deaerated supercritical water. The 316L SS samples coated with TiO_2 exhibited relatively better corrosion resistance than the samples coated with Al_2O_3 in supercritical water at 500 °C, 25 MPa for 80 h [12]. ZrO_2 ceramics exhibits good corrosion resistance in the aqueous aggressive sulfuric acid at 390 °C and 27 MPa [62].

Therefore, on the basis of good corrosion resistance in supercritical water, TiO_2 and ZrO_2 were selected as coating materials to inhibit metal matrix corrosion in MS SCWO environment in this work. The surface treatment of 316 stainless steel (316SS) was carried out by atmospheric plasma spraying to form a protective coating of ZrO_2 or TiO_2. Three samples (i.e., 316SS, ZrO_2/316SS, and TiO_2/316SS) were exposed to a typical MS SCWO environment (at 450 °C, 24 MPa and 1.2 of oxidation coefficient) in a batch reactor for 20–150 h. Corresponding corrosion morphologies, element

Table 6.4 Physical properties of two types of coating powders

Physical properties	Melting point (°C)	Rockwell hardness	Crystalline form	Density (g/cm^3)	Particle size (μm)
ZrO_2	2535	25	Monoclinic crystal system	5.3	10–53
TiO_2	1920	53	Tetragonal system	4.1	45–106

compositions and oxide components of oxide films, corrosion rates, and corrosion layer structures were evaluated by scanning electron microscopy, energy dispersive spectroscopy, weight measurement, X-ray diffraction, Raman spectroscopy, and X-ray photoelectron spectroscopy. The corrosion characteristic differences of ZrO_2-, TiO_2-coated and uncoated 316SS samples in MS SCWO with various exposure times are documented here for the first time. This information is valuable for guiding reactor material selection and corrosion control in MS SCWO.

6.3.2 Experimental Section

6.3.2.1 Materials

316SS was purchased from Shenzhen Metal Alloy Material Co., Ltd., and its elemental compositions (wt%) were C 0.08, Ni 9.69, Cr 16.87, Mo 1.13, Ti 0.12, Fe 69.21, Si 0.35, Cu 1.05 and others in trace. The 316SS sample (with the size of 80 × 20 × 2.5 mm) was polished to form a smooth mirror surface by the sandpapers of #400, #600, #1200 and #2000 in turn. It was cleaned by acetone in an ultrasonic cleaner and then dried in an oven (Electro-thermostatic blast) at 50 °C for 20 min. Next, one surface of the sample was coated by TiO_2 or ZrO_2 via atmospheric plasma spraying. The physical properties of these two types of coating powders are shown in Table 6.4. Herein, the mixture of argon and hydrogen was adopted as the plasma gas. After sprayed, the sample was dried in vacuum at 100 °C for 2 h. Eventually, three obtained samples (i.e., 316SS, TiO_2/316SS and ZrO_2/316SS) were weighted and kept in filter papers for subsequent corrosion tests in MS SCWO.

Municipal sludge was obtained from the Beishiqiao Sewage Treatment Center in Xi'an (China). The original MS contained a much high moisture content and large amounts of organic matters. Moreover, the used oxidant was hydrogen peroxide (30 wt%) produced by Tianjin Tianli Chemical Co., Ltd.

6.3.2.2 Experimental Procedures

A batch reactor (415 mL) was used to carry out corrosion tests in MS SCWO, and the reactor was made of 316SS and could withstand the limitations of 500 °C and 30 MPa. Before experiments, 316SS, ZrO_2/316SS and TiO_2/316SS samples were suspended in the reactor center. Certain amounts of original MS and hydrogen peroxide were then added into the reactor, and their addition amounts were calculated based on the reactor capacity, oxidant concentration and desired reaction conditions. After being sealed, the reactor was heated by a temperature controller up to 450 °C with 3.5 °C min^{-1} of heating rate, so its internal pressure increased accordingly. Hydrogen peroxide is easy to decompose into oxygen and water, and the actual oxidant in supercritical water was O_2 but not H_2O_2 in the experiments. After the desired exposure time (also including the preheating-up time and cooling-down time of the reactor), the heater was turned off and the reactor was cooled naturally to ambient temperature. Then, the reactor was opened and all tested samples were taken out. These samples were rinsed in ultrapure water for 6 min and then in ultrasonic acetone for 10 min to remove possible contaminants or deposits on their surfaces [15, 34]. They were further dried in a thermostatic oven at 50 °C for 2 h, and then weighed and kept in filter papers for following analyses. Notably, the experimental uncertainties of temperature, pressure and weight were 0.42%, 0.69% and 0.0038%, respectively.

6.3.2.3 Analytical Methods

The corrosion weight change of the tested sample was measured by an electronic scale with an accuracy of 0.0001 g. The corrosion rate of the sample was evaluated by weight variation before and after corrosion tests, based on the general corrosion hypothesis. The surface morphology of the corrosion sample was observed by a scanning electron microscope (Hitachi SU3500) equipped with an energy dispersive spectrometer (EDS). The elemental composition of oxide film was determined by the EDS analysis, and the line scanning of the sample cross-section was performed to detect elemental changes in the oxide film. A laser Raman spectroscopy (LabRAM HR Evolution) equipped with a powerful laser (1.5 W of maximum power) was used to identify sample surface oxides. The crystal structures of the oxides were analyzed by an X-Ray diffraction (XRD, D8 Advanced, Bruker AXS).The phase compositions of corrosion products were examined by an X-ray photoelectron spectrometer (XPS, ESCALAB Xi).

6.3.3 Results and Discussions

6.3.3.1 Coating Characterization

Figure 6.24 elucidates the SEM morphologies and line scanning images of coated samples (i.e., ZrO_2/316SS and TiO_2/316SS) before exposure to MS SCWO. Apparently, the coatings adhered tightly to the 316SS substrate surface, and their average thicknesses were separately around 105 μm (87–122 μm) and 121 μm (108–133 μm)

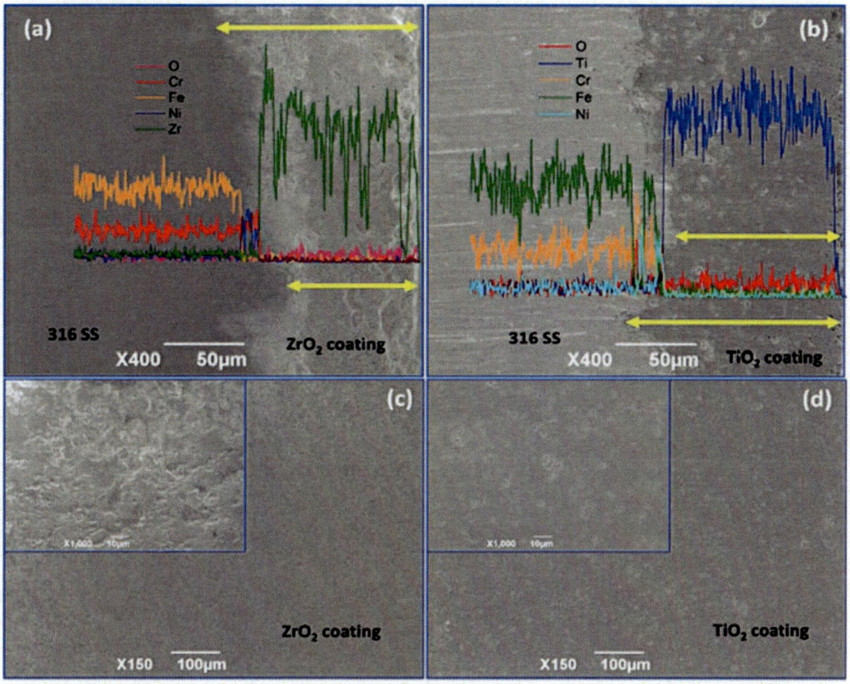

Fig. 6.24 SEM morphologies and line scanning images of ZrO_2/316SS and TiO_2/316SS before exposure to MS SCWO. **a** ZrO_2/316SS cross-section, **b** TiO_2/316SS cross-section, **c** ZrO_2/316SS surface, **d** TiO_2/316SS surface

for the ZrO_2- and TiO_2-coated samples (see yellow arrows in Fig. 6.24a, b). The coating elements were dominated by Zr and O for the ZrO_2/316SS sample, but Ti and O for the TiO_2/316SS sample. Notably, each coating surface was smooth and flat before corrosion tests (see Fig. 6.24c, d). However, there were some shallow micro-pits on the TiO_2/316SSsurface, which may have a negative impact on the coating protective effect.

6.3.3.2 Corrosion Morphology

Figure 6.25 illustrates the surface corrosion morphologies of 316 SS, ZrO_2/316 SS and TiO_2/316 SS exposed to the MS SCWO environment. Apparently, 316SS showed the most severe pitting corrosion morphology, while ZrO_2/316SS seemingly suffered from the lightest corrosion among the three samples. One possible explanation for the pitting corrosion phenomenon is that a large amount of dissolution loss of Fe-rich oxides occurs after oxide film formation, similar to the finding on the 316L SS oxide film in supercritical water [63]. TiO_2/316SS also took place severe surface cracks and even oxide scale exfoliation in the case of 150 h. However, Drews et al. [61] reported

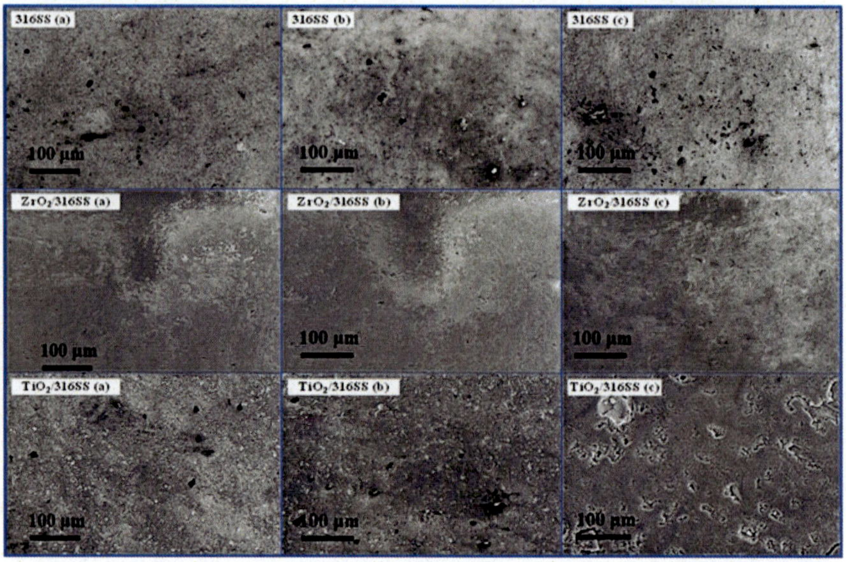

Fig. 6.25 Surface corrosion morphologies (×150) of tested samples exposed to MS SCWO for different exposure times. **a** 20 h, **b** 70 h, **c** 150 h

that TiO_2 coating could significantly improve the corrosion resistance of 316SS in sub/supercritical HCl solution. The inconsistence is likely due to the higher oxygen concentration and the existence of complex inorganic salts in this work. Overall, ZrO_2/316SS exhibited the best corrosion resistance among the three samples in MS SCWO. Additionally, prolonging exposure time accelerated each sample corrosion, and the unobvious change here can be attributed to the relatively short test time.

Figure 6.26 shows the high-magnification of SEM images of surface corrosion morphologies of three tested samples after exposed to MS SCWO for 150 h. Hence, the 316SS surface was covered with many amorphous and irregular oxides (see

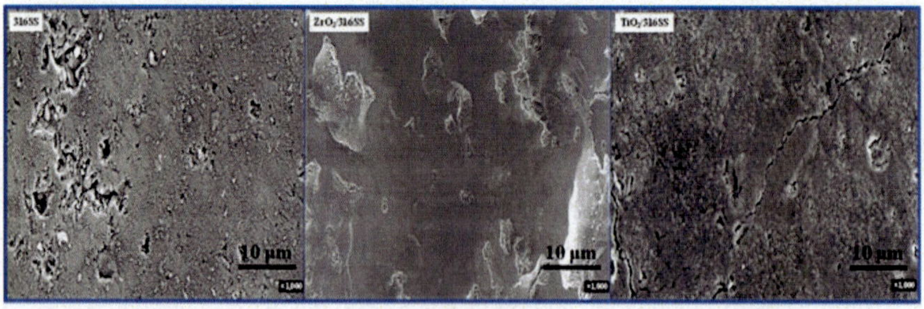

(a) (b) (c)

Fig. 6.26 High-magnification (×1000) of SEM images of surface morphologies of tested samples after 150 h exposure to MS SCWO. **a** 316SS, **b** ZrO_2/316SS, **c** TiO_2/316SS

Fig. 6.26a), meaning that severe corrosion occurs on the 316SS surface in MS SCWO. TiO_2/316SS displayed some remarkable surface cracks (see Fig. 6.26c), so it is difficult to form a protective continuous oxide film on the sample surface. Once the corrosion cracks continue to develop vertically and horizontally, high concentration of corrosion products will be generated near the corrosion pits. This may further aggravate corrosion and even causes the exfoliation of the oxide layer. In essence, the corrosion resistance of one material generally depends on the stability of its oxide film formed on the material surface [41, 64]. Notably, the ZrO_2/316 SS sample formed a relatively continuous and stable oxide scale, which helps to effectively prevent corrosive species (e.g., Cl^-) from contacting the 316SS substrate and thus provides a good protection effect. The findings regarding high-magnification of SEM images are consistent with those in Fig. 6.25.

6.3.3.3 Elemental Composition of Oxide Film

Figure 6.27 displays the major elemental compositions of oxide films of three tested samples exposed to MS SCWO for different times. Apparently, the Ni content significantly reduced while the Fe and O contents slightly increased for the 316SS sample with time increasing (see Fig. 6.27a). The phenomenon can be explained by the fact that Ni migrates outward and then dissolves in the MS SCWO medium, and more Fe oxides are formed on the 316SS surface in the presence of excessive oxygen as exposure time increased. The ZrO_2/316 SS corrosion products showed that the Zr content in the coating significantly decreased from 57 to 52 wt% (see Fig. 6.27b), while

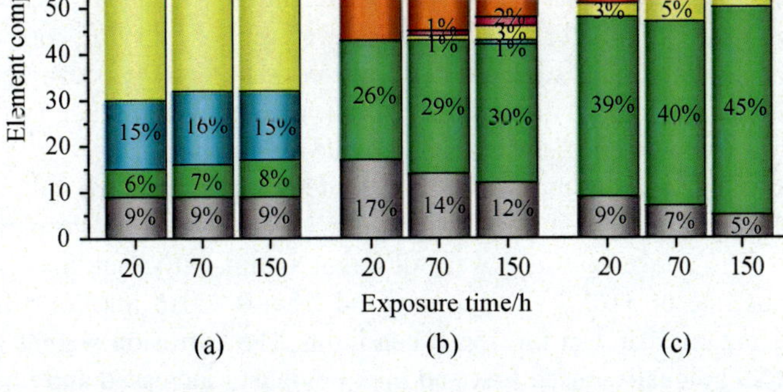

Fig. 6.27 Major elemental compositions of oxide films of tested samples exposed to MS SCWO for different exposure times. **a** 316SS, **b** ZrO_2/316SS, **c** TiO_2/316SS

the O content increased from 26 to 30 wt% with time changing from 20 to 150 h. The Fe and Ni contents increased with exposure time increasing. This suggests that ZrO_2 takes place dissolution or exfoliation to some extent in MS SCWO with time increasing, and corrosion products (e.g., Fe oxides and Ni oxides) are formed and then penetrate the ZrO_2 coating accordingly. Thus, although ZrO_2/316SS displayed good corrosion resistance (see Figs. 6.25 and 6.26), corrosion still occurred and developed after a certain period of time. Similarly, for the TiO_2/316 SS sample, the Ti content in corrosion products significantly decreased from 49 to 41 wt% (see Fig. 6.27b), whereas the O and Fe contents separately increased from 39 to 45 wt% and from 3 to 7 wt% with test time changing from 20 to 150 h. Ni first appeared and then its content increased as exposure time increased. The obvious drop of the Ti content implies that TiO_2 is easily converted into water-soluble ion states, and partial Ti in the coating may dealloy from the sample surface and dissolved in supercritical water [65] or take place exfoliation in MS SCWO.

The major elemental composition variations above further confirm that increasing exposure time accelerates the corrosion of each tested sample. Notably, compared with the TiO_2/316SS sample, the ZrO_2/316SS sample exhibited less Fe oxides on its surface after corrosion for the same time. This shows that ZrO_2/316SS had better corrosion resistance than TiO_2/316SS in MS SCWO, in good accordance with the findings in Figs. 6.25 and 6.26.

6.3.3.4 Corrosion Rate

Figure 6.28 provides the corrosion weight changes and corrosion rates of tested samples in MS SCWO for different exposure times, which help quantificationally evaluate the applicability of each coating process in MS SCWO. The lowest corrosion weight change seemingly occurred on the 316SS sample at the same conditions, but in fact this change was reflected by weight loss. This can be ascribed to the exfoliation of surface oxides and/or the loss of alloy elements. On the contrary, if formed oxides adhere on the alloy surface stably and uniformly, this helps reduce corrosion by preventing corrosive species from contacting the alloy substrate [48]. Thus, the weight loss here implies that 316SS has relatively low corrosion resistance. Different from 316SS, ZrO_2/316SS or TiO_2/316SS showed weight gain after corrosion tests at the same conditions. This suggests that some corrosion products can tightly adhere on the sample surface and these products have a higher weight than lost substances caused by metal element dissolution and oxide exfoliation.

As shown in Fig. 6.28, the corrosion weight loss or gain of each sample increased as exposure time increased (see bar graphs). However, their corresponding variation rates (i.e., corrosion rates) were different significantly (see line graphs). The weight loss rate of 316SS slowly increased from 0.00035 mg cm^{-2} h^{-1} at 20 h to 0.0011 mg cm^{-2} h^{-1} at the 150 h condition. The corrosion weight gain rate of ZrO_2/316SS gradually raised first and then obviously increased and further slowly reduced within 20–150 h, and the maximum value was 0.0097 mg cm^{-2} h^{-1} in the case of 70 h. Surprisingly, there was a time interval (i.e., 50–70 h), in which

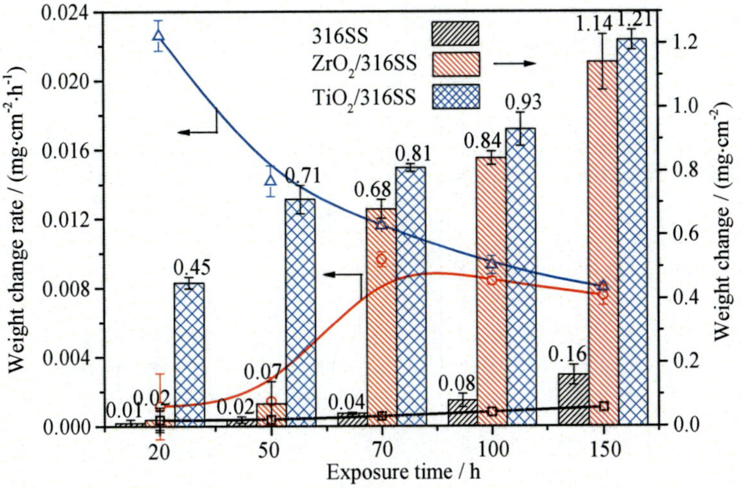

Fig. 6.28 Weight and corrosion rate variations of three tested samples exposed in MS SCWO for different exposure times (weight loss for 316SS (square), and weight gains for ZrO_2/316SS (circle) and TiO_2/316SS (triangle))

the corrosion rate of ZrO_2/316 significantly increased in MS SCWO. Hence, how to reduce the corrosion rate in the time interval should be explored in subsequent work. The weight gain rate of TiO_2/316SS was higher than that of ZrO_2/316SS totally at the same conditions. It had the maximum value of 0.23 mg cm^{-2} h^{-1} and presented a remarkable reduction tendency. One possible explanation is that the total exfoliation and dissolution rate of surface oxides and alloy elements is lower than the total formation rate of surface oxides, especially within 20–70 h. Overall, according to the weight changes and corrosion rates, it can be believable that the order of corrosion resistances of three tested samples in MS SCWO was ZrO_2/316SS > TiO_2/316SS > 316 SS, in agreement with the findings in Figs. 6.25 and 6.26.

6.3.3.5 XRD Analysis

Figure 6.29 indicates the XRD patterns of surface oxides of 316SS, TiO_2/316SS and ZrO_2/316SS after 150 h exposure in MS SCWO. Different peaks represent various crystal structures of oxides, and strong substrate peaks imply the thin surface oxide layers of tested samples. According to the XRD analysis, the oxide film of 316SS was composed of NiO, Cr_2O_3, FeO, Fe_2O_3, and Fe_3O_4. Consistently, the oxide layer of 316SS after exposed to supercritical water consists of oxides rich in Fe and Cr [66]. The Cr element plays an important role in improving the corrosion resistance of alloy material [67]. The much higher Fe content in the oxide layer can be explained by the fact that the Fe element has higher diffusivity and dissolution rates than other elements in supercritical water [68–70].

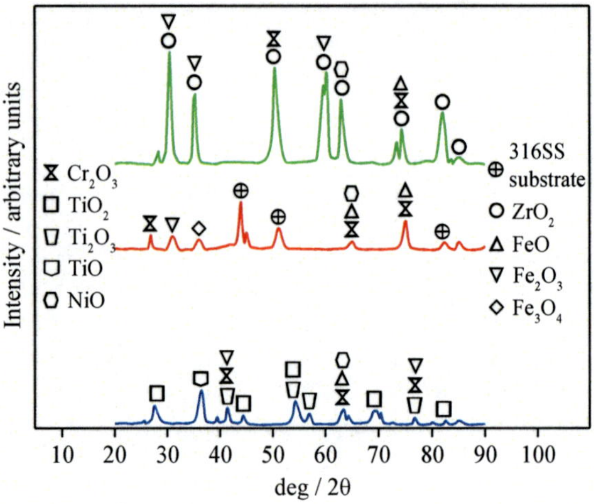

Fig. 6.29 XRD patterns of surface oxides of three tested samples after 150 h exposure in MS SCWO, and the lines from top to bottom represent ZrO_2/316SS, 316SS, and TiO_2/316SS, respectively

As shown in Fig. 6.29, the oxide film of ZrO_2/316SS after 150 h exposure in MS SCWO consisted of ZrO_2, FeO, Fe_2O_3, NiO and Cr_2O_3, in agreement with the findings in the EDS analysis (see Fig. 6.27b). Apart from the coating ZrO_2 peaks, the peaks of FeO, Fe_2O_3, NiO and Cr_2O_3 all exhibited relatively low strength. This shows that ZrO_2 was relatively stable and had a good inhibiting effect on the attack of corrosive species (e.g., Cl^- and O_2) in MS SCWO. However, some other alloy elements (e.g., Fe, Ni and Cr) still react with oxygen to form small amounts of metal oxides by diffusion. It is worth noticing that TiO and Ti_2O_3 were formed on the TiO_2/316SS surface after 150 h exposure in MS SCWO. This can be explained by the fact that Ti departs from the TiO_2/316SS surface of into the fluid phase and is converted into ionic states, which further react with supercritical water and dissolved oxygen in water [34]. Hence, the stability of the coating TiO_2 should be paid more attention if TiO_2 is selected as the coating material of 316SS in MS SCWO. Similarly, FeO, Fe_2O_3 and Cr_2O_3 were also detected on the surface layer of TiO_2/316SS, indicating that the Fe and Cr elements of 316SS diffuse outward to react with oxygen via oxidation reactions. The surface oxides of TiO_2/316SS after corrosion exposure were mainly composed of Cr_2O_3, FeO, Fe_2O_3, NiO, TiO, TiO_2 and Ti_2O_3.

6.3.3.6 Raman Analysis

Figure 6.30 displays the Raman spectra of surface oxides of three tested samples in MS SCWO for different exposure times. Seemingly, there were eight signal peaks in the Raman spectra of 316SS in the range of (100–700) cm^{-1}, i.e., at approximately (225, 297, 304, 424, 480, 506, 633 and 671) cm^{-1} (see Fig. 6.30a). According to previous literatures [71–73], the peaks at 225 and 297 cm^{-1} represented Fe_2O_3, the characteristic peaks of Cr_2O_3 were at 304 and 633 cm^{-1}, the peaks at 424 and

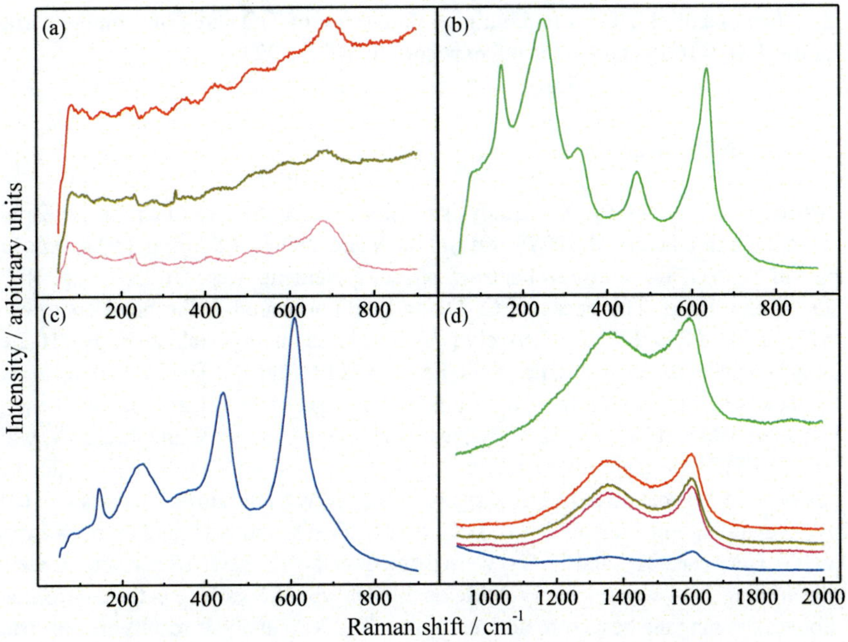

Fig. 6.30 Raman spectra of oxide films of three tested samples exposed in MS SCWO for different exposure times. **a** 316SS (the lines from top to bottom represent the results of 20 h, 70 h, and 150 h, respectively), **b** ZrO_2/316SS (150 h), **c** TiO_2/316SS (150 h), **d** three samples (the colors correspond to the lines in (**a–c**)

480 cm^{-1} were separately FeO and NiO, and the peaks at 506 and 671 cm^{-1} were consistent with the characteristics of Fe_3O_4. Notably, the peak signal of FeO (at 424 cm^{-1}) seems to have a higher strength at 20 h than that at 150 h, meaning the drop of the FeO content in the oxides. This suggests that increasing exposure time accelerates 316SS corrosion, because Fe is transformed from the low valence state of FeO to the high valence state of oxides (i.e., Fe_2O_3 and Fe_3O_4).

Figure 6.30b shows the Raman spectrum of the oxide film of ZrO_2/316 SS in the case of 150 h. Five characteristic peaks appeared at around (146, 271, 381, 473 and 636) cm^{-1}, representing the characteristics of ZrO_2 [74–76]. Surprisingly, the characteristic peaks of common oxides (i.e., FeO, Fe_2O_3, NiO and Cr_2O_3) were not found here, probably due to the much weak intensities of their corresponding sub-peaks. This further verifies that ZrO_2/316 SS had good corrosion resistance in MS SCWO tested in this work. In the Raman spectrum of the TiO_2/316 SS oxide film after 150 h (see Fig. 6.19c), there were four characteristic peaks at about (143, 279, 332, 426 and 611) cm^{-1}. According to previous reports [77, 78], the peaks corresponding to Ti_2O_3 were observed at 279 and 426 cm^{-1}, the peaks of TiO_2 appeared at 143 and 611 cm^{-1}, and the peak at 332 cm^{-1} could be attributed to TiO. It is worth noticing that TiO_2 had a strong signal (at 611 cm^{-1}) in the Raman spectrum, so suggesting that the content of TiO_2 is much higher than those of TiO and Ti_2O_3 on the surface of the TiO_2/316SS sample during exposure in MS SCWO.

6.3.3.7 XPS Analysis

Figure 6.31 indicates the XPS analysis results of valence states of Cr, Fe, Ni, Zr and Ti in the oxide layers of 316SS, ZrO_2/316SS and TiO_2/316SS after 150 h exposure in MS SCWO. The core level spectra of major elements were Cr 2p, Fe 2p, Ni 2p, Zr 3d and Ti 2p. The peaks of Cr_2O_3 and NiO appeared at the bond energies of 577.5 eV and 856.9 eV, respectively [79]. The bond energy peak at nearly 710.2 eV may consist of three oxides, i.e., FeO, Fe_3O_4 and Fe_2O_3 [24]. The Zr 3d bond energy at about 182.4 eV represented the characteristic peak of ZrO_2 [80]. The bond energy peaks of TiO and Ti_2O_3 separately located at around 455.9 eV and 462.2 eV, while those of TiO_2 were approximately at 464.4 eV and 458.8 eV [81]. Overall, the XPS analysis shows that after 150 h exposure in MS SCWO, the surface oxides of 316SS were mainly composed of Cr_2O_3, FeO, Fe_2O_3, Fe_3O_4 and NiO, and extra two types of Ti oxides (i.e., TiO and Ti_2O_3) were formed on the TiO_2/316SS surface. However, the coating ZrO_2 was relatively stable on the ZrO_2/316SS surface in the environment above, showing the best corrosion resistance. The XPS analysis results are consistent with those of SEM, XRD and Raman analyses mentioned before.

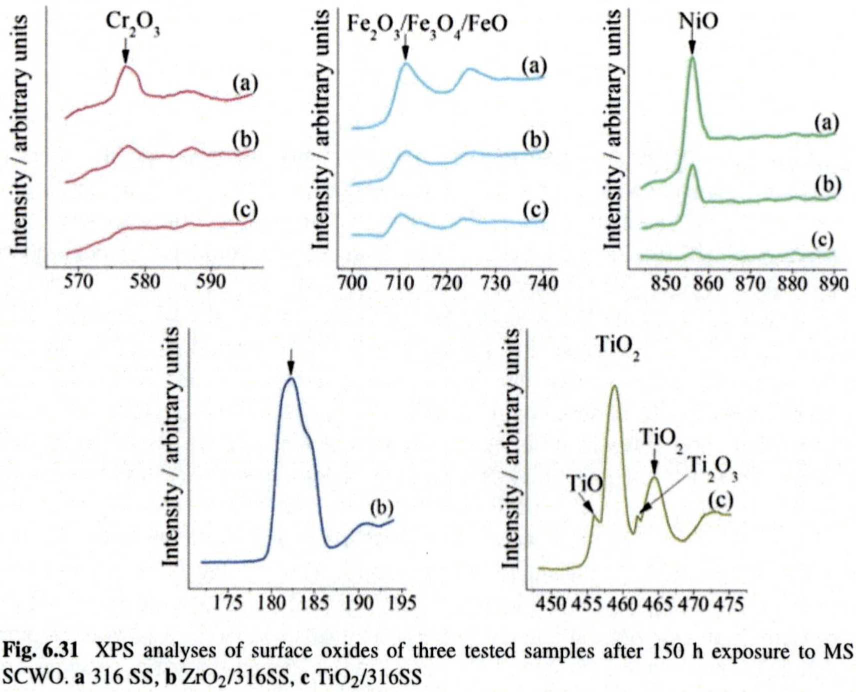

Fig. 6.31 XPS analyses of surface oxides of three tested samples after 150 h exposure to MS SCWO. **a** 316 SS, **b** ZrO_2/316SS, **c** TiO_2/316SS

6.3.3.8 Corrosion Layer Structure and Composition

Supplementary materials illustrate the SEM morphology, line scanning, and major element distribution of the 316SS cross-section after 150 h exposure in MS SCWO. Obviously, the thickness and structure of the 316SS oxide film can be observed from supplementary materials, and its oxide film was composed of an inner layer and an outer layer, based on major element content variations. Overall, the Fe and Cr contents significantly decreased while the O and C contents remarkably increased from the 316SS substrate to the interface of the inner and outer layers. This has good agreement with the element distributions shown in supplementary materials. It can be concluded from Figs. 6.18, 6.19, 6.20 and supplementary materials that the inner and outer layers of the 316SS oxide film were mainly composed of Cr_2O_3 and Fe-rich oxides (i.e., FeO, Fe_2O_3 and Fe_3O_4), respectively, in accordance with previous findings on 316SS corrosion in supercritical water [66]. The 316 SS surface showed severe Fe loss likely due to element dissolution and oxide exfoliation, which may significantly exacerbate corrosion and cause pitting corrosion (see Fig. 6.25).

Figure 6.21 further provides SEM morphologies and line scanning images of cross-sections of ZrO_2/316SS and TiO_2/316SS after 150 h exposure in MS SCWO. Apparently, in the coating transition layer, the major elements (i.e., Fe, Cr and Ni) derived from the 316SS substrate remarkably reduced, while the coating elements (i.e., Zr and O in ZrO_2, and Ti and O in TiO_2) obviously increased. These variation phenomena tended to be mild in the coating due to the diffusion effects of elements. This change region can be called as the diffusion layer of ZrO_2 or TiO_2 coating in corrosion exposure. Hence, the whole coating layer can be considered as the combination of the transition layer and the diffusion layer. The mass ratios of Zr/O and Ti/O separately reduced approximately from 2.8 to 2.3 and from 2.0 to 1.4 in the diffusion layer. This shows the inward diffusion of the O element, which likely destroys the protection performance of the ZrO_2 or TiO_2 coating. Similarly, the oxide film of ZrO_2/316SS or TiO_2/316SS could be divided into an inner layer and an outer layer, based on elemental composition changes. Therefore, the ZrO_2/316SS or TiO_2/316SS sample after 150 h exposure to MS SCWO consisted of the 316SS substrate, the coating layer (the transition layer + the diffusion layer), and the inner layer and the outer layer of the oxide film.

As indicated in Fig. 6.32a, in the inner layer of the oxide film of ZrO_2/316SS, the Zr and O contents first sharply reduced and then raised, whereas the Fe, Cr and Ni contents rapidly increased and then decreased accordingly. Different from the cases in the inner layer of the oxide film of ZrO_2/316SS, the Ti content obviously increased and the Ni content was almost constant in the inner layer of the oxide film of TiO_2/316SS. Based on the results in Figs. 6.29, 6.30, 6.31, 6.32 and supplementary materials, the oxides in the inner oxide layer of ZrO_2/316SS were composed of FeO, Fe_2O_3, ZrO_2, Cr_2O_3 and NiO, while those in the inner oxide layer of TiO_2/316SS consisted of FeO, Fe_2O_3, Cr_2O_3, TiO, TiO_2, Ti_2O_3 and NiO. Cr_2O_3 was rich in the inner layer other than in the outer layer of oxide film of each sample. The coating ZrO_2 was much stabler than the coating TiO_2 in the tested MS SCWO environment, which may be responsible for the better corrosion resistance of ZrO_2/316SS than that

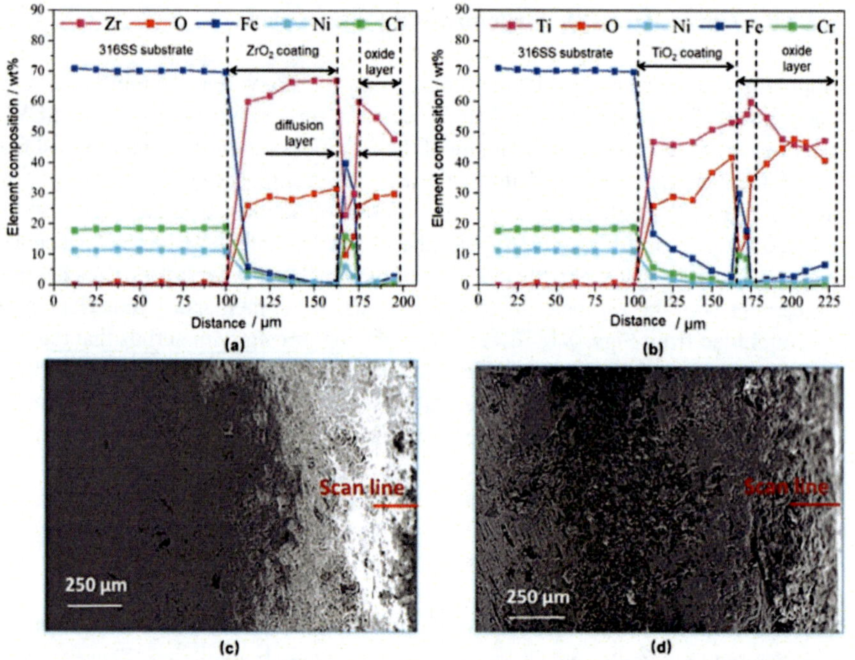

Fig. 6.32 **a**, **b** SEM morphologies and **c**, **d** line scanning images of cross-sections of ZrO_2/316SS and TiO_2/316SS after 150 h exposure to MS SCWO. **a**, **c** ZrO_2/316SS, **b**, **d** TiO_2/316SS, and the scales of EDS (**a**, **b**) match microscopic images (**c**, **d**), respectively

of TiO_2/316SS (see Figs. 6.25 and 6.26). This can be further verified by the findings that ZrO_2 was the most prevailing component in the outer layer of ZrO_2/316SS, whereas Ti oxides and other metal oxides were major components in the outer oxide layer of TiO_2/316SS, based on element content variations. The solid-state growth of metal oxides may lead to the formation of the outer oxide layer. However, it is still believable that the TiO_2 coating had a positive effect on the corrosion protection of 316SS in MS SCWO, according to the composition analyses of oxide films in Figs. 6.29, 6.30, 6.31, 6.32 and supplementary materials. Notably, the present work mainly compared the corrosion characteristics of ZrO_2-, TiO_2-coated and uncoated 316SS samples in MS SCWO systematically. However, the effects of MS conversion processes in SCWO on the corrosion characteristics of coated 316SS samples require to be deeply explored in subsequent work.

6.3.4 Conclusions

In MS SCWO at 450 °C, 24 MPa and 1.2 of oxidation coefficient, 316SS suffered from severe pitting corrosion, TiO_2/316SS took place surface cracks and even oxide film exfoliation in the case of 150 h, while ZrO_2/316SS showed the lightest corrosion. The order of corrosion resistances of three tested samples in MS SCWO was ZrO_2/316SS > TiO_2/316SS > 316SS. The corrosion rate of ZrO_2/316SS gradually raised first and then slowly reduced with exposure time increasing within 20 h to 150 h, and the maximum value was 0.0097 $mg \cdot cm^{-2} \cdot h^{-1}$ for 70 h. The ZrO_2/316SS sample after 150 h exposure in MS SCWO consisted of the 316SS substrate, the transition layer and diffusion layer of the coating layer, and the inner layer and outer layer of oxide film. The oxide film of ZrO_2/316SS consisted of ZrO_2, FeO, Fe_2O_3, NiO and Cr_2O_3, and the ZrO_2 coating can be considered as the coating material of a 316SS reactor for MS SCWO. However, a longer time of corrosion test and more comprehensive property evaluation concerning ZrO_2/316SS in MS SCWO are still required for its large scale application.

Supplementary Materials

The reactor temperature was measured by a K-type armored thermocouple with a measurement range from 0 to 800 °C, and the maximum value of the measurement deviation was $\pm 800 \times 0.25\% = \pm 2$ °C. The signal output of the thermocouple was transmitted onto the operation panel for display, and the absolute error of the temperature displayed on the panel was less than 0.5 °C. In the experiments, the minimum temperature of the fluid inside the reactor was 300 °C. Hence, the maximum relative uncertainty of the fluid temperature was

$$\frac{\delta T}{T} = \frac{\sqrt{2^2+0.5^2}}{300} = 0.69\%.$$

The pressure gauge (0–40 MPa) in the experimental system had a test accuracy of ±0.25%. Thus, the maximum value of the measurement deviation was $40 \times 0.25\% = 0.1$ MPa. In the experiments, the working pressure of the reactor was 24 MPa, so the relative uncertainty of the pressure was

$$\frac{\delta P}{P} = \frac{0.1}{24} = 0.42\%.$$

The accuracy of the electronic balance used in this work was 0.1 mg, and the minimum measured value in the experiments was 2.6×103 mg. Thus, the maximum relative uncertainty of the sample weight was

$$\frac{\delta m}{m} = \frac{0.1}{2.6 \times 10^3} = 0.0038\%.$$

References

1. E. Fauvel, C. Joussot-Dubien, P. Guichardon, G. Charbit, F. Charbit, S. Sarrade, A double-wall reactor for hydrothermal oxidation with supercritical water flow across the inner porous tube. J. Supercrit. Fluids **28**, 47–56 (2004)
2. C. Sun, R. Hui, W. Qu, S. Yick, Progress in corrosion resistant materials for supercritical water reactors. Corros. Sci. **51**, 2508–2523 (2009)
3. P.A. Marrone, G.T. Hong, Corrosion control methods in supercritical water oxidation and gasification processes. J. Supercrit. Fluids **51**, 83–103 (2009)
4. K.K. Mandapaka, R.S. Cahyadi, S. Yalisove, W. Kuang, K. Sickafus, M.K.P.S. Was, Corrosion behavior of ceramic-coated ZIRLO^{TM} exposed to supercritical water. J. Nucl. Mater. **498**, 495–504 (2018)
5. S. Bsat, X. Huang, S. Penttila, Corrosion behaviour of bare and NiCrAlY coated alloy 214 in supercritical water at 700 °C. J. Nucl. Eng. Radiat. Sci. **4**, 011017 (2017)
6. A. Selvig, X. Huang, D.J. Kim, D. Guzonas, Surface oxide formation on IN625 and plasma sprayed NiCrAlY after high density and low density supercritical water testing. Mater. Corros. **65**, 768–777 (2014)
7. X. Huang, Developing corrosion prevention coating solutions for the canadian SCWR concept. JOM **68**, 480–484 (2015)
8. F. Khatkhatay, L. Jiao, J. Jian, W. Zhang, Z. Jiao, J. Gan, H. Zhang, X. Zhang, H. Wang, Superior corrosion resistance properties of TiN-based coatings on zircaloy tubes in supercritical water. J. Nucl. Mater. **451**, 346–351 (2014)
9. M. Oksa, S. Tuurna, J. Metsäjoki, S. Penttilä, Oxidation performance coating for future supercritical power plants. J. Nucl. Eng. Radiat. Sci. **2**, 011018 (2016)
10. X. Zhang, Q. Yan, Y. Yang, Z. Hong, L. Zhang, C. Ge, Corrosion resistance of chromium-coated ferritic-martensitic steel in supercritical water. Corrosion **71**, 50–56 (2015)
11. F.J. Perez, S.I. Castaneda, TG-mass spectrometry studies in coating design for supercritical steam turbines. Mater. Corros. **59**, 409–413 (2008)
12. Y. Wang, F. Gao, J. Yang, Y. Zhu, C. Fang, S. Wang, G. Zhao, Comparative study on corrosion characteristics of Al_2O_3/316L and TiO_2/316L stainless steel in supercritical water. Int. J. Hydrogen Energ. **42**, 19836–19842 (2017)
13. K.M. Garcia, R. Mizia, Corrosion investigation of multilayered ceramics and experimental nickel alloys in SCWO process environments, vol. 94 (Idaho National Engineering Laboratory INEL, 1995), p. 0017
14. R. Hui, W. Cook, C. Sun, Y. Xie, P. Yao, J. Miles, R. Olive, J. Li, W. Zheng, L. Zhang, Deposition, characterization and performance evaluation of ceramic coatings on metallic substrates for supercritical water-cooled reactors. Surf. Coat. Technol. **205**, 3512–3519 (2011)
15. X. Tang, S. Wang, D. Xu, Y. Gong, J. Zhang, Y. Wang, Corrosion behavior of Ni-based alloys in supercritical water containing high concentrations of salt and oxygen. Ind. Eng. Chem. Res. **52**, 18241–18250 (2013)
16. X. Ren, M. Zhao, C. Wan, Y. Zheng, W. Pan, High-temperature aging of plasma sprayed quasi-eutectoid LaYbZr2O7-part I: phase evolution. J. Am. Ceram. Soc. **98**, 2829–2835 (2015)
17. P. Fauchais, Understanding plasma spraying. J. Phys. D: Appl. Phys. **37**, 86–108 (2004)
18. G. Bidron, S. Goutier, M. Vardelle, P. Denoirjean, P. Fauchais, Flattening behavior of micro- and nano-sized yttria-stabilized zirconia particles plasma-sprayed on smooth preheated (610 K) nickel substrate: part I. J. Phys. D: Appl. Phys. **52**, 1–9 (2019)
19. S. Mahboubi, Y. Jiao, W. Cook, W. Zheng, D.A. Guzonas, G.A. Botton, J.R. Kish, Stability of chromia (Cr_2O_3)-based scales formed during corrosion of austenitic Fe-Cr–Ni alloys in flowing oxygenated supercritical water. Corrosion **72**, 1170–1180 (2016)
20. X. Guo, K. Chen, W. Gao, Z. Shen, L. Zhang, Corrosion behavior of alumina-forming and oxide dispersion strengthened austenitic 316 stainless steel in supercritical water. Corros. Sci. **138**, 297–306 (2018)
21. M. Payet, L. Marchetti, M. Tabarant, J.P. Chevalier, Corrosion mechanism of a Ni-based alloy in supercritical water: impact of surface plastic deformation. Corros. Sci. **100**, 47–56 (2015)

22. Y. Li, S. Wang, P. Sun, J. Yang, X. Tang, D. Xu, Y. Guo, J. Yang, D.D. Macdonald, Investigation on early formation and evolution of oxide scales on ferritic–martensitic steels in supercritical water. Corros. Sci. **135**, 136–146 (2018)
23. Z. Zhu, H. Xu, H.I. Khan, D. Jiang, N. Zhang, Effect of exposure temperature on oxidation of austenitic steel HR3C in supercritical water. Oxid. Met. **91**, 77–93 (2019)
24. X. Zhong, E.H. Han, X. Wu, Corrosion behavior of Alloy 690 in aerated supercritical water. Corros. Sci. **66**, 369–379 (2013)
25. K.I. Choudhry, D.A. Guzonas, D.T. Kallikragas, I.M. Svishchev, On-line monitoring of oxide formation and dissolution on alloy 800H in supercritical water. Corros. Sci. **111**, 574–582 (2016)
26. Y. Li, S. Wang, P. Sun, D. Xu, M. Ren, Y. Guo, G. Lin, Early oxidation mechanism of austenitic stainless steel TP347H in supercritical water. Corros. Sci. **128**, 241–252 (2017)
27. Y. Behnamian, A. Mostafaei, A. Kohandehghan, B.S. Amirkhiz, R. Zahiri, W. Zheng, D. Guzonas, M. Chmielus, W. Chen, J.L. Luo, A comparative study on the oxidation of austenitic alloys 304 and 304-oxide dispersion strengthened steel in supercritical water at 650 °C. J. Supercrit. Fluids **119**, 245–260 (2017)
28. X. Huang, Y. Shen, Oxidation behavior of an 11Cr ferritic-martensitic steel after Ar-ions irradiation in supercritical water. J. Nucl. Mater. **461**, 1–9 (2015)
29. J. Yang, S. Wang, X. Tang, Y. Wang, Y. Li, Effect of low oxygen concentration on the oxidation behavior of Ni-based alloys 625 and 825 in supercritical water. J. Supercrit. Fluids **131**, 1–10 (2018)
30. X. Zhong, X. Wu, E.H. Han, Effects of exposure temperature and time on corrosion behavior of a ferritic–martensitic steel P92 in aerated supercritical water. Corros. Sci. **90**, 511–521 (2015)
31. Y. Behnamian, A. Mostafaei, A. Kohandehghan, B.S. Amirkhiz, J. Li, R. Zahiri, E. Aghaie, W. Zheng, D. Guzonas, M. Chmielus, W. Chen, J.L. Luo, Internal oxidation and crack susceptibility of alloy 310S stainless steel after long term exposure to supercritical water at 500 °C. J. Supercrit. Fluids **120**, 161–172 (2017)
32. S.R.J. Saunders, M. Monteiro, F. Rizzo, The oxidation behaviour of metals and alloys at high temperatures in atmospheres containing water vapour: a review. Prog. Mater. Sci. **53**, 775–837 (2008)
33. M. Sun, X. Wu, Z. Zhang, E.H. Han, Analyses of oxide films grown on alloy 625 in oxidizing supercritical water. J. Supercrit. Fluids **47**, 309–317 (2008)
34. X. Tang, S. Wang, L. Qian, Y. Li, Z. Lin, D. Xu, Y. Zhang, Corrosion behavior of nickel base alloys, stainless steel and titanium alloy in supercritical water containing chloride, phosphate and oxygen. Chem. Eng. Res. Des. **100**, 530–541 (2015)
35. Z. Zhang, Z.F. Hu, L.F. Zhang, K. Chen, P.M. Singh, Effect of temperature and dissolved oxygen on stress corrosion cracking behavior of P92 ferritic-martensitic steel in supercritical water environment. J. Nucl. Mater. **498**, 89–102 (2018)
36. D. Rodriguez, A. Merwin, D. Chidambaram, On the oxidation of stainless steel alloy 304 in subcritical and supercritical water. J. Nucl. Mater. **452**, 440–445 (2014)
37. H. Shi, Z. Gao, Z. Fan, Y. Ding, Y. Qiao, Z. Zhu, Corrosion behavior of alloy C-276 in supercritical water. Adv. Mater. Sci. Eng. **2018**, 1–6 (2018)
38. P. Kritzer, E. Dinjus, An assessment of supercritical water oxidation (SCWO) existing problems, possible solutions and new reactor concepts. Chem. Eng. J. **83**, 207–214 (2001)
39. M. Akgün, O.Ö. Sögüt, Supercritical water oxidation (SCWO) for wastewater treatment (Springer, 2014), pp. 261–298
40. P. Kritzer, Corrosion in high-temperature and supercritical water and aqueous solutions: a review. J. Supercrit. Fluid **29**, 1–29 (2004)
41. Y. Daigo, Y. Watanabe, K. Sue, Effect of chromium ion from autoclave material on corrosion behavior of nickel-based alloys in supercritical water. Corrosion **63**, 277–284 (2007)
42. P. Kritzer, N. Boukis, E. Dinjus, The corrosion of alloy 625 (NiCr22Mo9Nb; 2.4856) in high-temperature, high-pressure aqueous solutions of phosphoric acid and oxygen corrosion at sub- and supercritical temperatures. Mater. Corros. **49**, 831–839 (1998)

43. Z. Shen, K. Chen, X. Guo, L. Zhang, A study on the corrosion and stress corrosion cracking susceptibility of 310-ODS steel in supercritical water. J. Nucl. Mater. **514**, 56–65 (2019)
44. J. Yang, S. Wang, X. Tang, Y. Wang, Y. Li, Effect of low oxygen concentration on the oxidation behavior of Ni-based alloys 625 and 825 in supercritical water. J. Supercrit. Fluid **131**, 1–10 (2018)
45. H. Li, Q. Cao, Z. Zhu, Oxidation behaviour of super 304H stainless steel in supercritical water. Corros. Eng. Sci. Techn. **53**, 293–301 (2018)
46. J. Yang, S. Wang, D. Xu, Y. Guo, C. Yang, Y. Li, Effect of ammonium chloride on corrosion behavior of Ni-based alloys and stainless steel in supercritical water gasification process. Int. J. Hydrogen Energy **42**, 19788–19797 (2017)
47. Z. Zhu, H. Xu, H.I. Khan, N. Zhang, Oxidation behaviour of Nimonic 263 in high-temperature supercritical water. Corros. Eng. Sci. Techn. **53**, 617–624 (2018)
48. Z. Ma, D. Xu, S. Guo, Y. Wang, S. Wang, Z. Jing, Y. Guo, Corrosion properties and mechanisms of austenitic stainless steels and Ni-base alloys in supercritical water containing phosphate, sulfate, chloride and oxygen. Oxid. Met. **90**, 599–616 (2018)
49. Y. Behnamian, A. Mostafaei, A. Kohandehghan, B.S. Amirkhiz, D. Serate, Y. Sun, S. Liu, E. Aghaie, Y. Zeng, M. Chmielus, W. Zheng, D. Guzonas, W. Chen, J.L. Luo, A comparative study of oxide scales grown on stainless steel and nickel-based superalloys in ultra-high temperature supercritical water at 800 °C. Corros. Sci. **106**, 188–207 (2016)
50. Z. Zhu, H. Li, Q. Cao, Oxidation behavior of austenitic steels in supercritical water containing dissolved oxygen. J. Mater. Eng. Perform. **28**, 44–52 (2018)
51. X. Zhong, X. Wu, E.H. Han, Characteristics of oxidation and oxygen penetration of alloy 690 in 600 °C aerated supercritical water. J. Mater. Sci. Technol. **34**, 561–569 (2018)
52. X. Huang, Y. Shen, Oxidation behavior of an 11Cr ferritic–martensitic ateel after Ar-ions irradiation in supercritical water. J. Nucl. Mater. **461**, 1–9 (2015)
53. X. Huang, Y. Shen, J. Zhu, Influence of Ar-ions irradiation on the oxidation behavior of ferritic–martensitic steel P92 in supercritical water. J. Nucl. Mater. **457**, 18–28 (2015)
54. E.J. Opila, Volatility of common protective oxides in high-temperature water vapor: current understanding and unanswered questions. Mater. Sci. Forum **461–464**, 765–773 (2004)
55. M. Watanabe, K. Sue, T. Adschiri, H. Inomata, R.L.J. Smith, K. Arai, Control of methanol oxidation by ionic behavior in supercritical water. Chem. Commun. **21**, 2270–2271 (2001)
56. J. Yang, S. Wang, Y. Li, Y. Zhang, D. Xu, Novel design concept for a commercial-scale plant for supercritical water oxidation of industrial and sewage sludge. J. Environ. Manage. **233**, 131–140 (2019)
57. L. Qian, S. Wang, D. Xu, Y. Guo, X. Tang, L. Wang, Treatment of sewage sludge in supercritical water and evaluation of the combined process of supercritical water gasification and oxidation. Bioresource Technol. **176**, 218–224 (2015)
58. D. Xu, C. Huang, S. Wang, G. Lin, Y. Guo, Salt deposition problems in supercritical water oxidation. Chem. Eng. J. **279**, 1010–1022 (2015)
59. X. Tang, S. Wang, L. Qian, M. Ren, P. Sun, Y. Li, J. Yang, Corrosion properties of candidate materials in supercritical water oxidation process. J. Adv. Oxid. Technol. **19**, 141–157 (2016)
60. A. Agüero, K. Spiradek, S. Höfinger, M. Gutiérrez, R. Muelas, Microstructural evolution of slurry Fe aluminide coatings during high temperature steam oxidation. Mater. Sci. Forum **595–598**, 251–259 (2008)
61. M.J. Drews, M. Williams, M. Barr, The corrosion of sol-gel-coated type 316 SS in chlorinated SC water. Ind. Eng. Chem. Res. **39**, 4772–4783 (2000)
62. M. Schacht, N. Boukis, E. Dinjus, K. Ebert, R. Janssen, F. Meschke, N. Claussen, Corrosion of zirconia ceramics in acidic solutions at high pressures and temperatures. J. Eur. Ceram Soc. **18**, 2373–2376 (1998)
63. E. Asselin, A. Alfantazi, S. Rogak, Corrosion of nickel–chromium alloys, stainless steel and niobium at supercritical water oxidation conditions. Corros. Sci. **52**, 118–124 (2010)
64. K. Sue, N. Tsujinaka, T. Adschiri, K. Arai, Y. Watanabe, Relationship between corrosion rate and metal oxide solubility in supercritical water. Corrosion **58**, 225–231 (2002)

65. M. Atashfaraz, M. Shariaty-Niassar, S. Ohara, K. Minami, M. Umetsu, T. Naka, T. Adschiri, Effect of titanium dioxide solubility on the formation of $BaTiO_3$ nanoparticles in supercritical water. Fluid Phase Equilib. **257**, 233–237 (2007)
66. M. Sun, X. Wu, Z. Zhang, E.H. Han, Oxidation of 316 stainless steel in supercritical water. Corros. Sci. **51**, 1069–1072 (2009)
67. Y. Daigo, T. Isobe, K. Sugahara, Y. Watanabe, Compatibility of nickel-based alloys with supercritical water applications: aging effects on corrosion resistance and mechanical properties. Corrosion **62**, 174–181 (2006)
68. G. S. Was, S. Teysseyre, J. McKinley, Corrosion and stress corrosion cracking of iron and nickel-base austenitic alloys in supercritical water. NACE Int. 04492 (2004)
69. J.H. Yoon, K.S. Son, H.S. Kim, B. Mitton, R. Latanision, Y.R. Yoo, Y.S. Kim, Corrosion behavior of 316L stainless steel in supercritical water environment. Mater. Sci. Forum **475–479**, 4207–4210 (2005)
70. Y. Watanabe, K. Shoji, T. Adschiri, K. Sue, Effects of oxygen concentration on corrosion behavior of alloys in acidic supercritical water. Aust. J. Agr. Econ. **31**, 112–126 (2002)
71. J.H. Kim, I.S. Hwang, Development of an in situ Raman spectroscopic system for surface oxide films on metals and alloys in high temperature water. Nucl. Eng. Des. **235**, 1029–1040 (2005)
72. J.E. Maslar, W.S. Hurst, W.J. Bowers, J.H. Hendricks, M.I. Aquino, In situ Raman spectroscopic investigation of aqueous iron corrosion at elevated temperatures and pressures. J. Electrochem. Soc. **147**, 2532–2542 (2000)
73. J.E. Maslar, W.S. Hurst, J.H. Hendricks, M.I. Aquino, In situ Raman spectroscopic investigation of nickel hydrothermal corrosion. Corrosion **58**, 225–231 (2002)
74. V.A.G. Keramidas, W.B. White, Raman scattering study of the crystallization and phase transformations of ZrO_2. J. Am. Ceram. Soc. **57**, 22–24 (1974)
75. C. Li, M. Li, UV Raman spectroscopic study on the phase transformation of ZrO2, Y2O3-ZrO2 and SO42-/ZrO2. J. Raman Spectrosc. **33**, 301–308 (2002)
76. B. Zhao, X. Xu, J. Gao, Q. Fu, Y. Tang, Structure characterization of WO3/ZO2 catalysts by Raman spectroscopy. J. Raman Spectrosc. **27**, 549–554 (1996)
77. R.J. Nemanich, C.C. Tsai, G.A.N. Connell, Interference-enhanced raman scattering of very thin titanium and titanium oxide films. Phys. Rev. Lett. **44**, 273–276 (1980)
78. C.A. Melendres, A. Narayanasamy, V.A. Maroni, R.W. Siegel, Raman spectroscopy of nanophase TiO_2. J. Mater. Res. **4**, 1246–1250 (2011)
79. Q. Zhang, R. Tang, K. Yin, X. Luo, L. Zhang, Corrosion behavior of Hastelloy C-276 in supercritical water. Corros. Sci. **51**, 2092–2097 (2009)
80. S. Velu, K. Suzuki, C.S. Gopinath, H. Yoshidac, T. Hattoric, XPS, XANES and EXAFS investigations of $CuO/ZnO/Al_2O_3/ZrO_2$ mixed oxide catalysts. Phys. Chem. Chem. Phys. **4**, 1990–1999 (2002)
81. S. Hashimoto, A. Tanaka, Alteration of Ti 2p XPS spectrum for titanium oxide by low-energy Ar ion bombardment. Surf. Interface Anal. **34**, 262–265 (2002)

Chapter 7
Prospects and Challenges of Candidate Alloys Used in Sub- and Supercritical Water Environment

To date, it is evident that no one material or structural design can withstand corrosion damage under all conditions in SCW technologies. Although a considerable amount of corrosion work has been carried out in SCW and related environments, they are mostly limited to nearly purified SCW and supercritical aqueous systems containing certain salts. Experimental research in a real, complex environment with multi-parameter coupling is relatively scarce. For SCWO, the combination of material type, process design, reaction conditions, and process control under different operating conditions requires large amounts of experimental results in addition to a theoretical basis for interpreting and projecting those data. Hence, subsequent research should focus on actual SCWO environments. In addition, some researchers believe that phosphate can relieve the corrosion of Ni-based alloys in SCW. However, the feasibility of using phosphate as a corrosion inhibitor needs to be fully explored from the aspects of mechanism, applicability, environmental protection, and economics under various perceived operating conditions.

There are now some surface treatment processes that can improve the corrosion resistance of Ni-based alloys, but the laws and principles behind these processes (e.g., coating) in modifying the corrosion characteristics of alloys are not well-defined. For example, the further study of protection mechanisms of ceramics and the development of effective ceramic coating processes for use in SCW seem to be warranted. If ceramic coating can be applied to the alloy with poor corrosion resistance but good economics, it can provide a new material with good corrosion resistance or a combinatorial material that meets specific performance requirement.

Although several corrosion mechanisms for Ni-based alloys in SCW are described in literatures, most primarily focus on general corrosion and the formation of oxide films. There is a lack of research on the mechanisms and control of localized corrosion such as crevice corrosion, pitting corrosion, and intergranular corrosion. A large number of studies on localized corrosion have been conducted in ambient- or near-ambient temperature water and solutions and the theory models developed therein should be extended to interpret localized corrosion phenomena in high subcritical and

D. Xu and S. Guo, *Corrosion Characteristics, Mechanisms and Control Methods of Candidate Alloys in Sub- and Supercritical Water*,
https://doi.org/10.1007/978-981-16-5525-8_7

supercritical aqueous systems. Existing surface analytical techniques are capable of clarifying oxide compositions and qualitatively describing microstructural characteristics of oxide layers formed on many types of Ni-based alloys in SCW. Nonetheless, few researchers appear to concern about the properties and effects of these oxides on corrosion in SCW. The oxide formation and reaction processes mainly rely on material determination and theoretical examination after experiments. The properties of these oxides formed on the Ni-based alloy surface in SCW and their protective mechanisms need to be explored in detail. It is necessary to develop the quantitatively deterministic model for describing the growth of oxide films at the atomic levels.

The transport of oxygen is the rate-determining step for Zircaloy corrosion in low temperature water and high temperature steam. Many studies mention that the transformation of ZrO_2 morphology is the mechanism of advanced corrosion, however, some research indicate that the change of ZrO_2 morphology is the result instead of the reason for further corrosion. The corrosion mechanism of Zircaloy becomes complex in the presence of H_3BO_3 and LiOH. Existing information can not provide enough evidence to support whether the influencing mechanism of LiOH is dominated by oxide transition or other reasons such as the increase of hydrogen concentration and the synergistic effect between LiOH and substances in primary water, and relevant research is expected to clarity the effects of Li^+ and OH^- on corrosion. Besides, theoretical support for the opposite effect of the concentration change and specific action of solutes in corrosion behavior needs to be further explored. In steam, the heterogeneous transformation of zirconia leads to breakaway oxidation, the post-breakaway oxide is less protective and much thicker with cracks. Further investigations are required to understand the mechanism of breakaway oxidation. When Zircaloy is used as fuel claddings in nuclear reactors, in-reactor irradiation is a big challenge for its integrity. Irradiation influences element migration and surface defects deterioration. The redistribution of alloying elements by irradiation is regarded as a primary contributor to enhanced corrosion. At present, most studies on cladding corrosion are conducted in autoclaves which fail to support the deep investigation of irradiation on corrosion mechanism, so more research in experimental devices with irradiation sources is necessary in subsequent work.

The corrosion behavior of FeCrAl in high-temperature and high-pressure water chemistry has been studied, but most corrosion experiments are carried out in pure water where the water chemistry is less complicated than that in nuclear power plants. As a consequence, current results can only judge the corrosion resistance of FeCrAl unilaterally, but not accurately and comprehensively evaluate the material. FeCrAl cladding shows a negligible maximum thickness in hydrogenation water chemistry, which is relatively closer to the actual situation of reactors. However, high neutrons absorption, enhanced tritium release and uncertain extra behavior brings challenges to the development of FeCrAl. Based on the existing research, corrosion of FeCrAl alloys in reactor environment is worth studying as following. Oxide dispersion strengthening has been proved to be one of possible means to improve the oxidation resistance of FeCrAl alloys at high temperature. However, limited information reported in literatures about oxide dispersion strengthened-FeCrAl can be found. The state of Al_2O_3 will change at different temperatures, so the mechanism of its transformation needs to be further explored. In reactors, when FeCrAl alloys are used as fuel cladding, in addition to the influences of alloying elements and corrosive environment parameters, plastic deformation or annealing, the flow rate of coolant and cooling temperature will cause the surface oxide film to fall off. Therefore, the influence of single factor on the corrosion behavior of FeCrAl and the coupling mechanism of multi elements require more investigation. Besides, there is little research on FeCrAl alloys in corrosive ions presently. Referring to the engineering application at the same level, corrosive anions have great influence on alloy corrosion. For this reason, the corrosion research of FeCrAl alloys in corrosive anions should be promoted.

Overall, excellent fuel cladding should have creep resistance, good mechanical strength, high thermal expansion, satisfactory radiation resistance, ideal thermal hydraulic characteristics and enhanced neutron economy. Therefore, in terms of successful commercial applications of candidate fuel cladding, whether the improvement of corrosion resistance affects other characteristics needs to be comprehensively estimated.